Forensic Engineering for Academia

Most engineering curricula lack forensic engineering content, particularly the role of science in investigation and assessment, engineering failures and cause-and-effect analysis, and jurisprudence as it relates to engineering. *Forensic Engineering for Academia* addresses these and other important aspects of forensic engineering education. Mainly structured as two parts: (1) Forensic Engineering (FE) and Jurisprudence, and (2) FE Practice, the first introduces the practice of forensic engineering, presents investigation methods and analysis, describes the expert witness and how to become one, introduces the legal system regarding FE, and discusses reporting and communication liability. The second part covers a variety of practical applications for different FE disciplines through specific examples and experiences. Student solutions will be compared to industry practices with consideration for liability and loss prevention. Further, the book explains the importance of standard of care and performance evaluation as they relate to design and failure, and how to avoid becoming a defendant.

- Includes a summary table of concepts, terminology, and exercises, and practical case studies in each chapter.
- Discusses the importance of professional ethics in the field of forensic engineering.
- Includes Digital Learning Resources: adaptable lecture slides, test and quiz banks, and links to additional resources.

Forensic Engineering for Academia

Dennis M. Filler, P.E., PhD

CRC Press
Taylor & Francis Group
Boca Raton London New York

CRC Press is an imprint of the
Taylor & Francis Group, an **informa** business

Designed cover image: Dennis Filler

First edition published 2027
by CRC Press
2385 NW Executive Center Drive, Suite 320, Boca Raton FL 33431

and by CRC Press
4 Park Square, Milton Park, Abingdon, Oxon, OX14 4RN

CRC Press is an imprint of Taylor & Francis Group, LLC

ISBN: 978-1-041-03320-2 (hbk)
ISBN: 978-1-041-03325-7 (pbk)
ISBN: 978-1-003-62331-1 (ebk)

DOI: 10.1201/9781003623311

Typeset in Palatino
by KnowledgeWorks Global Ltd.

Access the Support Material: www.routledge.com/9781041033202

IN MEMORIAM

The Scientist and the Engineer

One plants a seed, the other harvests.
Visions of the unseen,
the Mad Scientist tinkers.
Imaginings strewn across the room,
yet organized in mind.
With zealous reverence,
the gold standard of testing.
And the ONE gets the glory.

Science guys,
relentless pursuit of truth.
The Engineer thinks in new ways,
information comes,
consilience the goal.
Report to settle;
truth shall set them free.
And the ONE gets the glory.

The world was a better place with the Mad Scientist in it.

Contents

Preface

Fundamentals for the study of engineering are mathematics and physics. Math is the tool students use to describe physical phenomena: the stress-strain relationship of materials, soil stress and behavior caused by the building above, forces on a rocket as it travels through Earth's atmosphere to reach space. Students of all disciplines must obtain a level of mastery of these fields to earn an engineering degree.

Engineering curricula are generally structured for design, and the world needs good designers who hold paramount the health and safety of users of all things engineered. But engineering practice is much broader than just design. Engineers develop, test, and maintain systems, structures, and technologies, within disciplines that have unique responsibilities and specializations. Engineers are also inspectors, quality controllers, investigators, researchers, and innovators whose patents and products must function with the protection of people. Further, scientific principles and methods are the underpinnings of engineering practices and advances, and failures are an inherent part of progress. Failures with the new, used, and maintained often result in damage and harm. Our students would benefit from knowing a little about how to apply scientific principles to their work, how to avoid becoming a defendant, and the growing role of forensics in engineering. The ultimate goals are engineering excellence and fewer failures.

Forensic Engineering will play a greater role in our products, services, and built environments of the future. Yet this subject is not yet taught in our engineering curricula. It will be taught through the Department of Civil, Environmental, and Construction Engineering at the University of Central Florida beginning Fall 2026. *Forensic Engineering for Academia* is for all disciplines and includes numerous multidisciplinary case studies for broad application.

Several imperatives support the need to incorporate forensic engineering in curricula:

1. Students learn about engineering design but have little exposure to standard of care and performance evaluation as they relate to design and implementation.
2. Students do not know what negligence is relative to engineering practice.
3. Students have no idea how engineers become defendants, nor how to navigate legal challenges.
4. Students know little about liability as it relates to their future employers.

5. Students need a better framework for relating science to engineering practice.
6. Industries are embracing prevention as a business practice.
7. Industries and insurers need high-quality engineering experts.
8. Engineering consultancy is moving away from conclusions in favor of informed opinions.
9. Lawyers rely upon engineers for expertise.
10. Students need to learn how they can become expert witnesses.
11. Ethics is at the core of the integrity and reputation of professional experts.

My generation of engineers owes it to the next to help prepare them for the future of our profession. This textbook is reference material for future forensic engineering courses we create and succession planning – for the next generation of engineers.

Dr. Dennis M. Filler, P.E., Professor,
Expert Witness, and Forensic Engineer

About the Author

Dr. Dennis M. Filler is a registered Professional Engineer with 35 years of diverse civil, environmental, and engineering science experience with residential, commercial, industrial, educational, and utility facilities. He is a principal with *Insight-Forensics, LLC* and practices forensic engineering related to infrastructure damage assessment, geo-environmental (soils and environmental) and hydrologic investigations, cold regions engineering (infrastructure impacts and permafrost investigations), fire and explosion cause and origin, engineered system failures and impacts (primarily underground infrastructure), product liability investigations, and limited traffic accident reconstruction, evaluation of mechanical systems, and structural failure analysis support. Professor Filler also has 26 years of university experience and is currently teaching and developing new courses for engineering curricula and the next generation of engineers at the University of Central Florida.

And Then There Were Three

Part I

Forensic Engineering and the Legal System

1

Introduction to Forensic Engineering

CONTENTS

CONCEPTS	KEY PHRASES	ADDITIONAL READING
History of Forensics	Forensics	Exercise 1
Forensic Engineering	Forensic science	Exercise 2
Forensic Investigation	Engineering codes	
Causes of Engineering Failures	Nature's entropy and decay	Case Study 1
Expert Engineers	Building safety audits	
Embrace Challenge	Milestone structural inspections	
Industry Needs Forensic Engineers		
The Age of Disinformation		

What will engineering be like by 2050? From Imhotep, the first engineer who designed the Step Pyramid at Saqqārah between 2667 and 2648 BC, to China's great wall and the magnificent Acropolis, Parthenon, and Colosseum from the seventh to fifth centuries BC, to the world's tallest building the Burj Kalifa at 830 m in 2010, and now SpaceX's Starship and tomorrow's Joby Aviation flying cars, technology is sure to bring amazing innovations and advances during the next 25 years. All the while we will continue to build and replace our habitats, commercial and industrial complexes, and machines as we have since the ancients. To err is human, there will be failures, and engineers and lawyers will sort them out. The lawyers will rely on the engineers, who know to look at the science that underlies every failure. Further, how might *contrived durability* and *planned obsolescence* figure into the liability of things gone wrong? Also, since the 9/11 attack of 2001, forensics is sure to play a key role in the rise of targeted infrastructure and vulnerability assessments. Such is the evolution of engineering.

DOI: 10.1201/9781003623311-2

Over the past century, numerous significant engineering failures (Table 1.1) have occurred. Dam collapses around the world, interstate highway I-35, I-10, I-40, and FIU bridge collapses in the United States, and building collapses have occurred in Russia, Bangladesh, China, Taiwan, North Korea, and the United States despite technological advances and improved safety standards and regulations. To err is human and faulty designs, design flaws, defective or substandard materials, deferred maintenance, corrosion, destructive

TABLE 1.1

A Quarter-Century of Modern Engineering Failures

Incident/Disaster	Year	Description	Incident/Disaster	Year	Description
VertiGo Amusement Park Ride Incident, USA	2002	Vortex shedding tower failure	Deepwater Horizon Oil Spill, Gulf of Mexico	2010	Multiple engineering and safety lapses led to rig explosion and failed blowout preventer
Charles de Gaulle Airport Terminal Roof Collapse, France	2004	Design flaws	Fukushima Nuclear Disaster, Japan	2011	Nuclear reactor meltdowns after earthquake and tsunami
Transvaal Water Park Dome Collapse, Russia	2004	Faulty design	Rana Plaza Building Collapse, Bangladesh	2013	Structural failure, design code violations, management and safety lapses
Katowice Trade Hall Roof Collapse, Poland	2006	Faulty design	FIU Pedestrian Bridge Collapse, USA	2018	Faulty design and oversight failure
I-35 Bridge Collapse, USA	2007	Design flaw, maintenance issues	Boeing 737 MAX Groundings (after two international flight crashes), USA	2019	Faulty MCAS (Maneuvering Characteristics Augmentation System) flight stabilizing system
Hornslet Wind Turbine Collapse, Denmark	2008	Deferred maintenance	Champlain Towers Condominium Collapse, Surfside, FL, USA	2021	Water intrusion and long-term concrete degradation and reinforced steel corrosion of basement level structural supports, deferred maintenance, corruption at the time of construction, and other contributing factors.
Korba Thermal Power Plant Chimney Collapse, India	2009	Faulty design, substandard materials	Derna Dam Collapses, Libya	2023	Corruption, neglect, lack of safety culture

vibrations, corruption, and other human and natural impacts will continue to fail infrastructure, products, and systems. Forensic engineers and lawyers will have their roles in sorting out the causes and consequences.

Table 1.1 highlights some engineering disasters that made the news. Large-scale catastrophes tend to be more sensational. Yet behind the scenes, there are thousands more incidents that do not make the news, and automobile accidents are daily occurrences in urban settings. Sometimes it is necessary to keep them out of the news, at least until after legal challenges and insurance claims are settled. The last thing we need is speculations, rumors, and innuendo to muddy the waters.

1.1 A History of Forensics

The word *"**forensic**"* originates from the Latin *"forensis,"* which means *"of or before the forum."* (Fig. 1.1). In ancient Rome, criminal charges were presented before a group of public individuals in a public square or forum. Both the accused and accuser would present their sides, and the *tribunal* would render a decision in favor of the one who presented the best argument.[1]

The term ***forensic science,*** which combines *forensic* – relating to courts of law – with *science* – systematic methods of obtaining knowledge, first appeared in print in 13th-century China. Song Ci (1186–1249), an official of the Southern Song Dynasty, authored *Collected Cases of Injustice Rectified* (Xiyuan jilu) in 1248, the first systematic treatise that summarized forensic science achievements in medicine and the resolution of crimes of the time.[2] Notably, *forensic entomology* – the use of insects found on corpses to help solve crimes – is a technique still used today in *criminology.* Medical practitioners of the sixteenth and seventeenth centuries began to catalog the systematic studies of causes of death, and ***forensic*** became an official English word in *Merriam-Webster*'s dictionary in 1659.

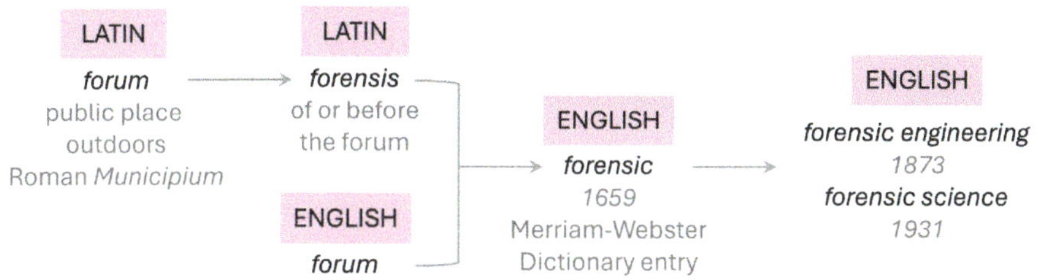

FIGURE 1.1
Forensic etymology.

Beginning in 1873, Dr. Henry Wigmore, an American lawyer and legal scholar, was a pioneer in the field of *Forensic Engineering* while teaching at Northwestern University. Renowned for his expertise in the law of evidence, his treatise on the *Anglo-American System of Evidence in Trials at Common Law* (1904), often called *Wigmore on Evidence,* was the forerunner to modern *Daubert* and *Frye*[1] evidence standards.[2]

Also, during the late 1800s to early 1900s, there was Dr. Edmond Locard (1877–1966), the French criminologist and pioneer considered the father of modern forensic science. Known as the *Sherlock Holmes of France,* he formulated *Locard's Exchange Principle:* "Every contact leaves a trace," a fundamental principle in criminology that extends to all forensic work.[3] *Locard's principle* – an extension of *Newton's Third Law of Motion* (i.e., the law of action and reaction) – stipulates that any two objects that come in contact will experience a transference of material between them and leave an indelible mark. First to establish a forensic laboratory in Europe (in 1910 in Lyon, France), Locard's greatest accomplishment was his seven-volume set *Traité de Criminalistique* (forensic treatise), the first formulation of forensic science methods and practices for the Western Hemisphere.

Next came additional developments with fingerprinting, blood typing, hair analysis, and ballistics, and *criminology* became synonymous with ***forensic science*** in 1931. In modern usage, *forensic science* refers to the application of scientific methods and principles to approximately 57 forensic disciplines (Table 1.2). This evolution makes a good case for *Forensic Engineering for Academia,* something lacking in engineering curricula. Four imperatives for the timing of this book:

1. The need to reduce failures and legal claims against designers and builders.
2. Engineering consultancy is moving away from conclusions in favor of informed opinions.
3. Engineering academia lacks forensic engineering education and training.
4. Technological advances and the need to update forensic engineering practices (Case Study 1 example).

Further, we need our next-generation engineers to be better than we are, and prevention requires a complementary understanding of the science behind how things work and fail.

Forensic engineering in North America has its roots in early investigations into bridge failures such as the Dee bridge and Tay rail bridge disasters of 1847 and 1879, and the term was first used by Dr. Henry Wigmore in 1873. Current definitions for forensic engineering have a 'failure investigation' connotation without reference to the underlying science.

TABLE 1.2

Forensic Disciplines in 2025

A Forensic Accounting and Investigation Standards by ICAI Forensic Anthropology Anthropometry Forensic Arts Forensic Astronomy Audio Forensics	*F* Forensic Facial Recognition Forensic Footwear Evidence Fractography	*P* Forensic Palynology Forensic Pathology Forensic Photography Forensic Podiatry Forensic Polymer Engineering Forensic Profiling
B Forensic Biology (DNA Forensics) Blood Stain Analysis	*G* Forensic Glass Analysis Gait Analysis Forensic Geology Forensic Geophysics	*R* Forensic Radiology
C Forensic Chemistry Criminal Investigation Forensic Colorimetry	*I* Forensic Identification Investigative Genetic Genealogy IoT (Internet of Things) Forensics	*S* Forensic Seismology Forensic Serology Forensic Social Work Forensic Software Engineering Forensic Statistics Stylometry
D Forensic Dentistry Digital Forensics Diplomatics	*L* Forensic Limnology Forensic Linguistics	*T* Forensic Tire Tread Evidence Forensic Toxicology
E Forensic Economics Edgeoscopy Forensic Electrical Engineering Forensic Engineering Entomotoxicology Forensic Epidemiology Environmental Forensic Science	*M* Marine Forensics Forensic Materials Engineering Forensic Meteorology Forensic Metrology Forensic Mycology	*V* Forensic Video Analysis *W* Wildfire Forensic Science

Forensic engineering today involves inspections, high-tech performance experimentation and corrosion testing, development of models and solutions to challenging problems, and can involve multidisciplinary studies, a much broader scope than the collection of evidence, measurements, and *exemplar* testing of failure investigations. Further, investigations may incorporate *Failure Mode and Effects Analysis* (FMEA) or *Fault Tree Analysis*[2] for structured examination of failures. The point is: engineering involves science.

An important distinction relative to liability is that engineers become defendants in *civil* courts, not criminal courts. When things fail and injury and/or property damage occurs, recompense is usually a civil matter. Only through intentional negligence might an engineer wind up in a criminal court.

Modern Pioneers

Some of the first organizations and professionals in the United States dedicated to forensic engineering include:

- National Academy of Forensic Engineers (NAFE), founded in 1982 by Marvin M. Specter, Paul E. Pritzker, and William A. Cox Jr.
- Investigative Engineers Association (I-ENG-A®), founded in 1991 as the first association of professional forensic or investigative engineering consultants.
- FTI Consulting, founded by Joe Reynolds in 1982. Early focus on investigating aviation and utilities accidents and failures to determine causes led to establishing prevention methods. It is noteworthy that FTI recognized *prevention* as a cost-effective practice a half-century ago.
- Randal K. Noon, Dressler Consulting Engineers, authored five textbooks on forensic engineering (1994–2009). As a mentor, his brilliance and expertise were inspirational, and his books benefit engineering.

Forensic Engineering Journals

At the core of forensics is the study of details, and science plays a significant role in engineering analysis and evaluation of cause and effect, corrosion, distress, impacts, fatigue, and failures. Further, our journals recognize *prevention* as necessary to future engineering practice.

- *Engineering Failure Analysis* (ScienceDirect)
  Scope: Analysis and prevention of engineering failures, emphasis on the investigation of failure mechanisms, identification of the root causes of failure, and prevention to avoid failures.
- *Journal of the National Academy of Forensic Engineers* (JotNAFE)
  Scope: Forensic engineering principles and methods; critical analysis of case studies; specific cases that demonstrate forensic engineering best practices or new technology; specific cases that consider ethics, standards compliance, investigative procedures, legal admissibility, and jurisdictional considerations.
- *Forensic Engineering* (Institution of Civil Engineers)
  Scope: Studies of constructed facilities and systems that fail to perform as intended, resulting in environmental, property, personal, or economic damage.
- *Journal of Performance of Constructed Facilities* (American Society of Civil Engineers, under the umbrella of its Technical Council on Forensic Engineering)
  Scope: Disseminate information on failures and performance deficiencies of constructed facilities, including investigation methods,

special techniques for failure investigations, infrastructure monitoring and maintenance, design and construction practices that lead to failure, risk management and failure prevention techniques, and ethical issues.

It is worth noting that the field of forensic engineering gradually evolved from various engineering disciplines applying their expertise to failure analysis and accident investigation. Engineering firms engaged in forensic work to some degree and the reporting of case studies led to the formal recognition of forensic engineering as a distinct field. Today, engineering investigations and studies are rooted in first principles of science,[4] and convergence of evidence leads to truth.

Engineers as Experts

In civil court, focus is on admissible evidence and provable facts. Honorable attorneys build cases on verifiable evidence, knowing the burden of proof relies upon the ability to demonstrate claims with credible evidence. And they rely on credible experts from many professions – often engineering – to support with facts rooted in science and standard practices. More about the *expert witness* in Chapter 3.

1.2 What Is Forensic Engineering?

Forensic engineering is the application of scientific principles and engineering methods to determine facts. Currently, we use it primarily to address failure, cause and effect, and cause and origin analyses. However, the costs of failures have been great, leading some industries to consider prevention as a necessary cost of doing business. FE will play a greater role in building science, building condition and vulnerability assessments, asset management, corrosion detection, cybersecurity, digital systems, land reclamation (mines, legacy landfills, and brownfield sites), environmental chemistry, and quality control.

1.3 Forensic Investigation (Overview)

Forensics involves the study of information, details, and evidence. ***Forensic investigation*** is a scientific approach to the study of failures and cause-and-effect scenarios. It involves *first principles* of science: reviewing relevant available information and evidence, hypothesizing about cause, testing to validate hypotheses, and determination and analysis of facts before assigning cause.[4]

Science implies observable and reproducible data with a *consilience approach*[3] to evidence. Investigation integrity demands unbiased and impartial work under the premise of *res ipsa loquitur*: "let the facts speak for themselves." And every good investigator possesses excellent attention to detail, a prerequisite to becoming an expert in the field.

Libel actions related to the *as-built world* arise from system operational problems, material defects, or failures with associated damages, expenses, or personal injury. Because causation is often the result of multiple factors, lawyers may involve designers, manufacturers, fabricators, and constructors in a legal action, and let subsequent *forensic investigations* sort out attributable causes and responsibility.

Engineers need some understanding of forensic investigation and to be able to identify reported conflicts and refute erroneous findings or flawed accusations as they relate to their (or their firms') involvement. Forensics is a deeper level of critical thinking and attention to detail that teaches us how to identify and evaluate causal interactions. *The devil is in the details,* and details are discoverable if we know what to look for.

1.4 The Nature of Failures

Design, Construction, or Acts of God

Things engineered are designed, fabricated, and constructed. Inherent with each is human control and responsibility for safety, professional standards of practice, and liability with failure. Only an *Act of God* may relieve engineers, manufacturers, and constructors of responsibility. As a legal concept, an *Act of God* refers to a natural hazard beyond human control. Earthquake, tsunami, tornado, and hurricane are familiar examples. Legal implications of such calamities can release parties from contractual liabilities or be an insured peril in insurance policies.

Engineering codes are books used to ensure *minimal* acceptable standards with design. Improved methods and failures lead to their improvement every few years. Standard of care and performance evaluation are additional functionality and operating environment considerations engineers employ to further ensure things work as designed for a period of useful life. Nevertheless, to err is human, and the forces of nature are constantly at work against the things we make and build. ***Nature's entropy and decay*** refers to the universal principle that things move toward a state of dysfunction and disrepair if not regularly maintained for service. Medieval castles, churches,

FIGURE 1.2
Reminders of nature's entropy and decay.

and cemeteries of Great Britain attest to nature's constant weathering and decay of structures (Fig. 1.2).

Engineering failures arise from design flaws, material fatigue, extreme conditions, maintenance issues, wear and tear, and human error. They can also occur with insufficient regulatory oversight, flawed organizational processes, mismanagement, and a lack of safety culture. All are preventable when we care to take care of what we produce.

Sometimes we engineer at the forefront of technology, a realm with a high probability of failure. At other times, basic mistakes are expensive reminders. What do the four mishaps below have in common?

Lost Mars Climate Orbiter (1999)

Gimli Glider (1983)

Tokyo Disneyland roller-coaster derailment (2003)

Columbus miscalculated his New World destination (1492)

They all resulted from unit conversion mistakes. A common issue with engineering students can just as easily happen to the most experienced professionals.

Advancing Humanity

Sometimes we fail when we try to advance. To move onward and upward means challenge, deep science, pushing beyond limits, and to engineer where no one has gone before. *No research is bad research* – we learn from what works and what does not – is an adage that embodies *try, and try again, until we figure it out*. Failure is sure to come with making new things and going to new places. To err is human, and we learn the most from our mistakes. May engineering students be allowed to make mistakes early, so they avoid failures later.

1440 LESSON

A rookie engineer was provided the opportunity to manage a first project. It was a ground-penetrating radar (GPR) survey of a cleared site planned for a slurry pond at a paper mill. Upon arrival, operators were leveling ground with heavy equipment, a situation not conducive to GPR surveying. The engineer had a conversation with the site foreman, who made a phone call, and heavy equipment moved away to accommodate the survey. Upon return to the office, the engineer was called to a supervisor's office. The conversation was about a $1440 charge levied by the paper mill client, the cost of heavy equipment operation time lost. *Communication can make or break a project* and the *1440 Lesson* are reminders of the importance of thorough logistics and Murphy's law* before heading out for field work.

Anecdote: The engineer had no idea if the company paid the $1440, and the amount was not deducted from his pay. Perhaps the valuable lesson learned – *pay attention to details* – was intentional training.

* Murphy's law is the idea that with any endeavor, anything that can go wrong will go wrong.

1.5 Embrace Challenge

Like NASA, sometimes engineers are asked to solve challenging problems NOW. What might an insurance adjuster on a golf course in Arizona know about an *icehouse* problem in Alaska?

ICEHOUSE INCIDENT

In early December, a Fairbanks, Alaska resident called his insurance provider and told the adjuster a fantastical story about a glacier forming around his house. "Please send me a photo," said the adjuster. As the problem grew, photographs were sent, and anxiety escalated with conversations over the ensuing weeks.

Fifty-four days after the initial homeowner call, the adjuster called an engineer: "Can you please go and take a look?" An hour later, the engineer reported to the adjuster: "The situation is an emergency. You have a runaway artesian well that has glaciated around and under the home, across the road and onto neighboring property, and the insured's residence will soon be a total loss. Further, artesian waters are a protected state resource and there is risk of neighboring property damages." "Can you fix this?" said the adjuster. "Maybe," replied the engineer. With the right approvals granted, the engineer went to work. All stakeholders understood this was unknown territory and no assurances were made.

Literature review provided scant information about Scandinavian and Russian accounts of similar situations but no solutions. Calls around the network did not identify expertise. Finally, a lead from a North Slope oil-industry colleague: a unique fabricator in Fairbanks happened to also be a well driller/installer. A series of troubleshooting meetings, whiteboard scribblings, and teleconferences with a variety of experts led to a plausible solution: *cryogenic freezeback* of the well, the moniker that stuck with the method being conceived.

Ice would be removed to expose the wellhead, the well cap would be cut off, a fabricated evaporator cylinder would be inserted deep down into the well, and liquid nitrogen (LN_2) would be carefully discharged into it at a controlled rate for 24 hours.

"This should freeze back the soil around the well, thereby reestablishing permafrost and blocking off the artesian flow ... in theory," said the heat and mass transfer expert. Then, we convert the evaporator to a thermosyphon and retrofit the top with a radiator, and then concrete-seal the thermosyphon in place inside the well casing as a perpetual heat-liberation device.

But first, amazing ice carvers with chainsaws that cut through ice like butter would create an intercept trench to temporarily stop ice growth around the house and buy time to perform the freezeback. After workers prepared the well in –40°C weather, and LN_2 began to discharge, water stopped flowing at the wellhead in nine minutes. Another 24 hours and 23,000 pounds of LN_2 provided freezeback integrity...in theory. The thermosyphon provided the necessary long-term assurance against a repeat incident.[5],[6]

Cryogenic freezeback of the former *icehouse* well has worked since March 2007, and a solution to the problem of runaway artesian wells is now in the literature for future use.

A *thermosyphon* is a device that circulates a special fluid within a cylinder using natural convection and conduction instead of a mechanical pump. Installed in the ground, and fitted with a radiator at the top, the liquid circuit liberates heat from the subsurface to the above-ground radiator for discharge to the atmosphere. The process preserves permanently frozen ground. Sections of the Trans-Alaska oil pipeline have benefited from thermosyphons since 1977.

Embrace challenge and become the *go-to engineer*! Wing suiters get a rush flying fast like birds. Investigating or engineering new territory can be just as thrilling for the forensic engineer. Read Case Study 1 and imagine the opportunity that comes with challenge.

1.6 Engineering Consultancy 2025 and Beyond

Conclusions, Opinions, and Liability

We live in a litigious society. In the early 1990s, one in five US engineering consultancies were sued within their first five years of existence. Today, that number is one in every two to three. Why? Generally, engineering engenders an excellent reputation and is highly regarded among the professions, and lawyers routinely rely upon engineering expertise. For now, the answer to why is *because they can,* and Chapter 4, Jurisprudence will provide students with the necessary understanding of our legal system as it relates to engineering. Suffice to say, engineering is moving away from conclusions in favor of informed opinions to reduce liability. Opinions are the near future of engineering findings and determinations, until the legal profession and tort reform improve the situation.

Concluding is the final step of investigation. Conclusions summarize results of an experiment or analysis and evaluate their relationship to the original hypothesis. Conclusions can also summarize findings, be statements of cause as determined by facts, or be exculpatory (to clear someone from blame or fault). By their very nature, investigation conclusions are considered firmly established through rigorous scrutiny of the evidence and hard facts, and their implications may carry great weight. In a court of law, conclusions may be challenged and must be able to withstand rigorous cross-examination. Further, with technical or engineered documents, faulty conclusions can carry significant liability when they misrepresent or lead to damage or harm.

Opinions are a dime a dozen, and everyone has one, including the uninformed. However, an *informed opinion* can be rooted in facts and reasoned like a conclusion (Fig. 1.3). When stated as professional viewpoints in the context of an investigation report with a limitations clause – e.g., "We make no guarantee that others will reach the same determinations." – informed

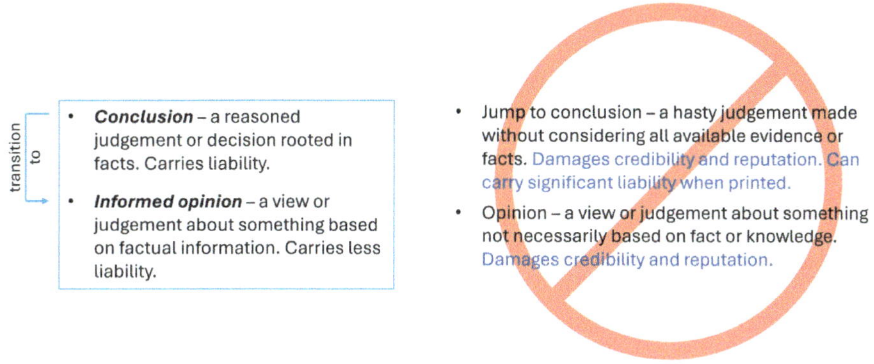

FIGURE 1.3
Liability attributed to conclusions, informed opinions, and unprofessional judgments.

opinions carry less liability. Trial courts frown upon attempts to sue for opinions. Finally, avoid jumps to conclusion and baseless opinions; they can be quickly disproved and tarnish credibility and reputation.

Forensic engineers currently have primacy with failure investigations and will play a greater role in the future of the insurance industry and infrastructure asset management.

Insurance Industry

The COVID-19 pandemic had interesting impacts on the national workforce. Major industries such as the commercial airlines experienced massive layoffs and attrition, as did restaurants, tourism, and insurance occupations. Large numbers of workers left their jobs in search of other opportunities during a two-year period of federal government subsidy. In response, insurance firms hired new recruits en masse, with Senior Adjuster titles and $25/hour pay as incentives. Consequences have been a substandard workforce, an increase in insurance claim challenges by the insured, and hits to the insurance industry's reputation. The insured are demanding engineers perform the damage assessments, and providers are complying. It is cheaper to hire an expert than risk litigation.

Infrastructure Asset Management

Forensic engineers will play an increasing role in effective *Infrastructure Asset Management*. Asset *Management* is a tool for maintenance planning and loss prevention. *Infrastructure Asset Management* (IAM) is a multidisciplinary approach to managing infrastructure assets through their useful life. The lifecycle of a physical or service asset encompasses planning, design, construction, operation or function, maintenance, upgrade, and replacement or decommission. Key components of IAM are condition assessment, maintenance planning, financial management (including maintenance, capital expenditures planning, cost optimization exercise), and risk management.

Also, the Surfside condominium collapse disaster of 2021 led to significant new legislation that directly impacts IAM. Florida enacted new legislation requiring ***Building Safety Audits*** (BSA) and ***Milestone Structural Inspections*** (MSI) for buildings three stories or higher. Inspections and repairs are underway all over the state. Also, for multistory residential condominium and apartment buildings, reserves studies now include funding for critical structural repairs and audit/inspection information must be shared with residents. Further, the National Institute of Standards and Technology is scheduled to complete its technical investigation of the disaster in 2025. Report findings are anticipated to further impact design codes and construction and oversight practices nationwide.

The new MSI must be performed by licensed professional engineers certified to assess building structural integrity. This new class of forensic engineers who perform technical investigations generates *building condition reports*. Loss prevention benefits insurers and property owners alike, and should improve *actuarial rates*, the premiums insurers charge. An ounce of prevention is worth a pound of cure, and IAM is evolving. Table 1.3 provides an example of how some infrastructure sectors are forward thinking.

TABLE 1.3

Examples of Infrastructure Asset Management Trends

Current Practices	Future Asset Management
Coastal Condominiums (≥ 4 stories) • **Reserve Study** • **Inspections** • **Deferred Maintenance** • **Expensive Insurance Coverage**	Multiunit Residential (≥ 3 stories) – NEW in FL 2023 • Reserve Study, Structural Repair Reserve, BSA, MSI • Asset Evaluation (Structural and Property) Report • Planned Maintenance and Risk Mitigation • Sensible Insurance*
Vulnerability Assessments • **Specifies which systems, components, applications, and environments require assessment.** • **Excludes some assets** • **Security focus**	Vulnerability and Integrity Assessments (VIA) • Assesses vulnerability assets for security • Assesses other infrastructure and property for integrity and longevity • Security, integrity, and loss prevention focus
Transportation Asset Management (TAM) • **Inventory and inspect** • **Economic Analysis** • **Asset Management Plan (TAMP) (maintenance scheduling, forecasting, and deferred maintenance)** • **Safety and viability focus**	Transportation Asset Management and Optimization (TAMO) • Smart Cities and multimodal systems (e.g., transport and energy production) • Sustainability Asset Management Plan (SAMP) • Economic viability with forecasting • Safety, security, and loss prevention focus • Minimal resourcing and environmental impacts

* The University of Central Florida launched its new *Institute for Risk Management & Insurance Education* in Daytona Beach, Florida in fall 2024. An inaugural MBA program will focus on risk management and development of pragmatic actuarial rates.

1.7 The Internet, Disinformation, and Pseudo Experts

Without doubt, the *Internet* is one of humanity's greatest achievements in providing more people access to information and each other than ever before. But it has a dark side. Built on open-source technologies and browsers that allow anyone to view, modify, and distribute information, we must contend with influencers and manipulators who falsify, mislead, and spread disinformation.

Further, the Internet of Things (IoT)[4] and artificial intelligence will present new ways of innovating and producing, and digital mishaps will be a new challenge for forensic engineers. Sensors are in more of our technology, infrastructure, and machines. Software is navigating our planes, drones, and rockets, and robots are coming. Was the failure physical, digital, robotic, or human error? With IoT and automation, does fewer human touches necessarily mean improved efficiency and functionality, or a whole new set of interactions we must make sense of with problems and failures?

Most troubling of all is what Tom Nichols calls *The Death of Expertise,* as a *Campaign against Established Knowledge.*[7] He makes an excellent point that mobile devices and quick and easy access to misinformation can lead to the rise of *pseudo experts,* people who think they know but do not. The forensic engineer must remain vigilant in upholding *first principles of science* and the ability to sort fact from fiction with investigations, studies, and assessments.

> The Internet lets a billion flowers bloom, and most of them stink.[7]
>
> **Tom Nichols, Professor of National Security Affairs,**
> *US Naval War College*

Notes

1. Evidence standards are discussed in more detail in Chapter 4, Jurisprudence.
2. FMEA is a procedural way to identify and address potential problems and their resulting effects on a system or process before an adverse event or failure occurs. Fault Tree Analysis (FTA) is usually a graphical model of how a problem moves through a system or process toward failure. FTA may involve statistics or mathematical algorithms to predict interventions, repairs, and replacements.
3. Evidence consilience is the principle that independent lines of information converge to support a cause hypothesis with reasonable correlation.
4. The Internet of Things (IoT) refers to the network of physical objects – "things" – that are embedded with sensors, software, and other technologies for the

purpose of connecting and exchanging data with other devices and systems over the Internet. These devices range from ordinary household objects to sophisticated industrial tools and machines. With approximately 7 billion connected IoT devices today, experts are expecting this number to grow to 22 billion by 2025. In this hyperconnected world, digital systems can record, monitor, and adjust each interaction between connected things.

References

1. Peter Garnsey (1987). *The Roman Empire: Economy, Society, and Culture.* University of California Press. 328 pp. ISBN 978-0-520-06067-8.
2. William J. Tilstone, Kathleen A. Savage, and Leigh A. Clark (2006). *Forensic Science: An Encyclopedia of History, Methods, and Techniques.* ABC-Clio, 307 pp. ISBN 978-1-57607-194-6.
3. Connie Fletcher (2006-07-25). *Every Contact Leaves a Trace: Crime Scene Experts Talk About Their Work from Discovery Through Verdict.* Macmillan, 386 pp. ISBN 978-0-312-34037-7.
4. Randall K. Noon (2001). Forensic Engineering Investigation. CRC Press, 447 pp. ISBN 0-8493-0911-5.
5. Dennis M. Filler and Rorik Peterson (2013). Cryogenic freezeback of uncontrolled artesian wells in permafrost. *J. Cold Reg. Eng.* 27(2):55–66.
6. Dennis M. Filler and Rorik Peterson (2011). Stopping runaway wells in permafrost: the cryogenic freezeback method. *Proc. IUGG 2011: Cold Regions Hydrology in a Changing Climate,* Melbourne, Australia, July 2011: 153–163 (IAHS Publ. 346, 2011).
7. Tom Nichols (2017). *The Death of Expertise: The Campaign against Established Knowledge and Why It Matters.* Oxford University Press, 249 pp. ISBN 978-0-190-46941-2.

Case Study 1: The Icehouse Affair

Freezeback of a Runaway Artesian Well

- Insurance Liability
- Property Damage
- Cryogenic Freezeback
- Hydrologic Control

A message to students:

Embrace challenge
It presents opportunity

DOI: 10.1201/9781003623311-3

Artesian Water

Artesian aquifers are nature's wellsprings of fresh water. They exist in lands around the world. As subterranean channels in limestone and other dolomite formations, and water-bearing layers sandwiched between impermeable clay and rock or frozen soil and rock, natural water under pressure greater than atmospheric is *artesian.*

We know springs to be nature's relief valves, holes eroded through confinement above that release water under positive pressure to the surface. Cold springs provide recreation and some of the best quality water money can buy in our shops and stores. Warm and hot springs soothe the chilled skier and snowboarder in mountainous geothermal regions. And wells installed into an artesian aquifer do not need a pump to provide homeowners and small communities with excellent-tasting drinking water.

But there is a downside to tapping nature's drink with an artesian well that loses control of the flow. Called a *runaway well,* damage can be great as well as the volume of pristine water lost. This case study tells the story about one such well, in a northern place where artesian waters are considered a protected natural resource.

Setting

Homeowners in a south-central Alaska permafrost valley tap into a high-quality artesian aquifer beneath frozen ground for their water supply. Well x1, the focus of this study, was a typical artesian well in the yard of a residential property in the small rural community (Fig. CS1.1). The two-story wood-frame home serviced by this well had an insured value of US $365,000. Raised four feet above ground with a post-and-pad foundation system, houses built on permafrost need a *crawl space* for air convection to prevent their heat from reaching the ground.

Rainfall and snowmelt from the surrounding hills infiltrate and percolate down through the porous frozen soils and into a schistose stratum that is underlain by bedrock (Fig. CS1.1(b)). The schistose is a highly cleavable and fractured rock of flaky micaceous mineral layers that makes for an excellent water filtration medium. From the head of the valley to its low end, six miles to the west, several hundred feet of elevation change serves to pressurize the water beneath its frozen confinement above. With the x1 well at the lower end of the valley, the pressure was high enough for water to rise above the ground surface, requiring hydrologic control.

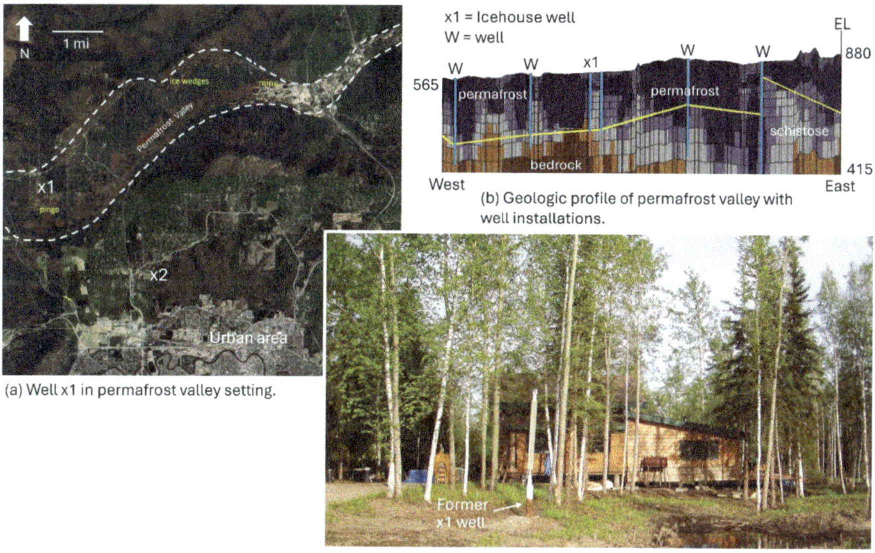

(a) Well x1 in permafrost valley setting.

(b) Geologic profile of permafrost valley with well installations.

(c) Artesian well setting post-cleanup. Note former x1 well retrofit with thermosyphon in foreground.

FIGURE CS1.1
Permafrost valley, artesian well setting.

Event Timeline

One winter morning in early December, the homeowner noticed a small pool of yellow-brown water-ice near a corner of the house. Within a week, the homeowner was on the phone with his Arizona insurance provider trying to explain the phenomenon of a *"glacier growing in the woods in front of his house."* After sending photographs and numerous conversations over weeks, the insurance adjuster called in an engineer to have a look. By then the *icing* had grown to a couple of acres in size, had encased the house in 4 feet (1.2 m) of ice, and was about to encroach the frontage road. The engineer reported the emergency and was given carte blanche approval to mitigate the problem by whatever means possible as soon as possible from the insurance company.

The event duration was seven months, from the beginning of icing through the Phase I and II emergency period, and then ice removal and structural foundation rehabilitation (Fig. CS1.2). A plan was devised to stop the runaway artesian well, save the home, prevent damage to adjacent properties, and placate the state Department of Natural Resources. Through considerable efforts by a team of oil and gas industry experts, an extraordinary fabricator and well installer, ice carvers, an expert contractor, and an engineer, a new method to stop runaway artesian wells is in the literature for future use.

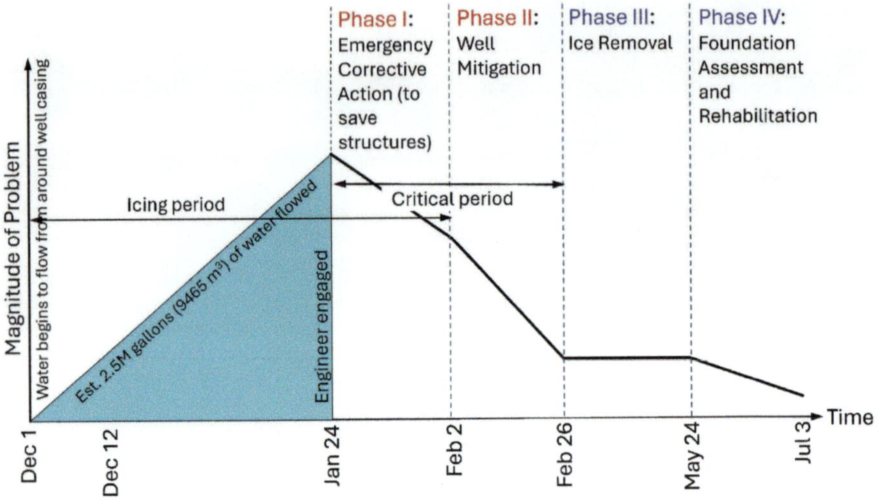

FIGURE CS1.2
Event timeline from estimated start to end of structure foundation rehabilitation.

Well Configuration

The x1 well comprised 144 feet (44 m) of 6-inch-diameter (15 cm) steel casing above a 12-foot (3.7 m) screened section, for a total length of 156 feet (48 m), the uppermost 4 feet (1.2 m) of which was the above-ground well head. Intermittent well heating to prevent *ice fouling* was accomplished with a pair of sheathed copper-wire heat tapes, one 15-foot (4.6 m) length for the upper wellhead section and the other 150-foot (46 m) for depth. Heat tapes were cycled one hour ON followed by 12 hours OFF with a timer (Fig. CS1.3(a)). In this way, the well was maintained free of ice blockages. The well was completed with a gasketed and bolted steel cap, fitted with Arctic-grade plumbing and wiring runs to the house, and protected with an insulated wellhead box, standard equipment for outside wells.

Loss of Hydrologic Control

Functional control of an artesian well in permafrost depends on proper heating. A 1-hr ON/12-hrs OFF cycle is standard. However, heating equipment (i.e., tapes, timer, and wire harness) can malfunction with age, and an interrupted cycle can render 24/7 heating of the well casing without the homeowner noticing until a high electric bill arrives in the mail. Long-term heating of a steel well casing will thaw surrounding permafrost, allowing pressurized artesian water to flow upward around the outside of the well casing (Fig. CS1.3).

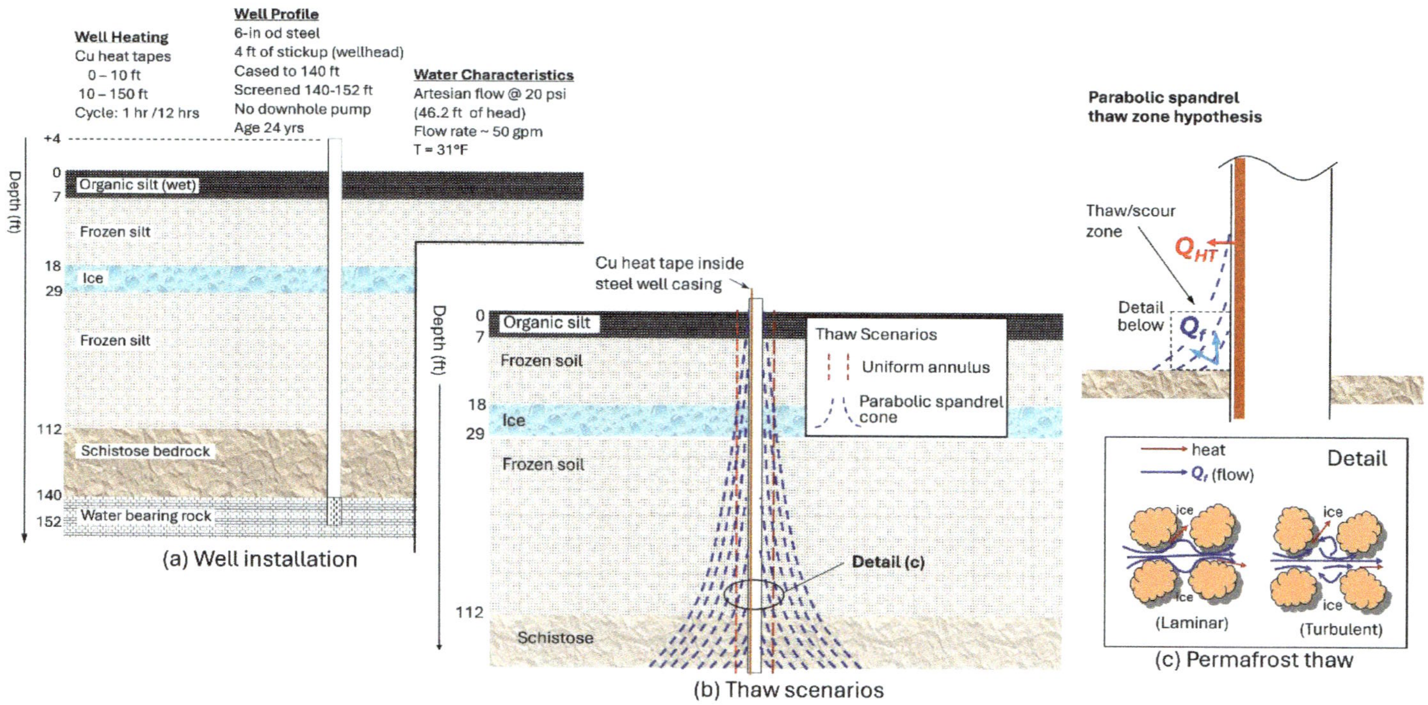

FIGURE CS1.3
(a) Well configuration, (b) thaw scenarios, and (c) the working hypothesis (parabolic radial thaw) for cryogenic freezeback design.

Two permafrost thaw scenarios were considered (Fig. CS1.3(b)). Uniform radial thaw implies a columnar annulus around the well for flow to the surface. The second is a parabolic spandrel thaw pattern that forms when well casing heat (Q_{HT}) and *sensible heat* from artesian flow (Q_f) work together to create thaw channels within soil at the schistose-silt interface (Fig. CS1.3(c)). Continued heating grows the thaw zone outward and upward within eroding permafrost until the seepage reaches the ground surface, where it then spreads to form *aufeis*[1]. Once surface spreading occurs, the well is considered *runaway* as nothing will stop the flow until the water supply runs out. To regain control, one must somehow stop the artesian flow. Prior to 2007, there were no published solutions to the problem of runaway artesian wells in permafrost in the scientific literature. Parabolic spandrel thaw, a worst-case scenario, became the working hypothesis for *cryogenic freezeback*, the method conceived for stopping the runaway x1 well.

Aufeis Icing

Aufeis, a German word that means "ice on top," refers to sheet-like layered ice that forms when spring water or groundwater seeps out of the ground and quickly freezes upon contact with subfreezing air in winter. Subsequent seepage overflows the newly frozen layer, and with successive overflows, an *icing* forms. The process continues until the pressurized water supply depletes.

What causes near-surface water to pressurize is ice blockage from soil freezing into ice layers in the ground. The pressure forces water toward the least path of resistance, usually upward through soil channels or fissures in an ice layer to reach the surface. Variations in minerals and mineral content in the water color aufeis.

Wells installed in permafrost that lose control of their artesian water when heating goes awry present a different way to create aufeis (Fig. CS1.4).

(a) House encased in aufeis

(b) Tree trunk conduit for icing growth

(c) Aufeis icing

FIGURE CS1.4
Aufeis icing from runaway artesian well x1.

The icing in frame (a) is about 4.5 feet (1.4 m) thick where it meets the wall of the house, having already engulfed the foundation system under the structure. An interesting phenomenon that occurs when aufeis encircles trees is shown in frame (b). Trees are obstacles that act to impede the freezing, slow-moving flow long enough to mound it. Successful layering forms cones around the tree trunks. Frame (c) depicts aufeis layering in an intercept trench cut around the house by ice carvers to save the structure.

It is noteworthy to remind that when water freezes, its volume increases by about nine percent. As aufeis engulfed post-and-pad foundations under the home, expansion lifted the structure off of its foundation system. The house was literally *riding on the icing* (Fig. CS1.4(a)). Intercept trenches interrupted aufeis progression for a time, long enough to stop the runway well and icing.

Cryogenic Freezeback Plan

Cryogenic freezeback uses liquid nitrogen (LN2) at –321°F (–196°C) to freeze off well flow and refreeze thawed soil beyond the well, thereby restoring permafrost. The well must be opened and prepared for LN2 injection at sufficient depth to create a massive ice plug inside the well and refreeze soil outside of the well casing up to the surface. The refreezing of water inside the well and soil beyond is a heat transfer problem with a mathematical solution that determines the LN2 quantity needed and the duration of its discharge into the well.

Freezeback design is a function of well, permafrost, and aquifer characteristics, an understanding of well heating and failure modes, and LN2 requirements based on heat transfer computations. The study considered

- Seepage analysis for plausible thaw scenarios based on well and subsurface characteristics.
- Thermal analysis to determine LN2 requirements for freezeback.
- LN2 evaporator design parameters.
- Well preparation for LN2 injection and thermosyphon retrofit.
- Development of a safe freezeback plan for the work area.
- Well decommission and conversion to a heat exchanger.
- Thermal monitoring inside and outside the former well over several years to ensure long-lasting freezeback.

Thermal analysis to determine LN2 requirements for freezeback had to be based on assumptions about permafrost soil characteristics for heat and mass transfer theory to quantify the LN2 requirement. Therefore, a high factor of safety for freezeback design was needed.

Thermal Analysis

An energy balance approach to cryogenic freezeback considered soil and ice properties for the upper 40 feet (12.2 m) of ground and well radius in estimating the LN2 requirement. The phase change interface between thawed and frozen soil *balances* the rate of heat conduction to the interface. This balance, expressed as the ratio of a temperature driving force, ΔT, and a thermal resistance, R_{th}, in cylindrical coordinates, is

$$q = \frac{\Delta T}{R_{th}} = \frac{T_{in} - T_{out}}{\frac{\ln(r / R_{in})}{2\pi k}} = 2\pi r L \frac{dr}{dt} \qquad [1]$$

With equation [1], T_{in} and T_{out} are the respective temperatures inside and outside of the well casing, r *is* the thaw radius from the center of well, R_{in} is the well inside radius, k is the soil thermal conductivity, and L is the volumetric latent heat of freezing. The term on the far right is the rate of phase change at the well-soil interface. *Sensible heat* effects can be neglected in this balance because the *Stefan number*[2] for freezing soil is small[1].

Equation [1] can be rearranged and integrated as an expression for the radial location of the freeze-thaw interface, R_{out}, as a function of time, t.

$$\int (T_{out} - T_{in})\, dt = I_f = \frac{L}{k}\left(\frac{R_{out}^2}{2} \ln \frac{R_{out}}{R_{in}} - \frac{R_{out}^2 - R_{in}^2}{4} \right) \qquad [2]$$

The time integral on the left-hand side of Equation [2] is often called the *freezing index*, I_f, because both the temperature difference and time have the same linear effect on the degree of freeze or thaw in these types of problems[2].

Equation [2] can be used to perform a conservative analysis for freezeback design. It was reasonable to assume a constant heat tape temperature when turned on for the inner boundary, and that of phase change at 32°F (0°C). Recall that frozen silt and massive ice define the confining stratum above schistose (Fig. CS1.3(a)). Latent heat values for the silt and ice were 4025 Btu/ft^3 (150,000 kJ/m^3) and 8937 Btu/ft^3 (333,000 kJ/m^3), respectively. The thaw radii around well x1 as a function of the freezing index are plotted in Figure CS1.5(a). From the plot, we see that the longer thawing occurs, the greater is the thaw radius from the center of the well, and the larger the thaw radius, the warmer is the seepage water.

Looking again at Figure CS1.3(b), soil nearest the ground surface was last to thaw and the diameter of seepage influence was the least for both thaw scenarios. Wellhead exposure confirmed this assertion, and the estimated water seepage rate was measured at $\sim 15\ gpm\ (3.4\ m^3/hr)$, and soil erosion was minimal. For conservative freezeback design, a freezing index based on

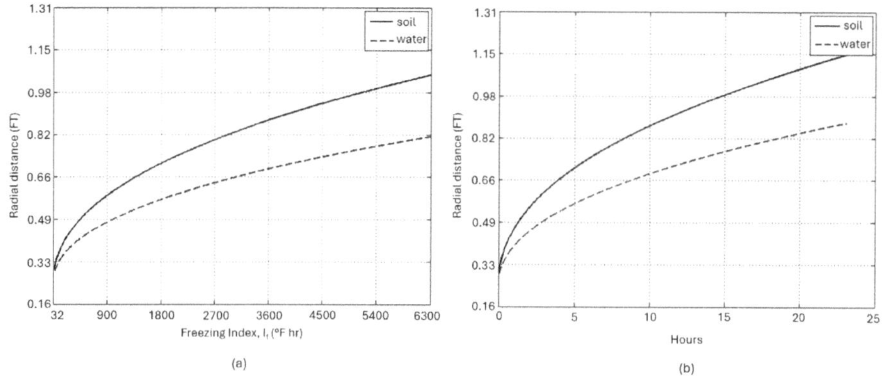

FIGURE CS1.5
Freezeback analysis: (a) Well thaw radius as a function of freezing index for silt and water/ice, and (b) freezeback radius as a function of LN2 injection duration.

36.5°F (2.5°C) water flowing for eight weeks gives $I_f = 6000°F{\cdot}hr$ $(3315\ °\mathrm{C}\cdot hr)$, for a minimum radius of freezeback of about 10.63 in (27 cm).

LN2 would first freeze off the water inside the well and in the annulus between the LN2 evaporator and casing, and then the soil beyond the well. Consequently, LN2 requirements were then computed based on heat removal rates for water and silt. For flowing water, the heat removal rate (q_w) is quantified as

$$q_w = \dot{m}_w c_p \Delta T \qquad [3]$$

Parameters in Equation [3] are: $\dot{m}_w$ is the mass flow rate of water, c_p the specific heat of water (1 $BTU/lb°F$ or 4.184 J/g°C), and ΔT the difference between the seepage water temperature and that of freezing, or 4.5°F (2.5°C).

The heat removal rate for flowing water was computed as $35{,}830\ Btu\,/\,hr\,(10.5\ kW)$. The minimum volume of soil required to be refrozen between the well casing and the maximum radius of thaw (or from 3 to 12.5 inches (7.6–30 cm) beyond the well), for a height (H) of 40 ft (12 m), was 128.5 cf (3.64 m^3). The energy required to refreeze this soil volume was calculated as

$$E = \pi\left(R_{out}^2 - R_{in}^2\right)HL = 502{,}343\ Btu\ (147\ kWh) \qquad [4]$$

for an estimated rate of $20{,}930\ Btu/hr$ for 24 hours. The heat removed by LN2 is expressed as

$$q_s = \dot{m}_{LN2} \cdot h_{vap(N)} \qquad [5]$$

Nitrogen's heat of vaporization is $\sim 86\ Btu\,/\,lb\left(200\,\frac{kJ}{kg}\right)$ and at an LN2 mass injection rate of $\dot{m}_{LN2} = 2.14\ gpm\ \left(0.49\ m^3/hr\right)$, the calculated rate of heat removal

FIGURE CS1.6
Well x1 preparation and LN2 freezeback.

for ground freezing was $68,242\ Btu/hr$ ($20\ kW$). With a factor of safety for freezeback design of 4, approximately $23,000\ lbs$ ($10.430\ kg$) of LN2 was required to achieve the desired freezeback radius. This conservative design would account for variable soil conditions and thaw profile uncertainties. Figure CS1.6 photographs sequence well preparation and the freezeback process.

Implementation and Well Conversion

A 40-foot-long (12.2 m) LN2 evaporator tube with welded *ears* was slipped through a rubber donut gasket before insertion into the well. Once installed, the tube's *ears* rested on the gasket to form a partial seal at the top of the well (Fig. CS1.6 (images 4 and 5)). The sheer weight of the assembly significantly reduced the artesian flow sufficiently for LN2 discharge. A 1-inch-diameter (2.54 cm) copper line was inserted down to the bottom of the evaporator and connected to the LN2 tanker truck supply. An expert operator discharged LN2 at the controlled design rate for rapid freezing of the well water in nine minutes, followed by 24 hours of permafrost freezeback. After freezeback, the well was then retrofitted with a *thermosyphon* for continued liberation of ground heat as a passive heat exchanger to ensure lasting freezeback.

Other Challenges

The emergency ended with well freezeback, which stopped icing growth. In one location, ice thickness reached 11.5 feet (3.5 m). However, ice abatement became the next challenge. How to thaw the ice under the home and reestablish the foundation without supersaturating the ground and damaging the structure required the expertise of one savvy contractor. A skirted crawl space, torpedo heaters, careful monitoring and measuring, and intermittent removal of cut-away ice chunks allowed the structure to slowly descend onto its 34 foundations. Residual ice melted away and air blowing eventually dried out the crawl space before the foundation system could be assessed. Eighteen of 34 post-and-pad foundations required rehabilitation (Fig. CS1.7). The photographs in this figure show examples of ice separation of foundation posts from pads, collapsed foundations, and ice-saturated insulation in the flooring above.

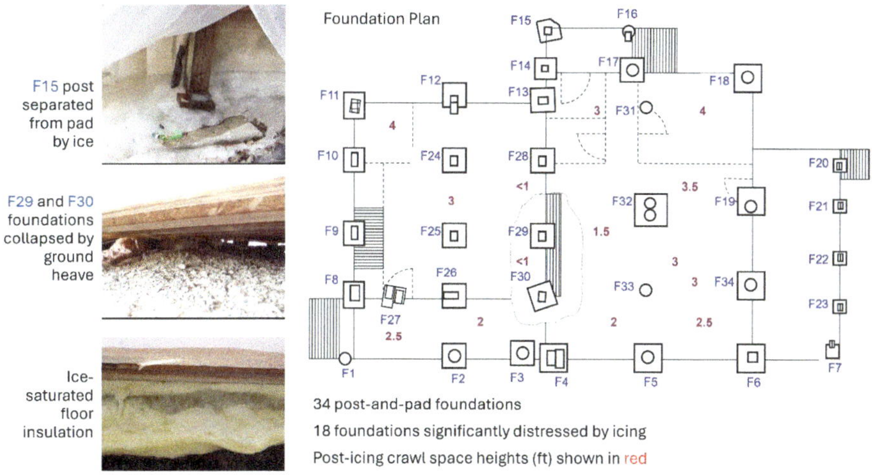

FIGURE CS1.7
Foundation distress caused by icing.

FIGURE CS1.8
Artesian well in the garage destroys a home.

Foundation rehabilitation required the services of an expert who specialized in jacking, moving, and rehabilitating foundations of large structures. The North Country has some exceptionally talented people who enjoy working on unique challenges.

Anecdote

Two years after the well x1 incident, artesian well x2 (Fig. CS1.1(a)) destroyed a home in a nearby valley. The resident was traveling abroad when the well heating system failed. Figure CS1.8 shows what a couple of weeks of 24/7 well heating can lead to with an interior well. It is not a good idea to install an artesian well inside a building.

Notes

1. Aufeis is a sheet-like mass of layered ice that forms from successive flows of upwelled groundwater during freezing temperatures. The overflow (or icing) continues to grow with an unabated supply of water through winter. Large icings can appear as glacial formations with interesting, mottled surface patterns.

2. Sensible heat is the heat that raises or lowers an object's temperature. For example, when water is heated or cooled, the increase in temperature is its sensible heat. The Stefan number is a dimensionless parameter that measures the duration of the solidification process. Defined as the ratio of sensible heat to latent heat, it quantifies as $e = c_p \Delta T / L$, where c_p is the specific heat of solid phase in the freezing process in this instance. The smaller is the Stefan number, the less significant is a soil's sensible heat with freezing.

References

1. Filler, D. M. and R. Peterson (2013). Cryogenic freezeback of uncontrolled artesian wells in permafrost. *J. Cold Reg. Eng.* 27(2):55–66.
2. Andersland, O. B. and B. Ladanyi (2003). *Frozen Ground Engineering* (Second Ed.). John Wiley & Sons, 384 pp. ISBN: 978-0-471-61549-1.

2

Investigation and Analysis

CONTENTS

CONCEPTS	KEY PHRASES	ADDITIONAL READING
Engineering and Science Relationship	Applied science	Exercise 3
First Principles of Science	First principles	Exercise 4
Consilience of Evidence	Consilience of evidence	
Modern Scientific Methods of Inquiry	Working hypothesis	Case Study 2
Inferences	Abductive, inductive, and deductive reasoning	
	Notorious drive	
	A priori problem	
	Perception of impropriety	

2.1 Engineering, Science, and First Principles

A student asked her professor: "Is engineering a science?" What an opportunity to engage! The perfect opener to relate science and engineering, and to discuss Forensic Engineering as the most relatable engineering discipline to science.

Strictly speaking, engineering is not a science. It is an *applied science* because of its systematic application of scientific principles. Science formulates hypotheses (theories or presumptions) and uses observations and experimental outcomes to describe natural phenomena. When repeated experiments produce similar outcomes, we can conclude a scientific fact. Engineers, on the other hand, create practical solutions to real problems using scientific principles. Therefore, we can say the two are closely related but differ in their methods. However, with FE, engineers learn to think like a scientist and apply *first principles of science* to get at the facts of what happened.

 DOI: 10.1201/9781003623311-4

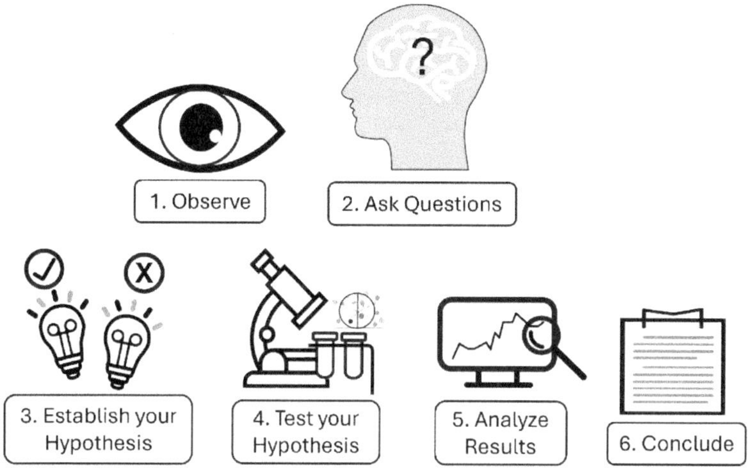

FIGURE 2.1
First principles of science.

Failure analysis is a kind of reverse engineering; reassembling the parts or reconstructing the rubble to see what fits and what no longer does. Working backward from broken to reassembled, with particular attention to distressed or deformed parts or members, the forensic engineer zeroes in on cause. Identifying the most obvious truths or facts about a failure, resolving the problem into its basic elements, and determining cause and effect relationships is applying ***first principles of science*** (Fig. 2.1). And integral to first principles is ***consilience of evidence***, the idea that independent sources of information converge to support a strong conclusion.

Reading *National Geographic* as a kid, maps, archaeology, and dinosaurs grabbed my attention. One particularly interesting series was about human migration and ancient civilizations. Anthropology was a new word, and I learned about how anthropologists go about deciding how different people groups lived. They considered historical and archaeological records, physical, environmental, and resource (food and water) influences, and of course safety was a concern. Information from different fields of study was considered before the anthropologist decided about culture (Fig. 2.2).

Fast forward to today, and we see forensic engineers doing something similar with failure investigations, drawing from at least three different sources of information to determine a cause. In both instances, the information had to be consistent, relatable, and converge toward reasonable inferences or conclusions. A coming together of knowledge is the essence of consilience.

Is there a unified theory of knowledge in disciplines that range from physics to biology, genetics, engineering, humanities, and the arts? In his book *Consilience: The Unity of Knowledge*, Edward Wilson proposes that principles

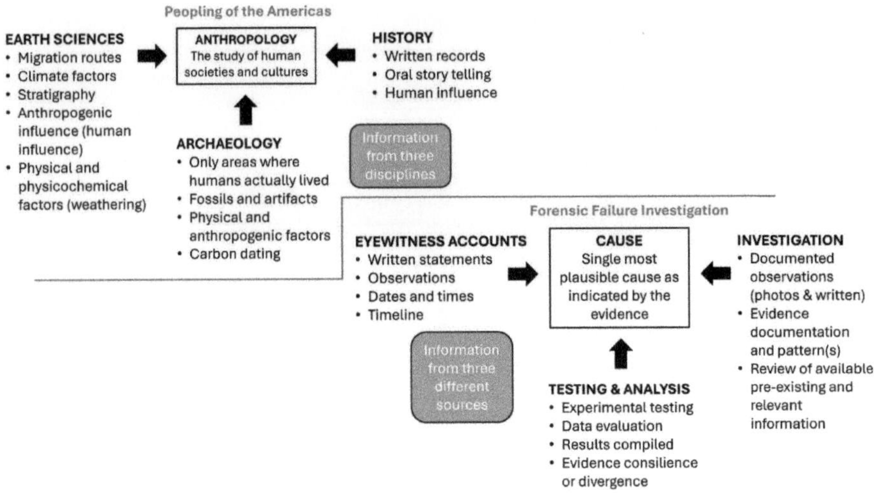

FIGURE 2.2
The principle of consilience applies to the anthropologist and forensic engineer alike.

from different disciplines must be *consilience-relatable* to define a comprehensive theory[1]. The laws of math and physics govern engineering practice, and with forensic engineering, multiple independent lines of evidence must converge for a definitive cause to emerge. I think the Pulitzer Prize-winning author and *Father of Biodiversity* is talking with Einstein and Newton about something profound!

> Consilience:
> Its central tenet, as Einstein knew, is the unification of knowledge.[1]
>
> **Edward O. Wilson**
> *Philosopher, Author, and Harvard Professor Emeritus*

> There is no one scientific method.
> We have more than one way to make knowledge.
> But there is a set of core principles that are fundamental to all Scientific Methods.

2.2 Modern Scientific Methods of Inquiry

Development of modern scientific methods has a rich history that spans thousands of years. Shaped by numerous philosophers, polymaths, and scientists across different cultures and time periods, we have a cornerstone of modern science, engineering, and forensic practices. From the ancient Greeks to the Islamic Golden Age and through the Scientific Revolution to modern perspectives, the story is best told with a timeline of the scientific method (Fig. 2.3). The story tells of a journey through scientific inquiry and discovery to get to ***modern scientific methods*** (emphasis on method<u>s</u> plural) we apply today. But at the core of them all is a set of immutable principles: *first principles of science* (or simply ***First Principles***).

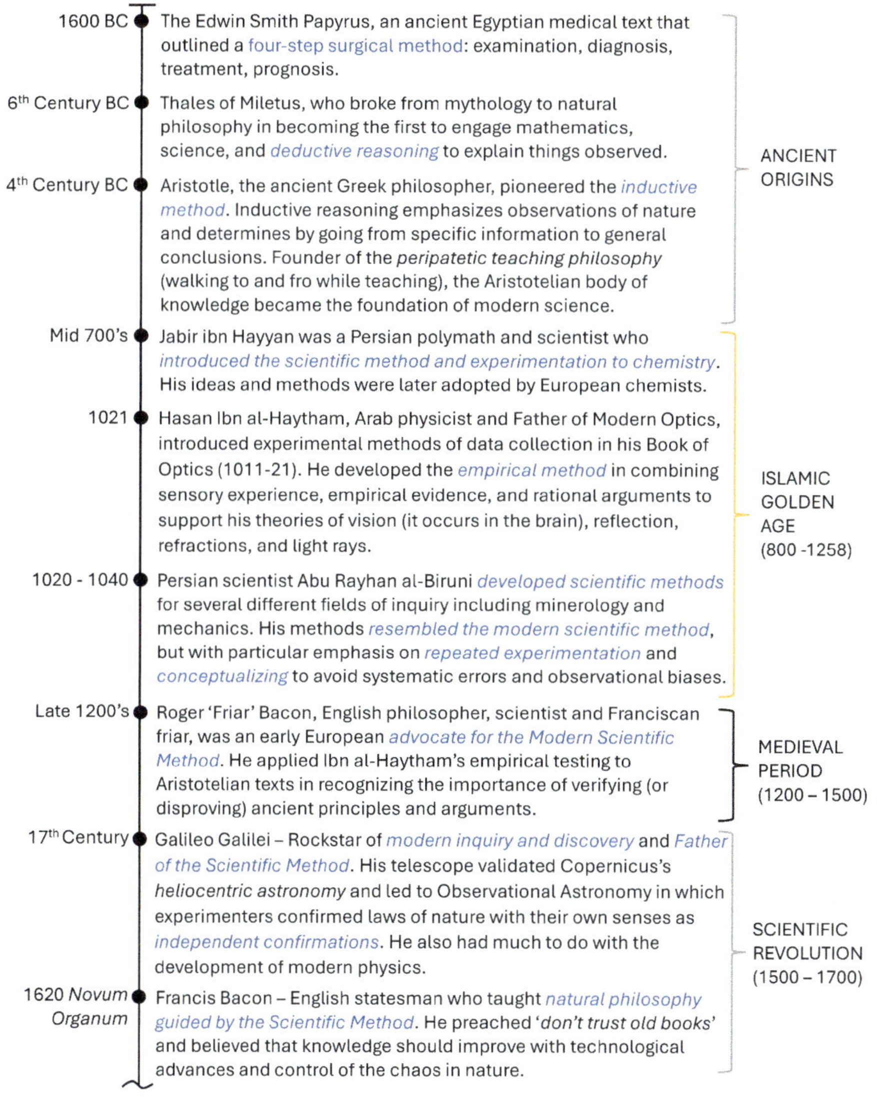

FIGURE 2.3
Timeline of the development of the scientific method.[2],[3],[4],[5],[6],[7] *(Continued)*

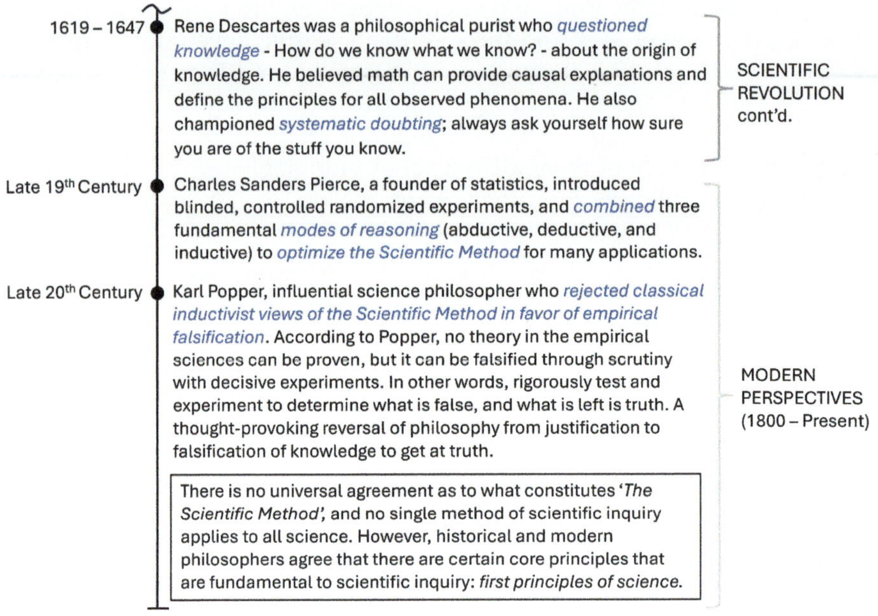

FIGURE 2.3 *(Continued)*

POP GOES THE CEILING

Servers and patrons in a golf country club restaurant periodically heard loud pops coming from above. A maintenance supervisor inspected the roof exterior first with a ladder, then a manlift for higher viewing; nothing out of the ordinary was observed. Over the next several months, the infrequent pops continued, and a drone survey and peek into the attic detected no issues with the roof system. An engineer was eventually called in to investigate the mystery.

The engineer first reviewed the drone videography which did not evidence depressions or missing Spanish barrel tiles. Next came an inspection of the attic space above the dining room. Initial observations discovered a couple of trusses with broken members; one had splintered. Deeper into the network were trusses with broken diagonals and a couple with sheared top or bottom chords. But most alarming was the sinuous alignment of a deformed 6-ply girder truss designed for supporting a heavy load. This bolted truss was 12 inches thick! The engineer hypothesized excessive roof load as the source of the problem.

Golf course country club

Spanish barrel clay tiles
Wt. 8 – 15 psf
Barrel vault tiles
Wt. 1.3 – 2 psf

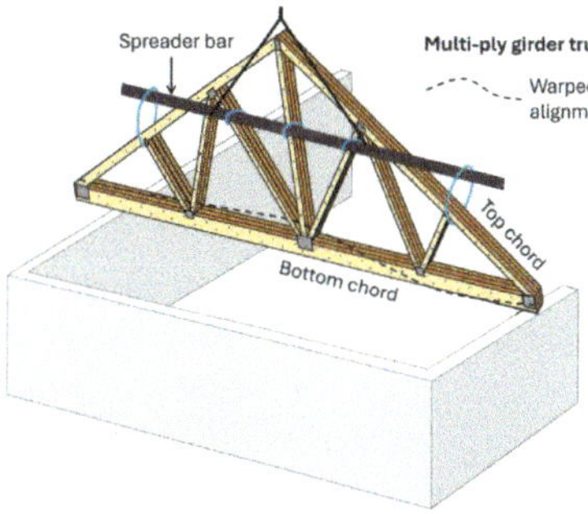

Multi-ply girders (MPGs) are trusses specifically designed to carry extra loads from structural framing members they support. MPGs are identical trusses fastened together to act as a single heavy-duty unit. ANSI/TPI 1 stipulates that up to three-ply thick girders can be fastened together with nails, girders more than three ply must be pre-drilled and bolted, and the maximum number of plies is five for girders that are supporting the extra load from only one side, and six if the load is distributed into both sides.

ANSI is American National Standards Institute
TPI is Truss Plate Institute

Review of the building plans and specs revealed two noteworthy details:

1. A change in roof tiles had been made. The Spanish barrel tiles on the roof weighed six to seven times more than the barrel vault tiles called for in the design documents.
2. Most stunning of all was a name on the golf country club design plans. An engineering disaster that occurred a decade before in a Midwestern city came to mind. We recall from the libraries of our minds.

ANECDOTE: KNOWLEDGE FINDS A WAY

Science and the scientific method advanced through medieval times and the Scientific Revolution despite state- and church-sponsored inquisitions to combat heresy, apostasy, blasphemy, witchcraft, and other *dangers*. Galileo, the Bacons, and Descartes had to repent and renounce or suffer consequences. As renowned and respected philosophers, they were spared execution in favor of periods of house arrest and restricted from teaching.

Giordano Bruno was the last person burned alive at the stake on orders from Rome in 1600.[5] A philosopher, alchemist, astrologer, and cosmologist, apparently his beliefs in *Copernicanism* (the Sun, not the Earth, is the center of the universe) and an infinite universe with innumerable inhabited worlds were more than the Catholic church could bear.

From Edwin Smith's ancient Egyptian papyrus to *Occam's razor*[1] (the simplest explanation is usually the best) of medieval times, and through reasoning philosophies (i.e., abductive, inductive, and deductive) and empirical methods of the Scientific Revolution, we see the essence of First Principles.

Abductive, inductive, and deductive reasoning are types of *inference* that differ in how they start and what they produce. Deductive reasoning uses general truths to reach specific conclusions. Inductive reason uses specific information to reach general conclusions. Abductive reasoning uses specific observations to reach a probable conclusion about cause and effect. An abductive conclusion is not a final truth, but a working hypothesis open to revision. For example, if we know that mineral soil comes from rocks, and clay is found in the ground, you can deductively conclude that clay also comes from rock. If you observe an alpine turquoise lake and know what glaciers do to underlying rock, you can use inductive reasoning to conclude that rock flour from the mountain above is responsible for the beautiful color of the lake. However, for situations with uncertainty or more than one possible cause, we need abductive reasoning as the logical process that lets observations lead to a working hypothesis that best explains the facts. With forensic investigations, abductive reasoning is more flexible and adaptable and lets the facts lead to conclusions.

With the papyrus, replace the words of the four-step surgical method with *observation, hypothesis, test,* and *conclude* and we have a four-step failure investigation method. Refine it with a healthy combination of Aristotelian science and Newtonian physics, Ibn al-Haytham's empiricism and Peirce's abductive reasoning, add a pinch of Galilean independent confirmation, a dash of Descartes doubting, and we have the making of an investigation that holds up against rigorous Popper falsification scrutiny. Or students could simply read about the modern approach to engineering investigation in the next section.

2.3 Forensic Investigation Process

If we take Figure 2.1 and apply *first principles* to a product liability scenario in which a machine failure injures the operator (Fig. 2.4a), we get an example of how the forensic engineer goes about investigating the incident as an exercise in *logic flow* (Fig. 2.4b). There are three possibilities for how this accident occurred:

1. Machine failure
2. Operator error
3. Intentional fraud (workers' compensation)

Starting with observations that lead to questions that lead to a working hypothesis, the engineer then goes about examination of the evidence and

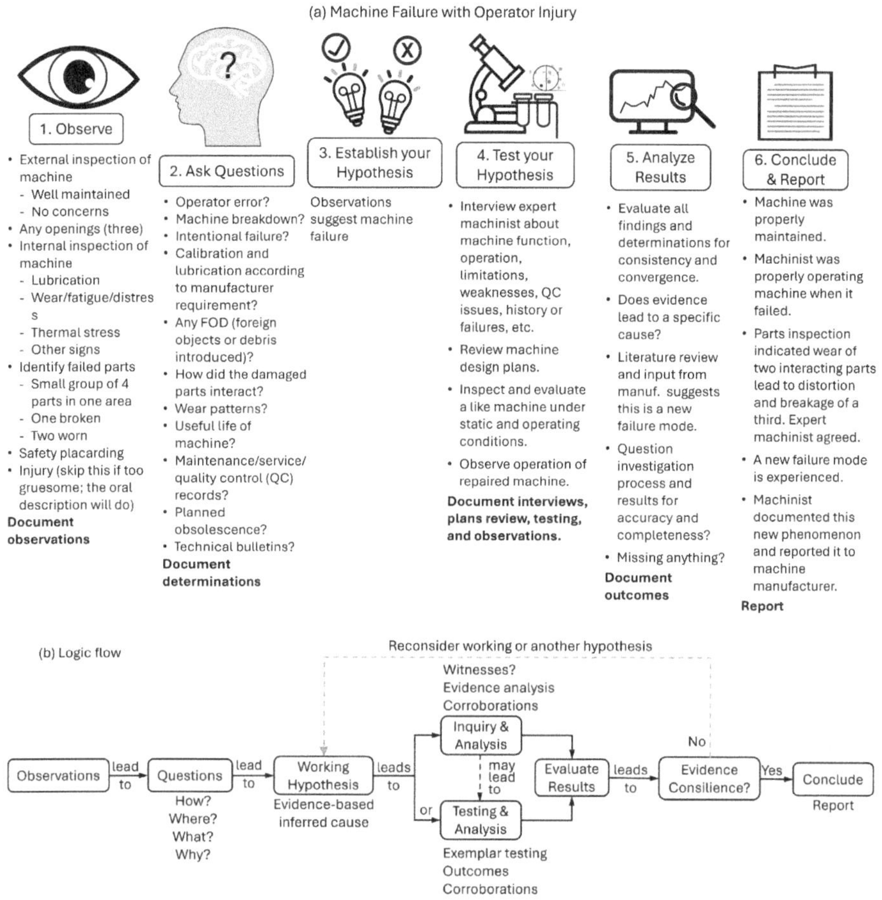

FIGURE 2.4
(a) First Principles applied to a product liability scenario, and (b) investigation logic flow.

may test an exemplar[2] machine to validate the hypothesis before concluding. The hope is that evaluation of all evidence and test results leads to consilience. At least three independent lines of converging evidence are necessary for concrete determinations with a high degree of confidence in the investigation process and outcomes. Students should thoroughly study Figure 2.4 to recognize the potential this logic flow process has for wide application to many failure scenarios.

A forensic investigation should be a truly independent exercise performed without assistance from anyone involved in the accident. Other than for initial interviews of key people directly involved with the mishap, the investigator works independently, is objective, and remains unbiased through the course of inquiry.

2.4 Example Applications

Let us apply first principles and logic flow to different scenarios to illustrate variants of the scientific method. First, a product liability scenario that involves exemplar testing of a commercial product to test performance claims. Second, an accident reconstruction (AR) scenario that involves a *notorious drive* where a truck backed over a pedestrian, causing serious injury. Lastly, an engineering disaster involving negligence that all engineering students should study.

Failed Elastomeric Roof Membrane (a Product Liability Scenario)

Elastomeric roofing membranes gained popularity in the United States in the 1970s during the Middle East oil embargo, which drove up the price of asphalt-based roofs and lowered the quality of available asphalt. This economic shift advanced elastomeric membranes as attractive alternatives to conventional built-up roofs. Over the past couple of decades, manufacturers have sought to simplify application procedures and improve roof-membrane performance with products that balance flexibility with durability. However, new products have their growing pains. Poor performances, application issues, underdeveloped technology, market limitations, and questionable marketing practices have tarnished the reputation of the elastomeric roofing industry.

For instance, an ethylene propylene diene monomer (EPDM) synthetic rubber product failed within a year of installation on a flat roof of a commercial building (Fig. 2.5). The client had researched and compared several products

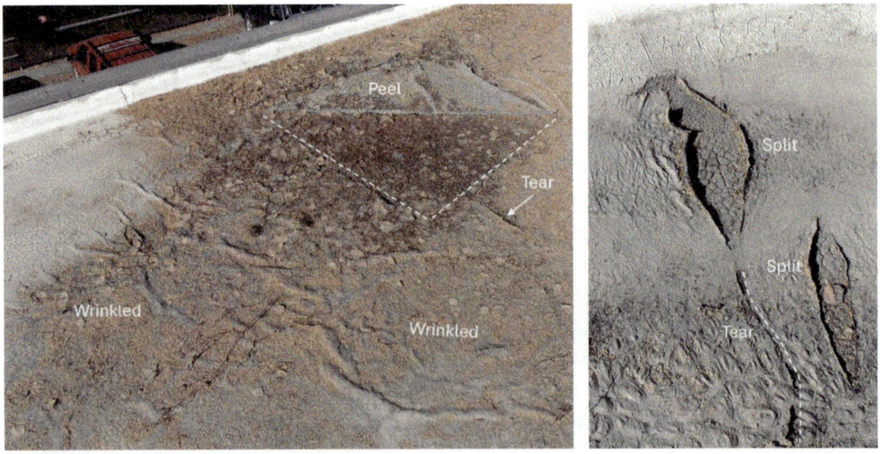

FIGURE 2.5
Failed elastomeric roof membrane.

before calling a distributor to discuss a chosen product and its viability with the building's weathered asphalt-based built-up roof. An experienced representative touted the merits of the EPDM product: durable, flexible, and able to withstand extreme temperatures, UV radiation, and weathering. In an email sent to the client, the rep stated, *"Our product is ideal for sealing old built-up roofs. It maintains good elastic properties within your [design] temperatures of -20°C to 35°C and can last over 50 years. Four 5-gallon canisters are all you will need to provide a 60-mil*[3] *waterproof cover for your roof."*

A building maintenance supervisor purchased the product and hired the contractor who installed it in accordance with manufacturer specifications. On each canister were step-by-step procedures with illustrations of roof preparation and elastomeric application during dry weather warmer than 10°C. The maintenance supervisor documented the roof pressure washing, drying, and air blowing (preparation), and membrane application process over a couple of days in early May. Photographs were taken of the newly installed light-gray roof cover, and again with subsequent roof inspections in August and November.

Then came March of the next year, and obvious signs of a failing roof cover. Numerous tears and splits were observed across the roof. Most significant were areas of wrinkled, split, and peeled-back membrane (Fig. 2.5) along the roof perimeters where water had periodically ponded. Water had also intruded the roof system. Distresses were documented, a claim was filed with the insurance carrier, and the carrier hired a forensic engineer to investigate.

A template of an investigation approach that involves materials testing for product liability cases is presented in Figure 2.6. We see application of first principles (Fig. 2.1), but without need for questioning. When a material fails, why and how did it fail are de facto questions. Also, sampling is a prerequisite to establishing a working hypothesis. This is because a material used with infrastructure has undergone rigorous standardized testing that produces its specific properties (i.e., specifications) that relate to its strength, expansion and contraction (flexibility), durability, temperature, permeability, etc. The forensic engineer knows that observations made through sampling and collection can narrow the focus to failure-related properties. Sampling is also a good idea for preserving actual conditions should the incident become a legal matter.

Steps 5 and 6 are often performed by a third party. It is routine to inquire of the manufacturer about its product. The engineer will want to see all relevant product information such as Material Safety Data Sheet (MSDS),[4] property specifications, test data, and literature posted on the manufacturer's website. In this instance, EPDM temperature rating and adherence properties are suspect. The engineer will look for information consistency, discrepancies, and unsupported grandiose claims.

Request for information (RFI) often requires the services of an attorney who can use subpoena power to get information not willingly disclosed by a

 1. Observe	• Distress patterns? • Distress types • Photo-document (macro & micro images) • Questions: Why? How ?
 2. Literature Review	• Client-distributor communications • Manufacturer product literature • Manufacturer website • Product reviews • RCMA (Roof Coatings Manufacturers Assoc.)
 3. Sampling	• Random sampling: pristine locations; failed locations • Membrane/substrate composites • Photo-document and log all samples • Chain of custody if transferring samples to 3rd party testing lab
 4. Establish your Hypothesis	• Performance Evaluation • Membrane cannot endure subfreezing temperatures • Membrane does not adhere well to weathered asphaltic substrate
 5. Request Information	• Request all product temp./adherence/ • permeance test data • Request MSDS • (Will likely need attorney to subpoena manufacturer for this info.)
 6. Lab Testing (test hypothesis)	• Exemplar canister • Prepare membranes (3) • Test membranes according to ASTM* membrane specific, property specific parameters. • **Let manufacturer do this at their expense if they engage a forensic expert.** *ASTM is American Society for Testing and Materials
 7 Analyze Results	• Analyze test results • Analyze defense expert report of test results (if manufacturer engaged testing)
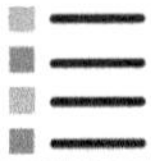 8. Findings	• List findings from observations, literature review, sampling and testing, and test results. • Categorize and assess consilience of evidence.
 9. Conclude & Report	• Conclude only what is necessary to support hypothesis • Conclude only facts • Issue report

FIGURE 2.6
Investigation approach for product liability with exemplar testing.

product manufacturer. This implies legal engagement at step 5. Once a subpoena is served, the manufacturer must provide the requested testing data or, if they cannot produce it, retain an expert of its own and test at their own expense. This is a good thing. When possible, let the manufacturer pay for the necessary testing to support their product claims.

We are back to first principles with steps 7 through 9. In step 8, it is important to organize results and determinations as independent lines of evidence and then evaluate for evidence consilience. The forensic engineer always hopes for at least three independent lines of converging evidence.

So, what have we learned with this product liability variant of the scientific method?

- It is applicable to material failures.
- It likely involves insurance.
- It has a sampling and testing component.
- It includes RFI and opportunity for exemplar testing.
- It affords multiple independent sources of evidence.
- It often requires the services of an attorney experienced in product liability law.
- A defense expert is likely to be involved.

Accident Reconstruction

A woman was walking on a trail talking to her father on her cell phone when she was backed over by a one-ton truck. The truck driver, having heard a scream over the din of radio and children's chit-chat in the back row, stopped the truck over the woman on the ground just ahead of the front tires. Emergency 911 was called, neighbors rushed to assist, and the ambulance quickly arrived. The woman was medevacked to a regional hospital, and the driver filed an insurance claim. Over the next couple of years, there were numerous surgeries, rehabilitations, and more than $1M in medical expenses, but the woman recovered to walk the trails again. However, her speech had slowed and her personality was different. The head trauma caused by the truck differential had taken its toll. *A notorious drive* regulators called it, the old dirt trail through the woods that had become a drive between two recently subdivided and developed properties.

The location of this horrific incident is denoted with the red *'x marks the spot'* circle with POI (point of impact) label in Figure 2.7. An aerial photograph superimposed on a county plat gives perspective as to what constitutes a notorious drive. The former trail, which became a drive, was coincident with a common property line and utility easement and joined a public access easement (the blue shaded flag stem[5]) that was surveyed and platted as the

FIGURE 2.7
Location of a truck-on-pedestrian collision on a dirt drive between two roads in a rural community where everybody walks their dogs.

common permitted driveway for the two properties shown above and below the drive in the figure. Further, the drive emanated from Bush Lane (road at right in the aerial), along a common property line between two additional properties. Four property owners had legal access to the drive, and many more dog walkers, bicyclists, and vehicles had traveled the drive between the two roads over the years.

At the core of this conflict are two former neighbors, now litigants, responsibility, a very large medical debt, and pending legal expenses. The truck litigant (defendant) sold everything and moved the family to another state far away within six months of the accident. The partner of the victim (plaintiff) was an honorable man, steadfast with care and love for his injured spouse. Bankruptcy and uncertain futures hung in the balance. The plaintiff's lawyer sought an expert to help sort the mess out, and a forensic investigation followed.

In comparing the AR investigation approach (Fig. 2.8) with that for the product liability scenario (Fig. 2.6), we see similarities, yet in a different

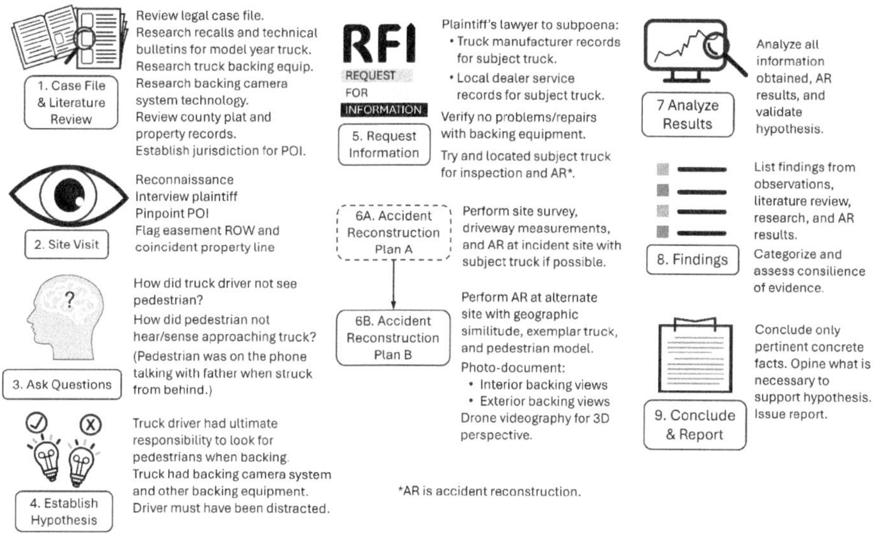

FIGURE 2.8
Investigation approach for accident reconstruction with exemplar testing.

order, with the tasks performed. If hired by an attorney representing a litigant, the first thing the forensic engineer requests is the *entire* (with emphasis) case file. This request seeks to avoid the ***a priori problem***[6] of only select information that benefits the attorney's client being transferred. The forensic engineer must always remain impartial, unbiased, and work dispassionately to avoid a ***perception of impropriety*** and a tainted investigation.

Another significant difference between the two approaches relates to exemplar testing and reconstruction of the incident. An exemplar can be the actual incident truck, a replica (same year model, make, and backing equipment), or a similar vehicle (different year, same model, chassis, and backing equipment) depending on availability. Reconstruction setup at the actual incident site is ideal, but an alternate geographically similar site with respect to terrain and visibility will suffice if necessary. Also, imagine the benefits from drone videography to truck-pedestrian model AR. Validation of backing trials and correlations with ground-level photography, i.e., driver views from backing mirrors and backing camera display screen, and exterior still images of the truck-pedestrian model approach sequence.

Next, we look at a third variant of the scientific methodology used in investigating structural failures. The incident was a well-publicized bridge collapse that was investigated by one of the world's preeminent investigation agencies, the US National Transportation Safety Board (NTSB).

2.5 NTSB: Distinguished Investigations

NTSB is an independent US government agency responsible for investigating civil transportation accidents and incidents. Civil transportation infrastructure encompasses aviation, highway, marine, railroad, and pipelines. The primary mission of the NTSB is to determine probable cause of transportation failures and improve transportation safety. Since 1967, with over 145,000 transportation incidents investigated, the agency has developed an exemplary reputation as an independent, objective, and authoritative voice in transportation safety. NTSB investigations are recognized worldwide for their expertise, thoroughness, accuracy, transparency, and freedom from conflicts of interest. Further, about 73%[8] of their recommendations have effectively improved safety and saved lives.

FIU Pedestrian Bridge Collapse

A partially constructed pedestrian bridge experienced a catastrophic failure and collapsed onto traffic on the afternoon of March 15, 2018 in Miami, Florida. Over the next 18 months, the NTSB conducted a forensic investigation of the *FIU Pedestrian Bridge Collapse.* The bridge before and after collapse is shown in Figure 2.9. Frame A depicts the point of origin of the collapse, a

FIGURE 2.9
FIU Pedestrian Bridge collapse. (Courtesy of NTSB.)

nodal connection of members 11 and 12. The collapsed bridge is shown in Frame B. The investigation approach undertaken to determine the cause of this complex structural failure was comprehensive (Fig. 2.10).

First, recognize that large-scale failures often require a team of experts, evidence preservation is essential to cause-and-effect determinations, and emergency response may be involved. Emergency responders will coordinate with NTSB to establish command and control (C2) of the incident scene that ensures a cooperative environment between the response parties.

Next, recognize first principles with Figure 2.10, and investigators forego hypothesizing about cause because design, construction, materials, equipment, sabotage, negligence, or any combination of these is possible with such a failure. Further, they know that the evidence will *speak for itself* in time.

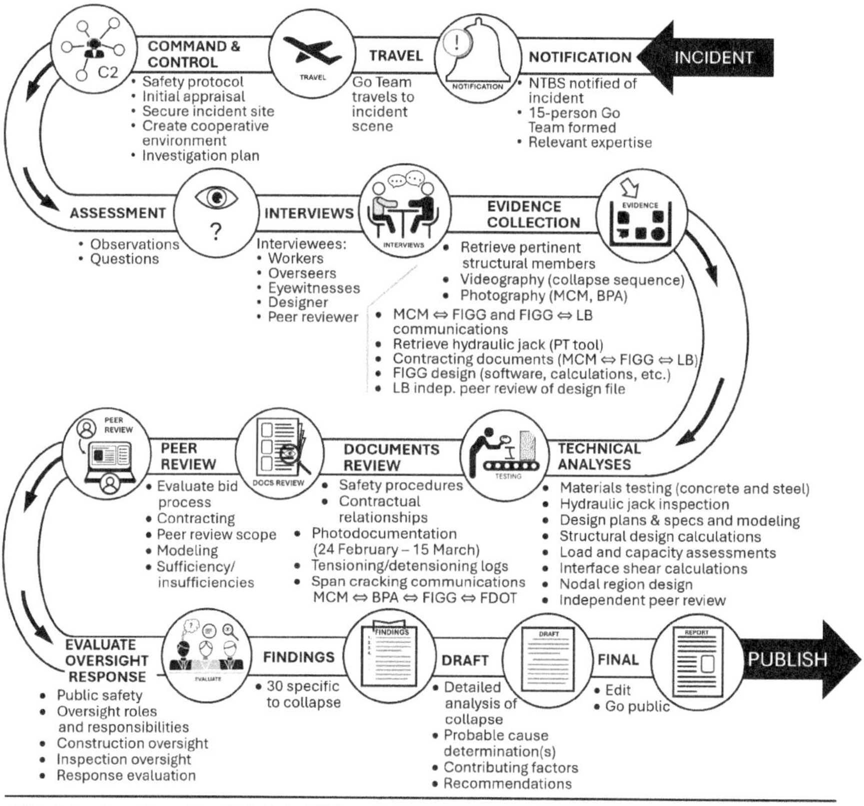

BPA is *Bolton, Perez & Associates*; FHA is *Federal Highway Administration*; FIGG is *Figg Bridge Engineers*; FDOT is *Florida Department of Transportation*; LB is *Louis Berger*; MCM is *Munilla Construction Management;* NTSB is *National Transportation Safety Board;* PT is post-tensioning.

FIGURE 2.10
Investigation approach for FIU pedestrian bridge collapse.

Case Study 2 at the end of this chapter provides the rest of the FIU pedestrian bridge collapse story. Students will learn about flawed design, gross negligence, how communication can make or break a project, and that the consequences of this engineering disaster far outweighed the cost of the bridge. A modern disaster that was totally preventable!

> "The investigation clearly highlighted basic design flaws and a complete lack of oversight by every single party that had responsibility to either identify the design errors or stop work and call for a safety stand-down, once it was clear that there was a massive internal failure."[9]
>
> **NTSB Vice Chairman Bruce Landsberg**
> *28 October 2019*

Notes

1. Attributed to William of Ockham, a 14th-century English Franciscan friar and philosopher, Ockham's (or Occam's) razor is a problem-solving principle that recommends searching for explanations based on the smallest set of factors. We know it today as *the simplest explanation is usually the best one.*
2. An exemplar is a replicate or like model of a thing central to the investigation.
3. A *mil* is a unit of measurement used to describe the thickness of a material. One mil equals one-thousandth of an inch, or 0.001 inch (or 0.0254 mm). A 90-mil elastomeric membrane is conservatively thick; 45- to 60-mil thicknesses are standard, and 30 mils is substandard for flat roofs.
4. MSDS is a material safety data sheet. It is a document that provides information about a product's ingredients, chemical properties, potential hazards, and how to use it safely.
5. Flag stem is a surveying term that refers to a narrow strip of land that provides public road access to a parcel of land located behind other properties. It is a common occurrence with the subdivision of large parcels into smaller lots within an area of few roads.
6. The *a priori problem* refers to a logical fallacy that starts with a preconceived notion, principle, or conclusion, and then searches for arguments to justify it.

References

1. Edward O. Wilson (1998). *Consilience: The Unity of Knowledge.* Knopf (Vintage), New York, 378 p. ISBN 978-0-679-76867-8.

2. Elizabeth Asmis (1984). *Epicurus' Scientific Method*, vol. 42. Cornell University Press. http://www.jstor.org/stable/10.7591/j.cttq45z9. Retrieved 12 June 2002.
3. Régis Morelon and Roshdi Rashed, eds. (1996). *Encyclopedia of the History of Arabic Science*, vol. 3. Routledge. ISBN 978-0-415-12410-2.
4. Peter Achinstein (2004). *General Introduction to Science Rules: A Historical Introduction to Scientific Methods*. Johns Hopkins University Press, 1–5 pp. ISBN 0-8018-7943-4. Retrieved 21 September 2002.
5. David C. Lindberg (1980). *Science in the Middle Ages*. University of Chicago Press, 21 p. ISBN 0-226-48233-2. Retrieved 23 August 2024.
6. Hugh G. Gauch (2003). *Scientific Method in Practice (Reprint ed.)*. Cambridge University Press, 1,45 p. ISBN 978-0-521-01708-4. Retrieved 7 July 2004.
7. Edward Grant (1996). *The Foundations of Modern Science in the Middle Ages: Their Religious, Institutional, and Intellectual Contexts*. Cambridge University Press, 164–7 pp. ISBN 0-512-56762-9.
8. *Annual Report to Congress* (2014). National Transportation Safety Board. Archived (PDF) from Original on September 5, 2015. Retrieved 11 November 2024.
9. National Transportation Safety Board (2019). Pedestrian Bridge Collapse Over SW 8th Street, Miami, Florida, March 15, 2018. Highway Accident Report NTSB/HAR-19/02. Washington, DC. https://www.ntsb.gov/investigations/AccidentReports/Reports/HAR1902.pdf

Case Study 2: A Bridge too Far

FIU Pedestrian Bridge Collapse

- Structural Failure
- Flawed Design
- Substandard Peer Review
- Negligence
- NTSB Investigation

Images courtesy of NTSB

Messages to students:

Welcome design reviews: they make engineers better
Public safety, health, and welfare first always

Acknowledgment

First and foremost, I thank the National Transportation and Safety Board (NTSB) for publishing the results of their excellent investigation of this disaster in the public domain. They know we learn the most from our failures, and

 DOI: 10.1201/9781003623311-5

the importance of teaching future engineers about them. An exhaustive complex investigation 18 months in the making, their work is exemplary forensic engineering. NTSB photographs and graphics used herein are greatly appreciated and credited with "Courtesy of NTSB."

> A novel pedestrian bridge was under construction when it collapsed onto traffic below on the afternoon of March 15, 2018. *For the twelve….*

A Unique Design-Build Bridge Project

The *FIU Sweetwater-University City Pedestrian Bridge* (pedestrian bridge) was a concrete truss bridge designed to look like a cable-stayed bridge (Fig. CS2.1). Designed by FIGG Bridge Engineers (FIGG), the structure comprised a 175-foot-long (53 *m*) main span over Southwest 8th Street and 99-foot (30 *m*) back span over the Tamiami Canal, both supported on piers. A single truss down the centerline of the deck and overhead canopy provided open-air walking appeal, and the central pylon with splayed high cables gave the structure the cable-stay look. Intended to connect the campus with student housing in dramatic fashion, the bridge would showcase FIU's leadership with Accelerated Bridge Construction (ABC) methodology. The bridge was also advertised to withstand a Category 5 hurricane and last 100+ years.

With its single centerline truss of asymmetrical triangular supports, the pedestrian bridge was classified a Category 2 Structure[1] because it was a posttensioned concrete bridge that used a novel *"reinvented I-beam"* design concept, components, and construction techniques with a history of less than five years in Florida. Category 2 structures are also required to undergo independent third-party peer review of design. Further, concrete truss pedestrian bridges are rare, ten times heavier than comparable steel truss bridges, and considered high risk by the Federal Highway Administration (FHA)[1].

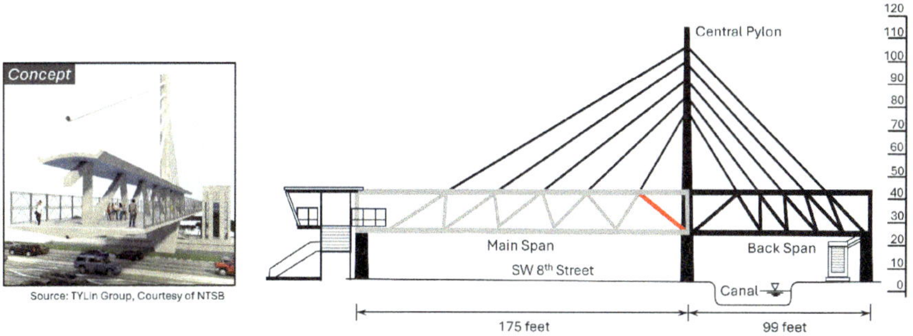

FIGURE CS2.1
FIU pedestrian bridge concept. The plan view at right highlights critical truss member 11 (red) within the 175-foot-long (53 *m*) main span (gray) that collapsed.

In 2013, FIU was awarded a US $19.4 million Transportation Investment Generating Economic Recovery (TIGER) and Transportation Alternatives Program (TAP) grant[2] from the US Department of Transportation (USDOT) to design and construct the pedestrian bridge. The FHA was the funding agency and the Florida Department of Transportation (FDOT) the state contracts manager. A Local Agency Program (LAP) agreement between FDOT and FIU established a management plan for using the federal funds, and FIU hired the T.Y. Lin Group in San Francisco, pioneering architects in the design of prestressed concrete structures, to provide the conceptual design. Munilla Construction Management (MCM) established a design-build relationship with FIGG to build the bridge, and FIGG hired Louis Berger US (LB) for the independent peer review of design[2]. This design-build-peer review relationship became a focal point of the post-collapse investigation (Fig. CS2.2).

The bridge would be constructed in sections in a casting yard located immediately adjacent to its planned location (Fig. CS2.3, upper right). First, support piers were installed. Then the main span was built and pretensioned (see *Design* section for further discussion) for the move to the piers. Self-propelled modular transporters (SPMTs), as shown in Figure CS2.3 (lower right), were used to move the span in place. This ABC method meant that bridge spans could be installed in a single night with minimal interruption to SW 8th Street traffic.

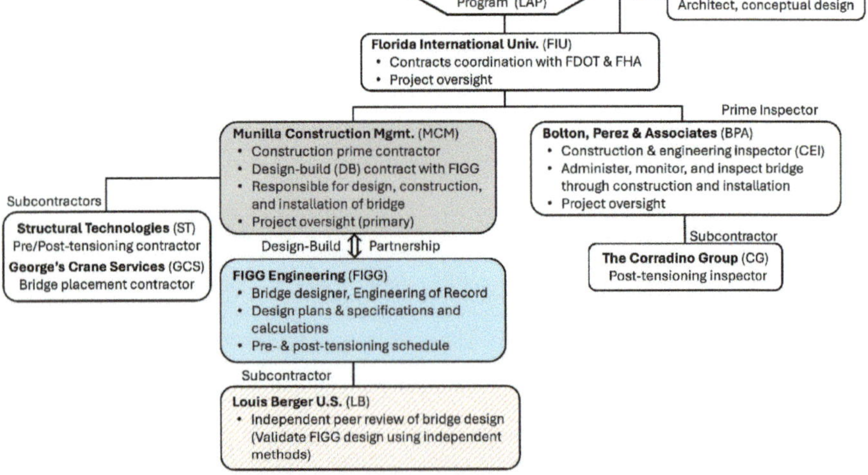

FIGURE CS2.2
Project organizational chart.

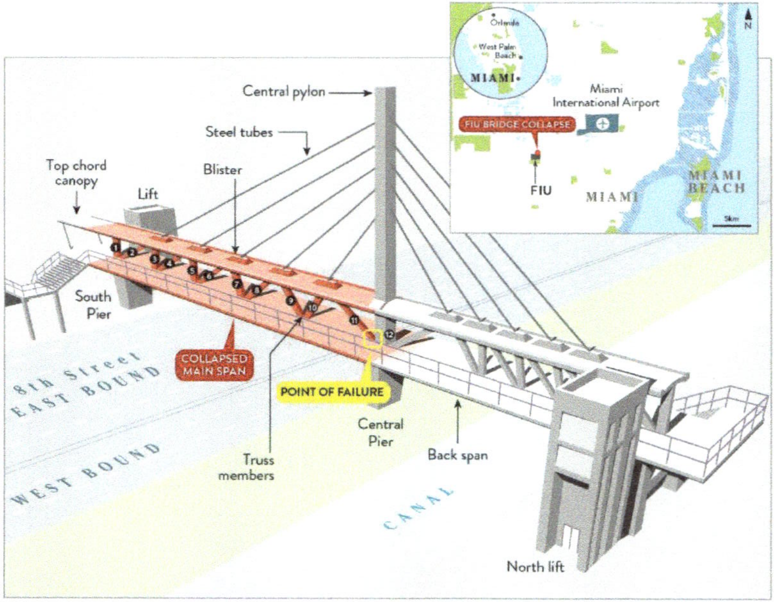

Nonredundant, single-load-path, pre-stressed concrete truss bridge designed to look like a cable-stayed bridge. (Courtesy of NTSB)

Casting Yard

- Accelerated bridge construction methods
- Preassembled bridge components
- Post-tensioning rods pre-tensioned for the move.
- Bridge main span installed in a single night with self-propelled modular transporters.

The Move

Self-Propelled Modular Transporters. (Source: MCM; Courtesy of NTSB)

FIGURE CS2.3
Portrayal of accelerated bridge construction of the main span.

After installation on the piers, workers on top were detensioning the main span when it collapsed onto traffic below. The rest of the bridge was never built.

Investigation

Structural failures that occur during construction or installation are rare in the United States. Failure of this novel bridge being built using ABC methods was unique; experience reminds us that infrastructure failures usually result from a combination of factors, and the families of those that perished deserve to know exactly what happened. No matter how challenging the situation, trained investigators leave no stone unturned in their quest to identify all contributing factors. We must learn and improve from failures.

The investigation of the pedestrian bridge was organized using a phased approach:

Phase I	Phase II	Phase III
Pre-investigation Command and Control (C2)	Technical Analysis	Findings
Observations and Questions	Documents Review	Determinations
Evidence Collection	Evaluation of Design Peer Review	Conclusions
Interviews	Oversight Evaluation	Recommendations

Initiated with a phone call to NTSB, cooperative C2 was immediately established. The NTSB team of a dozen experts was then assembled, and investigators traveled to the disaster site to set about observing, photo-documenting, and preserving evidence. Phase I established a pristine investigation site and acquired storage space sufficient for securing, examining, and packaging evidence. NTSB also collected all available videography and interviewed employees from the firms and organizations involved with bridge design, construction, inspection, and oversight.

Phase II involved extensive technical analysis and reviews of all aspects of design, peer review, and oversight from a safety focus perspective. The scope is outlined below:

	Technical Analysis	
Safety Focus	**Design**	**Other**
Design errors	Structural design calculations	Study video recordings of collapse
Construction errors	Load and capacity assessments	Inspection photo-documentation
Unique bridge	Interface shear calculations	Materials testing (concrete and steel)
Lack of redundancy	Nodal region design	Equipment evaluation (hydraulic jack used for tensioning PT rods)
Failure mechanisms	Design analytical models	Design documents (plans and specifications)
Independent peer review of design		Peer review of design
Oversight response		Witness interviews*

* NTSB investigators interviewed employees from FIGG, MCM, BPA, ST, LB, FDOT, CG, and GCS.

The bridge was safety-compromised from the beginning with its lack of redundancy. A single truss to transfer load means a nonredundant bridge (Fig. CS2.4, at left). If any member failed, bridge collapse can occur. Bridges with two or three parallel steel trusses (i.e., redundant structures) are less likely to collapse should a member fail.

Materials and equipment were tested and analyzed with Phase 2. The hydraulic jack used with tensioning PT rods in the bridge at the time of collapse was inspected and found to be fully functional. PT rod and other steel reinforcement samples were tested, as were concrete samples. The tested materials all met strength and quality standards.

Also with Phase II, review of design plans and specifications, calculations, and computer modeling, and independent peer-review documentation was performed. A truss equilibrium calculation shows the magnitude of shear demand at the 11/12 nodal region (Fig.CS2.6(b)). More about this later in the design section of this digest. Suffice it to say, document review is a tedious process that requires great attention to detail and patience.

Once testing and analyses were completed; results were compiled and evaluated to develop a list of findings and determinations. Findings are direct outcomes from testing; determinations arise from cause-and-effect understanding. When allowed to *speak for itself,* evidence usually converges toward cause and validates or invalidates hypotheses. Sometimes, reevaluation of the evidence and analysis with new eyes and focus leads to an alternate hypothesis.

Phase III of the investigation was about summarizing the data and evaluating results to reveal *the tell.* The tell is the most probable cause of

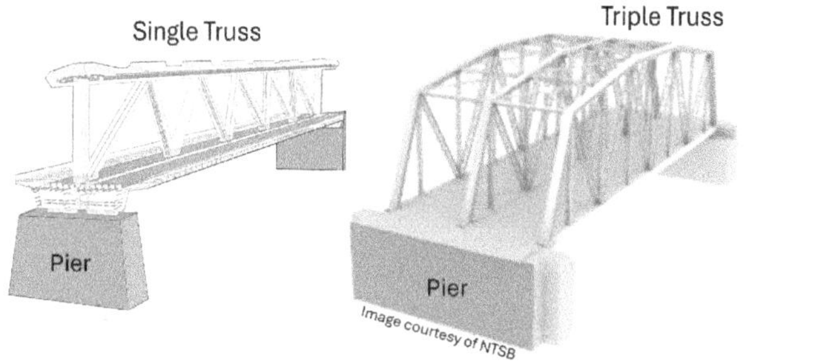

Nonredundant, single load-path FIU reinforced concrete pedestrian bridge (L) and exemplar redundant triple-load-path steel bridge (R). With only a single load-path to maintain stability, the FIU bridge could collapse if only one member failed. The triple truss bridge can withstand failure of multiple members without collapsing.

FIGURE CS2.4
Load path redundancy.

what happened. Rooted in scientific methods of investigation, truth will come forth through evidence convergence as it always does with forensic inquiry. Convergence of evidence for this bridge failure is depicted in Figure CS2.5. Note six independent lines of evidence point to *probable cause*. A minimum of three is required to substantiate probable cause as a rule of thumb.

A minimum of three independent lines of converging evidence is required to substantiate probable cause.

Probable cause is a high standard based on objective circumstances and evidence, *"cause"* usually refers to a single source of failure, and forensic engineers avoid using the term *"most probable cause"* because *"most"* is subjective, especially with multiple-cause situations. The last thing an engineer wants to be involved with is ranking cause-and-effect and contributing factors, which is a subjective exercise best left to arbitrators.

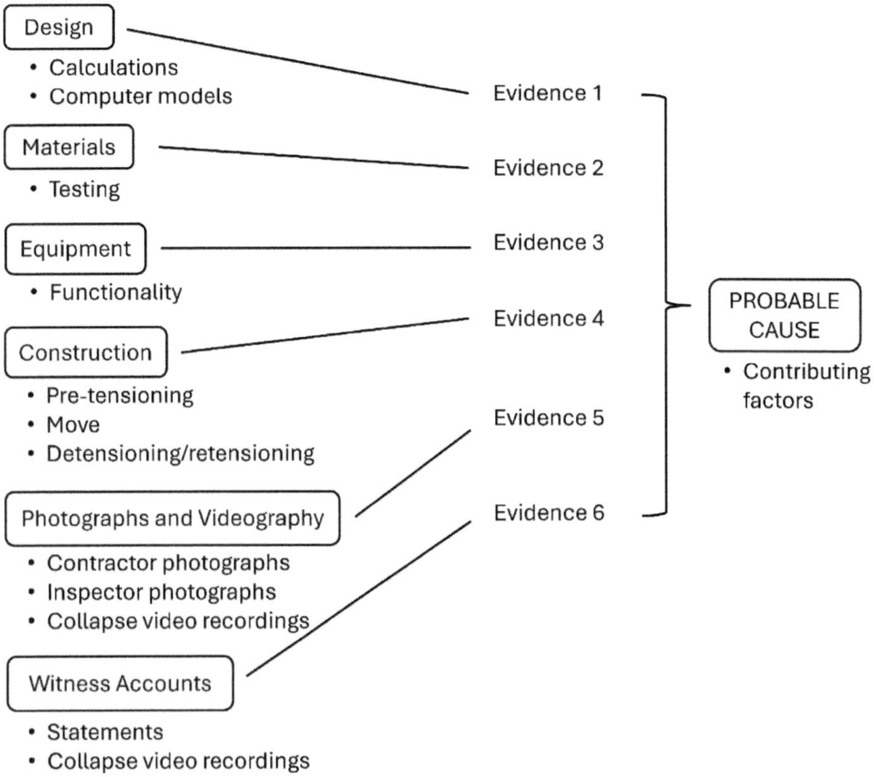

FIGURE CS2.5
Convergence of evidence.

Design

A flawed design, specifically main span load and resistance calculations, is at the core of why this pedestrian bridge failed. MCM and BPA photo-documentation evidenced progression of cracking in nodal region 11/12 between February 24 and March 15, through the period of main span transport with the SPMTs (self-propelled modular transporters) and installation on the piers. This suggested node 11/12 had been actively dislodging from the deck. Dislodgement implied underestimation of load forces (demand), overestimation of capacity to resist demand, or both; the *"design calculation error"* hypothesis. Further, an in-vehicle video captured the concrete blowout of node 11/12 that initiated the collapse. Therefore, we examine what went wrong with *demand-capacity analysis* with focus on critical nodal regions[3] at the ends of the main span. The results portrayed here reflect a primary metric among the design flaws that led to bridge collapse. Bolded statements indicate design errors.

Governing design codes for this novel pedestrian bridge were

- *American Association of State Highway and Transportation Officials, Load and Resistance Factor Design 2015* (AASHTO LRFD) and
- FDOT *Structures Design Guidelines* (FDOT 2015b).

FIGG design calculations were the basis for plans that specified construction materials, components, and procedures MCM used to construct the bridge.

The AASHTO LRFD was the principal design guideline for the FIU pedestrian bridge. This standard uses a reliability-based approach to ensure adequate structural capacity to resist load forces (demands) on the structure. Design typically uses a two-step process whereby the designer first analyses the forces on structural components due to applied loads and then designs the components and connections to withstand the forces with a desired reliability against failure. The desired reliability is a two-fold factor of safety concept: (1) amplify the expected loading on the structure by a statistically based multiplier called a load factor, and (2) reduce expected structural capacity by a statistically based multiplier called a resistance factor. In amplifying expected loading and reducing anticipated capacity, a level of confidence is achieved that sufficient capacity will be available to resist the loadings (demands). A check to ensure reduced capacity is greater than the factored load effects completes the design.

Forces acting at nodal regions where diagonal truss members connected to the deck are shown in Figure CS2.6(a). Nodal regions 11/12 are the focus here since it is where the bridge collapse initiated. Phased construction resulted in cold joints between concrete pours. For this bridge, a cold joint is a discontinuity at the deck, which hardened before the concrete truss was cast. In accordance with AASHTO LRFD, cold joints were roughened to improve resistance to shearing. Member 11 in axial compression has a downward

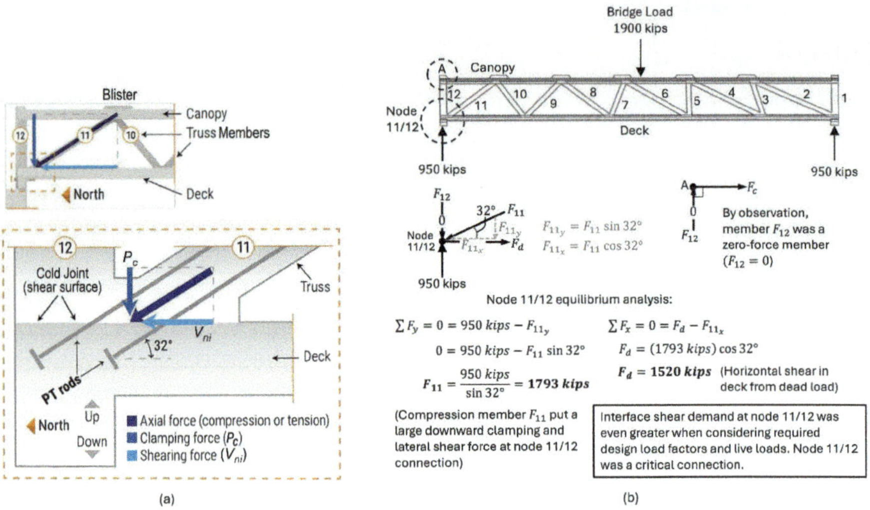

FIGURE CS2.6
(a) Typical depiction of axial force and its clamping and shearing components on a nodal region cold joint interface shear surface between deck and truss. (Annotated. Source (force diagrams): FHA; courtesy of NTSB). Also, (b) equilibrium analysis of node 11/12 provides an estimate of interface shear demand caused by the bridge dead load.

clamping component force that increases resistance to sliding caused by the horizontal shearing component. The equilibrium calculation (Fig. CS2.6(b)) provides estimates of these forces.

The AASHTO LRFD groups different load combinations into *strength limit states*, conditions beyond which a bridge or its components cease to function as designed. In other words, strength limit states relate to the safe load-carrying capacity of a structure. There are five strength limit states in AASHTO LRFD:

- Strength I: Normal vehicular use without wind
- Strength II: Special vehicles or permit vehicles without wind
- Strength III: High wind velocity (exceeding 55 mph)
- Strength IV: Very high dead load to live load ratio
- Strength V: Normal vehicular use with 55 mph wind velocity

Since the FIU structure was a pedestrian bridge, FIGG design appropriately used the Strength I limit state to evaluate interface shear demand on the truss nodal regions.

Design Calculations (Demand)

Discussion here pertains to FIGG design calculation errors that led to inaccuracies with *demand-capacity analysis* of the main span. FIGG used the following load factors to determine the Strength I limit state applied force effects (demand) on the truss member nodal regions:

- Dead load factor (of structural components and nonstructural attachments), DC = 1.25
- Pedestrian live load factor, PL = 1.75
- Posttensioned tendon[4] force factor, PT = 1.0
- Force effect due to uniform temperature factor, TU = 0.5

The AASHTO LRFD defines the dead load factor, DC, as 0.90 to generate minimum demand and 1.25 for maximum, but does not specify what DC to use when calculating interface shear capacity. FHA strongly recommends the lower value when determining the permanent net compressive (or clamping) force normal to the shear plane, P_c (Fig. CS2.6(a)) to obtain a minimized interface shear capacity value. Also, the highest justifiable DC for determining P_c according to FHA is 1.0. **FIGG used 1.25**.

Modeling and Analysis

FIGG's design used LARSA and LUSAS software for evaluation of three critical construction stages. The stages and corresponding models are illustrated and described in Figure CS2.7. However, the FIGG design level of detail

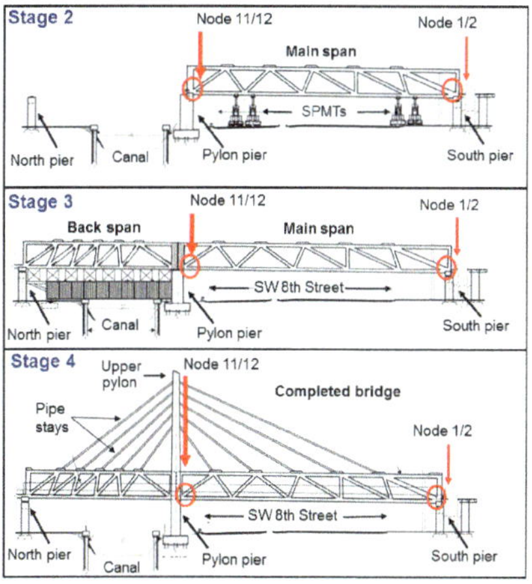

Construction Stages

Stage 1: Casting yard construction of bridge substructure elements and main span.

Stage 2: SPMT transport of main span from casting yard to placement on south and pylon piers in a *simply supported* condition. Also includes detensioning of truss members 2 and 11.

Stage 3: Back span construction, supported on the pylon and north piers.

Stage 4: Connection of the main span, back span, and pylon pier; construction of the upper pylon; and installation of steel pipe stays.

FIGG Analytical Models for Final Design Calculations:

LARSA Main Span Erection Model: 2D model of main span supported at SPMT locations during transport (Stage 2).

LUSAS Simple Support: 3D model of main span simply supported at south and pylon piers (Stage 2).

LARSA Longitudinal Model: 2D model of completed two-span, continuous structure (Stage 4).

LUSAS Fixed Pylon: 3D model of main span simply supported at south pier and fixed at pylon pier, adjusted to replicate main span performance in its completed condition (Stage 4).

LARSA and LUSAS are the proprietary names of software packages used for modeling.

FIGURE CS2.7
Construction stages and corresponding FIGG analytical models used for final design calculations. (Annotated. Illustrations courtesy of NTSB.)

varied with the stages, the fully completed two-span bridge having the most thorough documentation, and that of the simply supported main span condition (its state of collapse) the least. FIGG did not provide a rationale for selective use of modeling results that led to consistent underestimation of interface shear demand at many truss member nodal regions. Therefore, FHA performed post-collapse analysis using the same models FIGG used as necessary to demonstrate correct use of the AASHTO LRFD design guide.

Figure CS2.8 shows the nodal region interface shear demand results generated for the FIGG bridge design and underestimation of shear demand at critical nodal regions as determined by FHA. Note that V_{ni} is nominal interface shear resistance (or nominal capacity) as a function of P_c for dead load (weight of bridge and its components), and V_{ui} is the factored interface shear force (or nominal resistance) due to total load (dead load and live load) based on the applicable Strength I limit state load combination. There were significant variations in the factored interface shear demand at the nodal regions for the models (Fig. CS2.8(a)). The LARSA longitudinal model (for completed bridge superstructure) generated the largest forces at critical nodal regions of the ends of the main span (vicinity of pier supports). **However, FIGG exclusively used LUSAS model results for design of all main span nodal regions.** Figure CS2.8(b) depicts the FIGG and FHA comparative modeling results for nominal interface shear demand based on dead load. **FIGG's use of only the simple support and fixed pylon (LUSAS) model resulted in the following underestimations of interface shear demand**:

- Nodal region 11/12, underestimated by 850 kips (or 46%)
- Nodal region 10/11, underestimated by 1360 kips (or 91%)
- Nodal region 1/2, underestimated by 1100 kips (or 49%)
- Nodal region 2/3, underestimated by 1100 kips (or 43%)

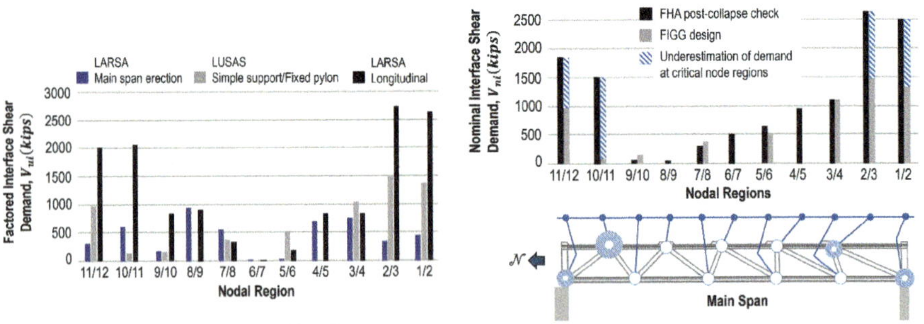

(a) Modeling shear demand results for FIGG bridge design. (Annotated. Source: FHA, Courtesy of NTSB).

(b) Comparison of FIGG and FHA modeling results for underestimation of shear demand at critical nodal regions. (Annotated. Courtesy of NTSB).

FIGURE CS2.8
Interface shear demand (a) computer modeling results generated for FIGG bridge design, and (b) underestimation at critical nodes.

The FIGG calculation errors led to too little steel rebar in the concrete between the deck and truss members at crucial nodal regions. For example, recalling our truss calculation for demand at nodal regions 11/12 (Fig. CS2.6(b)):

$$F_{11(clamping)} = 1793\ kips(\sin 32°) = 950\ kips = P_{c(11/12)}$$

This P_c value is significantly lower than that shown in Table 1 for nodal region 11/12. Recall FIGG erroneously used a dead load demand factor of 1.25 for their P_c design values. The higher is the clamping force P_c, the less reinforcement is needed (i.e., A_{vf} decreases) and the greater is A_{cv}, the area of concrete engaged in interface shear transfer. This relationship is depicted with the AASHTO LRFD governing equation [1] that determines the nominal connection capacity, V_{ni} as

$$V_{ni} = c\overset{\uparrow}{A_{cv}} + \mu\left(\overset{\downarrow}{A_{vf}}f_y + \overset{\uparrow}{P_c}\right) \qquad [1]$$

where $c\ (ksi)$ is the cohesion factor (0 in this case since no cohesion effects)

$A_{cv}\ (in^2)$ is area of concrete engaged in interface shear transfer (varies by node)

μ is a dimensionless friction factor that indicates [truss member-deck] interface surface condition (1.0 for clean concrete roughened to amplitude of 0.25 *in* for this bridge)

$A_{vf}\ (in^2)$ is area of interface shear reinforcement crossing shear plane within A_{cv} (varies by node)

$f_y\ (ksi)$ is yield stress of reinforcing steel that crosses the interface plane, not to exceed 60 *ksi* (60 *ksi* for this bridge)

$P_c\ (kips)$ is permanent net compressive force (or clamping force) normal to shear plane (varies by node)

Further, the nominal resistance calculated with Equation [1] must be reduced by a resistance factor multiplier such that

$$V_{ri} = \varphi V_{ni} \geq V_{ui} \qquad [2]$$

where φ is a dimensionless resistance factor (0.9 for normal weight concrete with this bridge)

V_{ui} is the factored interface shear demand as determined by the models (Fig. CS2.8(a))

The FIGG interface shear design calculation results are summarized in Table CS2.1.

Not only was nodal region 11/12 constructed with insufficient reinforcement, but concrete voids also (pipes) diminished effective A_{cv} and capacity to resist shear at this node. Figure CS2.9 depicts the punching shear

TABLE CS2.1

Nodal Region Interface Shear Design Calculation Summary (Courtesy of NTSB)

Nodal Region	$A_{cv}(in^2)$	$A_{vf}(in^2)$	$P_c(kips)$	$V_{ri}(kips)$
1/2	1210	8 - #7 = 4.80	1275	1407
2/3	1218	14 - #6 = 6.16	1298	1501
3/4	1798	18 - #7 = 10.80	579	1104
4/5	FIGG did not calculate for this nodal region			
5/6	1664	12 - #7 = 7.20	230	596
6/7	FIGG did not calculate for this nodal region			
7/8	1428	12 - #7 = 7.20	449	793
8/9	FIGG did not calculate for this nodal region			
9/10	1407	12 - #7 = 7.20	703	1021
10/11	1218	6 - #6 = 2.64	1210	1232
11/12	882	8 - #7 = 4.80	1233	1369

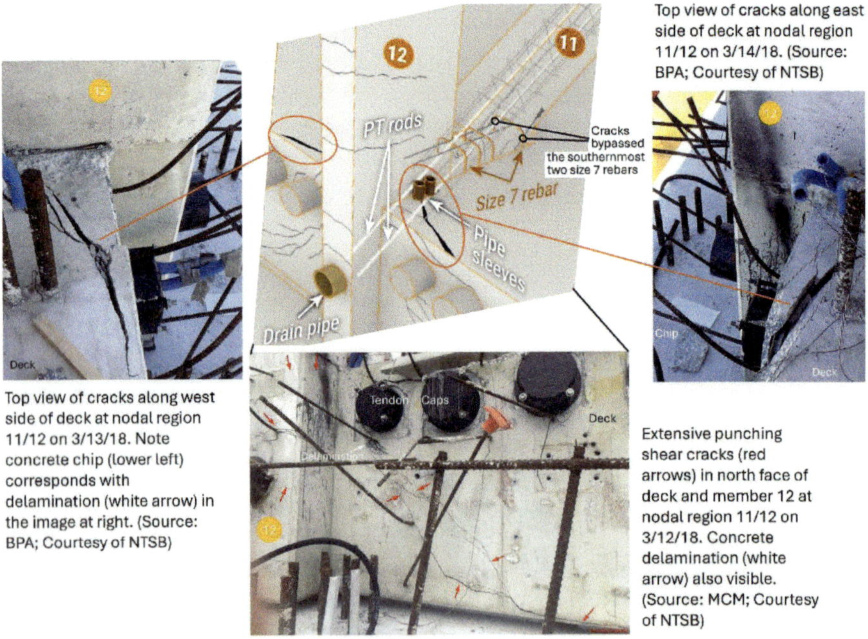

Top view of cracks along east side of deck at nodal region 11/12 on 3/14/18. (Source: BPA; Courtesy of NTSB)

Top view of cracks along west side of deck at nodal region 11/12 on 3/13/18. Note concrete chip (lower left) corresponds with delamination (white arrow) in the image at right. (Source: BPA; Courtesy of NTSB)

Extensive punching shear cracks (red arrows) in north face of deck and member 12 at nodal region 11/12 on 3/12/18. Concrete delamination (white arrow) also visible. (Source: MCM; Courtesy of NTSB)

FIGURE CS2.9
Insufficient reinforcement and concrete voids (pipes) at nodal regions 11/12 (L) resulted in the significant shear cracking visible in the deck photographs. (Annotated. Images courtesy of NTSB.)

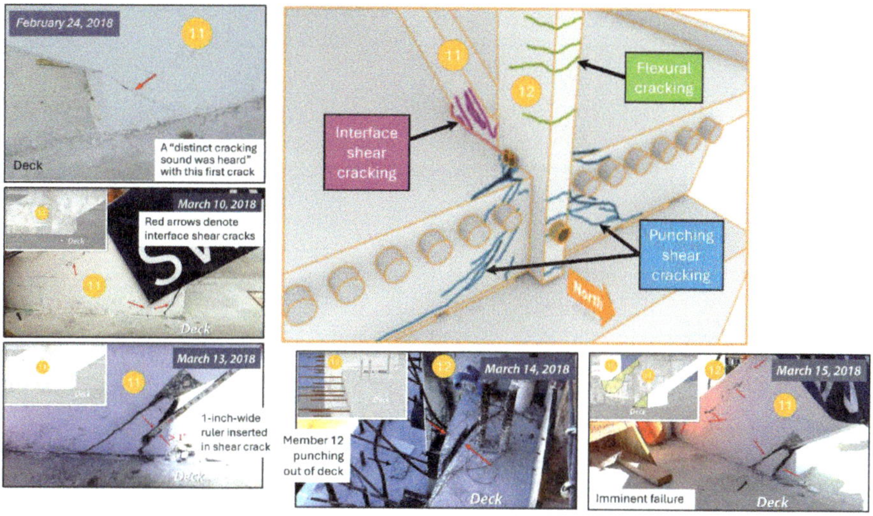

FIGURE CS2.10
The look of failure at nodal regions 11/12. Photographs represent 3 weeks of crack progressions. The graphic at upper right illustrates the different types of cracks associated with the punching shear failure. (Annotated. Photographs source: MCM; Illustration courtesy of NTSB.)

consequence that triggered the bridge collapse. Note the significant shear cracks in the side photographs that developed during the last days before collapse. Like the truss size 7 rebar comment in the upper middle graphic, these deck cracks also bypassed reinforcement.

Different types of stress cracks developed at nodal region 11/12 over 3 weeks before punching shear failure on March 15, 2018 (Fig. CS2.10). The extent and magnitude of cracking was alarming. Why this was not recognized as a problem and safety concern is mystifying.

The Move

A tensioning plan had been engineered for the span move from the casting yard onto piers. Main span PT tensioning was performed to counter cantilever effects during the move. In Figure CS2.11, white arrows denote the direction of cantilever pull and blue arrows denote induced tensile forces in critical truss members 2 and 11. The temporary pretensioning of members 2 and 11 was designed to keep the span level during the move. Once stabilized on the piers, the members were then de-tensioned.

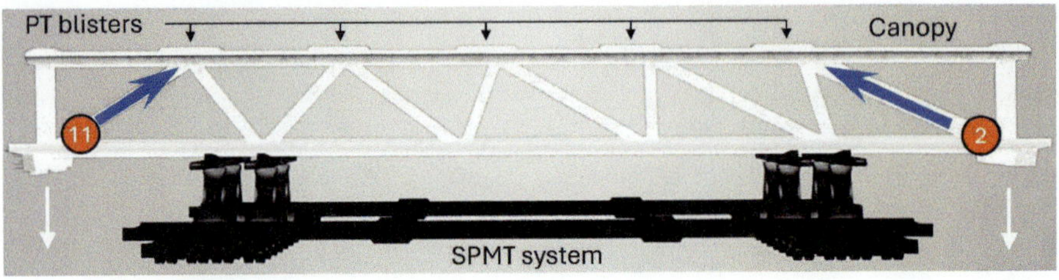

FIGURE CS2.11
Illustration of temporary pretensioning plan for the span move. (Annotated. Image courtesy of NTSB.)

Last Resort

While waiting on a red traffic light on the morning of March 15, an FIU employee heard a loud "whip crack" sound from the bridge span[3]. Nearby at the casting yard, FIGG lead engineers were meeting with FIU, FDOT, MCM, and BPA representatives. The EOR stated that the cracks were not a safety concern and the span's integrity was not compromised and recommended re-tensioning diagonal member 11 PT bars to stabilize the span on the piers. The representatives deferred to FIGG's recommendation to re-tension despite this step not having been peer-reviewed. FIGG also stated that no crack repairs should occur until after span stabilization on the piers. To further compound the urgency, the posttensioning inspection subcontractor was unavailable on short notice to monitor the re-tensioning.

Immediately after the meeting, the FIGG engineers went back to Tallahassee, and the tensioning crew began to re-stress truss member 11 PT rods to 280,000 pounds each[2]. Upon completion, with equipment attached and the crew still on the canopy, the bridge span collapsed. The NTSB concluded the re-tensioning of member 11 was a last-resort attempt to mitigate further cracking and stabilize the span[2].

Design Peer Review

Lapses in FIGG's design review process failed to detect erroneous calculations and conformance with AASHTO LRFD. Independent peer review of this bridge design was established through a manipulated bid that resulted in an unqualified reviewer and a truncated review. A separate case study could be written about the FIU pedestrian bridge peer review process and emphasize the importance of peer review for unique and challenging structures.

Further, design engineers work in government funding agencies. Internal design review should be a requirement with every infrastructure project awarded before construction begins. This point was well made in an

Inspector General's audit which criticized the FHA's Florida Division, "... *which knew the doomed bridge was a high-risk project but did not have design calculations reviewed by their own expert bridge engineers.*"[2]

Consequences

Eighteen civil lawsuits were filed against 25 businesses, 23 of which settled for undisclosed amounts. OSHA (Occupational Safety and Health Administration) cited five businesses for safety violations[1]. The designer, builder, design peer reviewer, and construction and engineering inspector were all found negligent by the courts. Table CS2.2 summarizes consequences of this structural failure.

How does a designer deal with a $19.4 million failure? Or an attorney who spends $50K of his/her client's money building a defense, only to find out the evidence does not support it? One might imagine the attorney has the easier time telling the client the bad news. Structural failures are expensive, and one might imagine the designer struggling mightily with failure. Overestimation of success, complacency, fear of failure, denial, and the psychology of failure affect us all differently. If all one knows is success, how might they deal with failure? A designer builds a thing 100 times. Magnificent landmarks; testaments to success. What happens when the novel 101st one fails?

> Failure and innovation are inseparable twins.[4]
>
> **Jeff Bezos**

TABLE CS2.2

Economic impacts of the FIU bridge collapse.

Consequences
Bridge builder: • settled for $42M • decertified by FDOT • lost ~$200M in future contracts • filed for Chapter 11 bankruptcy reorganization
Bridge design firm: • disagreed with some NTSB findings • lost ~$1B in bridge contracts in three states • suspended from FHA contracts for 10 years • filed appeal in federal court and asked for temporary restraining order on FHA suspension
Bridge EOR: • surrendered FL P.E. license • retired without admitting culpability
Project CEI: • firm acquired by large Canadian company

Anecdote: The Psychology of Failure

Notes

1. Chapter 16, section 1604.5, Table 1604.5 "Risk Category of Buildings and Other Structures" in the International Building Code assigns structures to one of four categories based on risk to human life in the event of a structural failure. Risk Category 2 structures represent "...a lesser hazard to life because of fewer building occupants and smaller size compared to those that are considered Risk Category III." Category 2 structures are essentially low risk to occupants.

2. The Rebuilding American Infrastructure with Sustainability and Equity (RAISE) transportation discretionary grant, Better Utilizing Investments to Leverage Development (BUILD), and Transportation Investment Generating Economic Recovery (TIGER), or *RAISE-BUILD-TIGER*, was a national infrastructure investment program initiated with the 2009 Recovery Act. Grants provided discretionary funding for investments in road, rail, transit, and port projects to achieve national objectives. Since 2009, Congress has provided 15 rounds of competitive grants totaling more than $14.4 billion for ~ 1100 projects in all 50 states, the District of Columbia, the Virgin Islands, Puerto Rico, and Guam.

3. A node is where a diagonal truss member connects with the deck or canopy.

4. Diagonal truss members, or *tendons*, had two to four PT rods that could be tightened or loosened in accordance with engineered design.

References

1. Wikipedia Contributors (2024, December 4). Florida International University Pedestrian Bridge Collapse. In *Wikipedia, The Free Encyclopedia*. Retrieved 8 December 2024, from https://en.wikipedia.org/w/index.php?title=Florida_International_University_pedestrian_bridge_collapse&oldid=1261167134

2. National Transportation Safety Board (2019). Pedestrian Bridge Collapse Over SW 8th Street, Miami, Florida, March 15, 2018. Highway Accident Report NTSB/HAR-19/02. Washington, DC. https://www.ntsb.gov/investigations/AccidentReports/Reports/HAR1902.pdf

3. "FIU Employee Heard a Loud Crack Crossing Under the Bridge. Hours Later, it Collapsed" (http://www.miamiherald.com/news/local/community/miami-dade/article205432799.html). *Miami herald*. Archived (https://web.archive.org/web/20180316043518/http://www.miamiherald.com/news/local/community/miami-dade/article205432799.html) from the original on March 16, 2018. Retrieved 8 December 2024.

4. Taylor Soper. (2016).' *Failure and Innovation are Inseparable Twins*': Amazon Founder Jeff Bezos Offers 7 Leadership Principles, Available at: https://www.geekwire.com/2016/amazon-founder-jeff-bezos-offers-6-leadership-principles-change-mind-lot-embrace-failure-ditch-powerpoints/, Accessed 1 December 2024.

3

The Expert Witness

CONTENTS

CONCEPTS	KEY PHRASES	ADDITIONAL READING
Expert Witness	Lawsuit (civil suit, civil litigation)	Exercise 5
The Golden Rule	Plaintiff and defendant	Exercise 6
Neutrality and Advocacy	Litigants (parties)	
	Answer, demurrer, motion to strike	Case Study 3
Credentials	Discovery, interrogatories, and motions	
Science in Everything	Ethics and excellence	
Reports	Hired gun	
Faces of Testimony	Keep it simple students (KISS) Communication liability	

DEFENSE ATTORNEY: ***MR. NOON, ARE YOU AN EXPERT IN ACCIDENT RECONSTRUCTION?***

Choose the best response:

a. Yes
b. That is for you to decide. In my 25 years of forensic engineering practice, I have investigated more than 200 collisions involving automobiles, motorcycles, bicycles, and pedestrians, for plaintiff and defense attorneys involved in both civil and criminal cases.
c. I have a master's degree in mathematics, am ACTAR* certified, and have 500+ accident reconstruction cases under my belt. I think that qualifies me as an expert.

DOI: 10.1201/9781003623311-6

d. The word *"Expert"* is highly subjective depending on credentials. I prefer *Ingénieur Extraordinaire.*
e. There are 57 forensic disciplines, and our firm practices most. A majority of our cases involve AR, so you could say that I – as a Sr. Engineer and Partner in the firm – am an expert in that discipline.
f. At three Franklins an hour, I'd better be.

Beware of the loaded question? It will come early in a deposition or testimony in a court of law. Avoid calling oneself an expert, especially in front of a jury. Let others decide that.

Mr. Noon would choose b). Wise humility is a badge of honor.

DEFENSE ATTORNEY: ***DR. SAGE, HOW MUCH ARE THEY PAYING YOU FOR YOUR TESTIMONY?***

This one is a classic: Beware of the *hired gun* trap.

* ACTAR is the *Accreditation Commission for Traffic Accident Reconstructionists,* an independent organization that provides accreditation to professionals involved in traffic accident investigation and reconstruction. The primary goal of ACTAR is to ensure that accident investigation professionals meet established minimum training standards to enhance their credibility within the legal system.

3.1 What Is an Expert?

Perhaps the best way to answer is to first explain why experts are needed. Experts play a crucial role in our legal system. Lawyers, judges, and juries rely on the specialized knowledge experts apply to the facts of a case to explain how, why, or what caused an event.[1] A case refers to a civil litigation (or civil lawsuit), and whether an accident, damage, breach, or catastrophe, the engineer as expert witness is called upon to explain the science and cause-and-effect relationship(s) of the event and rule out other noncausal factors. Let us explain in more detail using Figure 3.1.

Figure 3.1 shows the general timeline of a civil litigation, wherein the sequence of steps above the horizontal line is grouped into phases. A lawsuit is considered by a ***plaintiff*** (individual, business, organization, or government agency) against a ***defendant*** (also an individual, business, organization, or government agency) usually through a *Prelawsuit Initial Conference* with an attorney. Together, the plaintiff and the defendant are the ***litigants*** (or ***parties***) involved in a lawsuit. The plaintiff initiates a *civil suit* with formal *lawsuit*

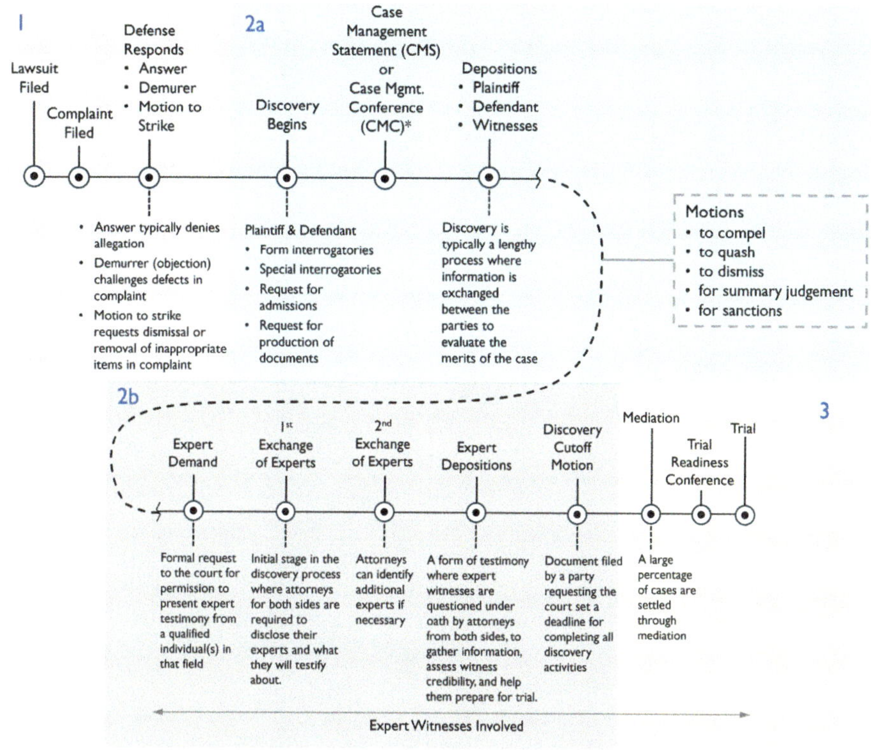

*A CMS filed in civil court serves to update on status and progress of a case, and helps the court determine readiness for trial or other resolution method. A CMC is an actual meeting between the judge, attorneys, and litigants to do the same. Scheduling and questions about procedural matters may also be discussed. Which – CMS or CMC - depends on jurisdiction, and litigants may be required to attend, or attorneys may attend on their behalf.

FIGURE 3.1
Civil lawsuit timeline.

and ***complaint*** filings with an appropriate jurisdictional court as Phase 1. The defendant, usually through an attorney, responds to the court with an *answer, demurrer,* or *motion to strike* (defense response). An ***answer*** denies the allegation, ***demurrer*** objects to defects in the complaint, and ***motion to strike*** seeks to omit parts of or deny the case altogether. A judge will decide the merits of the case and whether it moves forward to discovery (Phase 2).

Phase 2 involves the discovery of information pertinent to the case. Subdivided as ***Initial Discovery*** (2a) and ***Expert Discovery*** (2b) for the situation involving the need for expert witnesses (EWs), initial discovery seeks information through *interrogatories, admissions,* and *productions* from the parties. ***Interrogatories*** refer to information gathered through plaintiff and defendant written answers to written questions posed by the opposing attorneys. ***Admissions*** refer to admitting or denying specific statements, and ***productions*** are documents and/or tangible evidence. Next, litigants and witnesses (if any) are ***deposed***, where they provide sworn testimony outside of court.

A ***deposition*** is an oral form of information gathering. At a defendant deposition, for example, the defendant must answer questions under oath posed by the plaintiff's attorney. The defense attorney, also present, mostly listens but is ready to object *for the record* when legal clarity is needed, or interrogation protocol is violated. The same process is accorded to the plaintiff so that both sides of the event story are recorded. A stenographer also in the room records the depositions with a shorthand or stenotype machine, from which she/he later transcribes as a formal ***deposition transcript***. Generally, with civil cases, litigants are afforded the opportunity to review and correct their own deposition transcript through a formal ***errata process***. This completes the early part of discovery, and the attorneys and judge again evaluate the merits of the case. When a case requires expertise to explain aspects of the event in dispute, EWs are retained for Phase 2b discovery.

Need to involve experts arises when the attorneys and initial discovery information cannot sufficiently explain the event. *Expert demand* is the formal request for permission of the judge to present expert testimony about specific information. Usually, the defense identifies first, followed by the plaintiff's attorney. This process identifies EWs, exchanges their *civil disclosures* (e.g., resume, qualifications, summary listing of trial experience, and compensation rate) between the attorneys, and stipulates exactly what the EWs will testify to. In this way, both sides know what to expect, and no attorney has an advantage in the case. Judges do not like surprises and uphold the ***doctrine of fairness*** to the parties.

This ***First Exchange of Experts*** usually leads to the sharing of expert reports and reviews by the experts. If both reports are somewhat agreeable with ***findings*** (or determinations), settlement is the likely next step and the case avoids trial. Should report determinations significantly differ, be diametrically opposed, or be inconclusive with regard to certain important aspects of the case, the EWs will be deposed. Once again, testimonies lead to transcripts with errata opportunity. This opportunity is especially important to the EW, as statements made under duress, through inexperience, and stenographer mistakes can muddle testimony. Also, strategies and the way attorneys pose questions – statements, declarations, or incomplete phrases that are not questions – sometimes confuse or mislead to unintended responses.

When the first round of expertise does not answer everything or new information comes to light through discovery, a ***Second Exchange of Experts*** (need for additional experts) may be necessary. Eventually, sufficient expert participation and all available pertinent information lead to a ***Discovery Cutoff Motion***. This signals the end of discovery and the beginning of Phase 3 ***resolution***.

A civil case resolves through settlement, mediation, arbitration, or trial. When the evidence overwhelmingly supports one side, ***settlement*** is most sensible. A ***mediated agreement*** allows both sides to reach an amicable solution and saves time and costs. ***Arbitration*** relies on a neutral third party (often an experienced attorney, retired judge, or business professional with expertise in a particular area) to resolve the dispute. A civil lawsuit only goes to trial when it cannot be resolved out of court. Trial requires a jury selection process, usually involves a series of pretrial motions and a readiness conference of two, and in-court testimonies by EWs, litigants, and witnesses. Put simply, a trial is expensive and the outcome uncertain. Most parties prefer to avoid the time, cost, and uncertainty of a trial.

Now that we know why experts are needed, who they are, and how one becomes a credible ***expert witness***. Figure 3.2 summarizes the *credentials* lawyers look for when choosing experts, and the *qualities* jurors appreciate. One way to prioritize is to start at the top of CREDENTIALS and traverse clockwise around the circle. Level of education, training, professional licensure, and an ethically solid reputation are essentials of expert credibility.

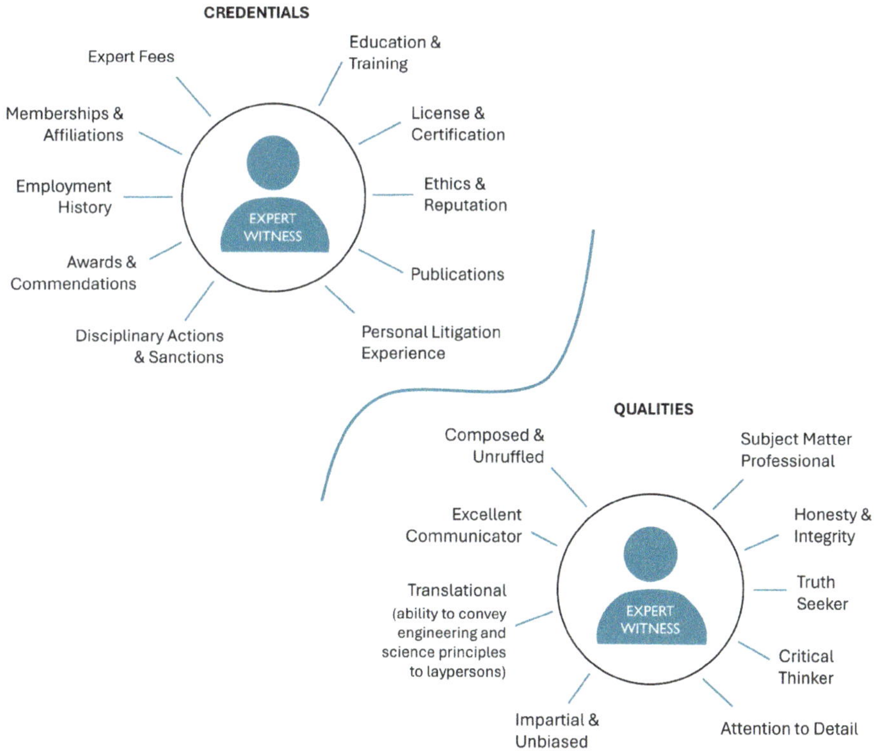

FIGURE 3.2
Attributes of the expert witness.

Publications, although a testament to *some* areas of expertise, may be limited to funded research of the academic expert or nonexistent. Not all forensic engineers publish. Generally, lawyers value deposition and trial experience that comes with a summary of personal litigation experience on a resume. And of course, they will want to know about disciplinary actions/sanctions associated with professional licensure. The rest of the left side of CREDENTIALS provides additional information usually included in a resume. The circle of QUALITIES is respected by everyone involved in a legal matter.

The journey to become a ***forensic engineer*** is somewhat different than the traditional engineering path. Most engineers study a discipline at university, practice that discipline at an engineering firm, and take a discipline-specific professional engineering exam (the PE) to obtain a specific professional license. "*Specific*" is the operative word in this statement. However, with forensic engineering, failure of something engineered or an in-depth study of a subject often requires multidisciplinary knowledge. Product liability, gas explosions, chemical spills, accident reconstruction, structural distress, materials fatigue, and corrosion... the list goes on. One gains talent across disciplines by studying more subjects and practicing alongside people with other knowledge. Experience is the best knowledge, and lifelong learning is an engineering mantra.

These days we have forensic engineering firms, something few and far between a couple of decades ago, and the best way to become an authoritative FE is to go to work for such a firm. Be the apprentice who expresses desire to learn, embraces challenge and work across disciplines, and enjoys a fascinating and enjoyable FE career.

An expert attains

- Comprehensive knowledge of subject matter supported by extensive practice and research.
- Excellence with attention to detail, cause-and-effect understanding, and technical writing proficiency.
- Ability to understand and solve complex problems and is adaptive to new situations.
- Deliberate continuous improvement and refinement of skills.
- Recognition by peers as an experienced authority with credentials, higher education, and published works.

3.2 The Golden Rule

Journalism has a code of ethics that encourages journalists to avoid political and other outside activities that may compromise impartiality, integrity, and credibility. ***Social responsibility ethics*** teaches that individuals and corporate leaders should act to benefit society in balancing economic and environmental needs with the welfare of people. Modern business ethics teaches core principles of integrity, fairness, accountability, respect, and sustainability. One description of business ethics found in the *Encyclopedia of Applied Ethics* suggests our ethical origins stem from Aristotelianism, Utilitarianism, and Kantianism principles and frameworks.[2] This book contains 3464 pages of entries related to all sorts of fields: bioethics, computer and information management ethics, economic and business ethics, environmental ethics, medical ethics, etc. Do politicians have a code of ethics? I don't recall ever reading a book on ethics. But I did learn *The Golden Rule* at an early age, taught by caring parents, in church, and by peers. The seed took root and matured with time.

Perhaps John Maxwell says it best with his *KISS* (keep it simple students) maxim:

> There is no such thing as business ethics. Ethics is ethics. If you are ethical, you decide with one standard rooted in honesty, integrity, and The Golden Rule.[3]

In everything, *do unto others as you would have them do unto you*. Treat others as you want to be treated. Our duty with everything is to do everything for each other. No matter how we state it, ***The Golden Rule*** is universal and immutable in all major religions and world philosophies. And it embodies one simple practice: *reciprocity*, our social norm of rewarding kind actions.

Institutions and some corporations require employees to take ethics refresher training annually. Try as they might to make their short courses entertaining and engaging with situational enactments and cartoon skits, they do not remind of *The Golden Rule*. Sometimes the simplest things are so hard to imitate with modern educational tools.

How powerful the message:

> DO UNTO OTHERS AS YOU WOULD HAVE THEM DO UNTO YOU

As a shingle hanging in the lobby of the firm one is proud to work at. What is your moral compass?

3.3 Neutrality Versus Advocacy

A lawyer is an *advocate* for his client. Unlike lawyers, experts owe no specific allegiance. An *expert witness* is supposed to be an impartial source of knowledge and opinions that are based on evidence, science, and facts. Therefore, an expert must always remain *neutral*.

Neutrality means not being aligned with or obliged to support any side or position in a controversy. This neutrality can be challenging as an expert hired by one party in a dispute. In litigation, neutrality should not depend on which party retains an expert, but rather should be a professional practice. It involves the expert maintaining a distance from the emotional aspects of the case and the desire of counsel to win. The expert must be impartial in the quest to uncover provable facts and irrefutable evidence. In this capacity, the expert provides an invaluable service by thinking through the complex details without bias. Further, the court is best served when an expert is free to report and testify about the facts without pressure to do otherwise.

However, the lines of advocacy and neutrality can sometimes get blurred. For example, a plaintiff's expert may be asked to assist counsel in preparing for a deposition of a defense expert. In this role, the expert reviews the case information and the opposing expert report and drafts questions. A planning meeting with the attorney is typically part of this process. At the meeting, the expert is exposed to the thoughts and strategies of the attorney, and the concerns of the client. At this point, it is incumbent on the expert to point out any flaws in the logic associated with these views. An open and honest exchange should be encouraged at this critical juncture, with the role of the expert clearly defined and respected.

Sometimes disagreements arise between members of the litigation team. For instance, a strategy dispute, incompleteness of information, treatment of evidence, conflicting expert reports, or underlying money matters. Consider weak evidence, for example. An expert may feel that the evidence does not overwhelmingly support the client's position, or enough strategy points. The basis for this position must be explained and time allowed for discussion and debate. Perhaps some evidence is circumstantial, or more evidence and documentation are needed to buttress an argument.

ADVOCACY SCENARIO

An accident reconstructionist is contacted by a defense attorney to discuss a traffic accident. The discussion is framed around selective information, weak evidence, and multiple plausible scenarios. The expert is hired if the attorney recognizes support for the strategy. The *advocate* meets with the attorney to discuss key elements of the case and strategy. Built on disavowing certain evidence, discrediting investigators and/or investigation work, and analysis, the strategy is designed to elicit reasonable doubt and/or comparative fault. The expert develops an analysis using industry-standard software, invalid or unsubstantiated assumptions, and math and physics to support the strategy. The more scanty or weak the evidence, the greater the potential to hypothesize fiction. Only hard science can refute the fiction.

When did the expert cross the line and become the advocate?

In another example, an attorney may be heavily invested in a client's position after having spent much time (and client money) developing a strategy, only to recognize how weak their case is after reviewing opposing expert reports. A meeting is called to discuss the new information and how to deal with it. A conflict of interest arises, and the expert's unwillingness to testify can be a signal to counsel not to depose witnesses, and settlement is in the best interest of the client.

NEUTRALITY ALTERNATIVE

An expert accident reconstructionist is hired by a plaintiff's counsel to determine the cause of a traffic accident. The expert meets with the attorney and client to discuss the key elements of the case. They talk about what is expected, as well as the expert's background, credentials, and any potential conflicts of interest. Full disclosure of case information and evidence is afforded, along with documents, time frames, and other administrative requirements. The expert reviews and opines about the quality of the evidence, the merits of the case, and the ability to support the case with sound science. The team (attorney/client/expert) communicates in an open forum on key positions, facts, damage theories, etc. All viewpoints are respected and evaluated. By doing so early on, the expert is better able to maintain a sense of independent objectivity. This leads to more effective testimony and less vulnerability to attack from opposing counsel, who will likely try to impeach the objectivity and independence of the expert. The expert applies hard science and develops evidence-based opinions about causation through the pursuit of the truth.

Expert independence and objectivity within the legal system must be respected. An expert witness is supposed to be an impartial source of knowledge and opinions that lead to facts. Jeopardizing the role of the expert comes at the risk of branding the ***hired gun***.[1] Slanted opinions that are not based on empirical data and theory are subject to vicious attack on cross-examination and potential dismissal of the expert by the court. ***Expert dismissal*** can negatively impact the reputations of the expert and legal team and does not serve the client well.

Testimony and written reports offer opinions and conclusions that should be factual and free of bias. A benefit of expert reports can be the timely resolution of a case without the need for trial. A well-written report based on facts, sound science and analysis, and independent lines of evidence that converge to the same or similar conclusions is highly desirable to those who respect truth. In contrast, a report that is based toward a position of advocacy and a predetermined outcome will not stand up to sound scientific scrutiny.

In criminal law, this is a primary reason why some attorneys now request their experts not write reports. Some examples of advocate reporting to avoid:

- A single-page introduction about the event (or incident) followed by four pages of conclusions and half a dozen references listed without indication of how they were used. (This is an unprofessional report that lacks substance and support for stated conclusions.)
- An analysis and results section that uses selective imagery to bolster a position that tells only part of the story or implies a misleading scenario. (Imagine opposing experts both investigating the same failed roof-cover scenario discussed in Chapter 2, and one selectively uses photographs of the best-performing areas in support of a position that implies poor installation rather than a bad product.)
- An ACTAR-certified nonengineer riding on the coattails of an engineer father who never reports about hundreds of AR investigations. With every difficult question, this pseudo-expert goes to a computer just out of view of the recording camera to supposedly look something up to help answer during a virtual deposition (mandated during the COVID-19 pandemic). One might imagine the disbelief of those witnessing this travesty in court.

> Experts need to own their opinions and hold each other accountable.[4]
>
> **Tom Nichols, U.S. Naval War College**

3.4 Credentials

Creds, the left side of Figure 3.2, are essential ***credentials*** of the expert. The forensic engineer should seek higher education such as a master's degree in a specialized field, or an interdisciplinary PhD, which usually comes with an in-depth study of a multidisciplinary research project. A PE license is the gold standard for the forensic engineer. Like the bar exam, lawyers know what it means to earn a PE, what it takes to maintain it, and the professional reputation it engenders. The point is: higher education and PE licensure are the two most important credentials of the forensic engineer. Obtain both and doors will open.

There is no substitute for experience. We have two hands and a brain, the basic tools needed to obtain experience. Get as much investigation experience as possible during the apprenticeship. Skills sharpen, critical thinking and attention to detail improve, and cause and effect become more apparent sooner. We also learn which tools best lead to the answers.

> What I hear, I forget. What I see, I remember. What I do, I understand.
>
> **Xunzi, Chinese philosopher, second century BC**

3.5 Science in Everything and Patience

Soil, water, and biota are ecosystems. The interrelationships and interactions can be complex, dynamic, intermittent, and spatially variable. Materials experience forces, angular acceleration, inertia, rotation, friction, wear, corrosion, stress, and fatigue. Explosions come from leaks, sparks, or impacts. Space is a place with lots of unknowns. If we want to go there, it requires novel designs with high factors of safety and only earthly means to test and simulate what we think it will be like. The laws of physics apply to everything as we currently understand them. Things we engineer require maintenance, get damaged, and people sometimes get hurt using them. When we endeavor with things in various situations, much critical thinking and effort go into minimizing failure.

Failure results in evidence. Examination of evidence leads to cause and effect. Evidence is relational for any single event, and the convergence of relational evidence leads to facts. Therefore, *relationship* is the key to understanding the science of failure, and applying scientific methods of inquiry is the best way to figure out why a thing failed. And sometimes the simplest explanation is the right one – *Occam's Razor.* With forensic investigation, failure is rarely unique, and its root cause is usually simple to understand once we find it. But the finding often requires patience.

One of the best ways to develop infrastructure damage assessment skills is to audit a functioning structure that has been around for a while. Take a building, for example. It is on the ground, has an exterior and interior, utilities run in and out of it, people work in some spaces, machines and electrical equipment in others, and people go up and down in it using elevators and stairways. There may also be equipment on the roof and in ground-level mechanical rooms, and an interior parking garage.

An audit requires a checklist, field log, camera, magnifying glass, tape measure, color tape and pencil for marking, and a video camera just in case. The objective is to document all signs of wear, distress, problems, and maintenance issues. From smallest to largest, document everything observed that suggests a potential issue and that is out of the ordinary. Cracks in the walls, foundation, columns, and rooftop parapets, wear on steps of the stairways and machines, odd noises and vibrations with mechanical equipment, elevator and fire and life safety system irregularities, utility leaks and signs of water intrusions, and so on.

Start with the outside of the building, looking at all sides from the ground up. Then go inside, floor by floor, room by room, again from the ground (or garage) up. Document all observed issues. Once the physical inspection is complete, then review the maintenance records. Maintenance records document schedules, repairs, and pending repairs. Such records might lead to something missed with the physical inspection. The inspection tools – tape measure, color tape, and pencil – are for measuring cracks, flagging issues worthy of photographing or that warrant a second look. Number the flags to correlate with photographs and location and log this information on the checklist and field book as appropriate. The patient mentor oversees, reminds, encourages, and answers every question throughout the process. Imagine what one can learn from auditing three different types of infrastructure during that first year as a forensic engineer.

3.6 Excellence – the Highest Standard

Excellence as applied to forensic engineering means leaving no stone unturned with investigation, searching until you know, and well-written technical reports. It is an exemplary standard of practice that builds an impeccable reputation. Do everything to excellence and always be in demand.

"What matters is only what I can prove." A quote from the character Helen Rodin, played by Rosamund Pike, expressed a significant sentiment in the film *Jack Reacher.* It emphasizes the importance of evidence and proof beyond a doubt. In forensic investigation, a minimum of three independent lines of evidence must converge for proof beyond a doubt. Physical evidence, videography that captured the event, and eyewitness accounts are a pretty good set of three, for example.

Excellence is not perfection. Perfection is unattainable and always debatable. Excellence implies a high standard of achievement. Therefore, strive for excellence and become an excellent engineer. There is no room for average in forensic engineering.

> It takes less time to do a thing right, than it does to explain why you did it wrong.
>
> **Henry Wadsworth Longfellow**

3.7 Reports

What the engineer writes and testifies to can be used against him or her in a court of law. Therefore, *KISS* and ***communication liability*** are essential tools of the report writer. Figure 3.3 outlines an example forensic investigation report. Six primary sections, followed by references and a limitations statement, is a KISS format for a narrative report. This template has broad application with investigations of different types of subjects, with minor adjustments. Content in later chapters and case studies will demonstrate its use.

Applied to the completed draft report, *communication liability* is about eliminating report liability. Condensing paragraphs, omitting unnecessary text

References

1. Stephen E. Petty, Editor. (2022). *Forensic Engineering: Damage Assessments for Residential and Commercial Structures,* 2nd Ed., CRC Press, Boca Raton, Florida, p. 740.
2. J.D. Rendtorff (2012). *Encyclopedia of Applied Ethics,* 2nd Ed., Elsevier, ISBN: 9780123736321, p. 365–72.
3. Borrowed from John C. Maxwell, 2003. *There's no such thing as "Business Ethics". There's Only ONE RULE for Making Decisions.* Warner Books, 142 pp.
4. Tom Nichols (2017). *The Death of Expertise: The Campaign against Established Knowledge and Why It Matters.* Oxford University Press, 249 pp. ISBN 978-0-190-46941-2.

Case Study 3: Real Estate Predicament

Forensic Engineer Reviews Conflict with Case Facts

- Real Estate
- Consumer Protection
- Insurance

Message to students:

> Sometimes people get involved in real estate predicaments that lawyers need an engineer to help sort out. When hired as an expert, focus on the engineering and remain neutral and unbiased through the entire legal process.

This case involves a real estate transaction between realtors that involves multiple engineering inspections, a history of home freeze-up events and repairs, and fraud accusations. The buyer filed a lawsuit against the seller claiming known problems with the home were not disclosed prior to the sale

DOI: 10.1201/9781003623311-7

as required by law. Through their first year of ownership, the new homeowners started an addition and experienced water intrusions that resulted in significant interior damage during winter. An attorney hired a forensic engineer to investigate and identify problems with the home. The FE performed an independent investigation free from interactions with the litigants. All names of people and businesses herein are fictional. The case takes the reader through the legal process from the plaintiff's complaint and defense answer, through expert discovery to demonstrate the role of the FE as an expert witness to help the court resolve the matter.

Background

The subject property was developed in 2005 with a large single-family split-level wood-frame house with a basement on three acres within a new upscale subdivision of Dodge, Utopia. First owners split time in two states and used Spartacus Property Management (SPM) company to manage their property through September 2013 when they sold the home to Rick Shyster, a local real estate broker and owner of Dream On Real Estate Services, LLC (Dream On). Mr. Shyster collaborated with the original owners to effect repairs prior to his purchase.

Meantime, Mr. and Mrs. Holiday moved to Dodge in 2013. Mrs. Holiday, a licensed professional realtor, became employed with Dream On. The Holidays rented the subject property from Mr. Shyster in fall 2013, and subsequently entered into an agreement to purchase the property in April 2014. Mr. Shyster hired a contractor to make a few minor repairs, and both seller and buyer hired separate engineers (Engineer A and Engineer B) to provide independent home inspections. The engineers identified deficiencies and made recommendations for repairs; the sale was contingent upon satisfactory completion of the repairs. Mr. Shyster coordinated the additional repairs, which were completed in Mar-Apr 2014. The Holidays then purchased the property in June of that year.

The Holidays experienced water intrusion and damage inside the home during the following winter of 2014–2015, and subsequently filed a claim with their insurance company (State Ranch) in March 2015. State Ranch hired Engineer C to investigate the claim. Engineer C determined that the primary cause of water intrusion was an insufficient roof system and made recommendations for repairs. Engineer C also noted that a single-story addition was underway at the back of the house. The Holidays hired a contractor to effect roof repairs in accordance with Engineer C's recommendations and hired Engineer D for the postmonitoring inspection of the home per State Ranch requirement. Engineer D confirmed that repairs had been made but recommended additional improvements to the roof system. Further, Engineer D suggested an improperly installed *cold roof* [1] was the cause of the water problem but would not guarantee the recommended upgrades would stop the moisture problem.

In June, while attending a home improvement workshop, Mrs. Holiday coincidentally met the contractor (Tudor Construction) that Mr. Shyster had used to make repairs to the subject home prior to the property sale. This contractor told a story about how Mr. Shyster told him to fix some things but ignore water damage issues, as he (Mr. Shyster) would fix those himself. Mrs. Holiday requested and obtained a letter from the contractor that retold this story in writing. At this point, the Holidays believed they had been defrauded, and subsequently retained an attorney and filed a complaint with the proper jurisdictional court in September 2015. A lawsuit was effectively initiated.

Complaint and Answer

Complaint	Defendant's Answer to Complaint
COME NOW Mr. and Mrs. Holiday, pursuant to State Rule of Civil Procedure #, by and through attorneys of record, the law office of Alius, Smith & Jones, and amend the complaint and allege the following against the defendant:	All defendants jointly answer plaintiff's complaint as follows:
1. Mr. and Mrs. Holiday are husband and wife and residents of Dodge, Eleventh Judicial District, State of Utopia.	1. Admit
2. Rick Shyster is a resident of the Eleventh Judicial District, is a licensed Real Estate Broker with the State of Utopia and owns Dream On Real Estate Service, LLC (Dream On).	2. Admit that Rick Shyster is a licensed Real Estate Broker and owns a portion of Dream ON. Deny legal import of Dream On d/b/a Millenium 31. Dream On's involvement as the entity did not exist at any relevant time herein.
3. Rick Shyster and/or Dream On are owners of the Millenium 31 Real Estate Corporation (Millenium 31).	3. Admit that Mr. Shyster owns the local Millenium 31 franchise. Deny remaining allegations.
4. Lori Kiter is a resident of the Eleventh Judicial District, State of Utopia, is a licensed Real Estate Agent with the State of Utopia, and is an owner of Dream On.	4. Admit. Lori Kiter owns a portion of Dream On.
5. Millenium 31 is a Utopia corporation in good standing and subject to the jurisdiction of this court.	5. As this allegation is directed at a nonanswering defendant, it is denied.
6. On or about April 9, 2014, Mr. and Mrs. Holiday entered into a Consensual Dual Agent Agreement consenting to Lori Kiter of Dream On acting as a Dual Agent to facilitate the Holiday's purchase of real estate and improvements located at 2001 Odyssey Lane, Dodge, Utopia. *See* Exhibit A.	6. Admit

(*Continued*)

Complaint	Defendant's Answer to Complaint
7. The real property and improvements were owned by Rick Shyster, a licensed Real Estate Broker in the State of Utopia at the time. *See* Exhibit A.	7. Admit
8. Pursuant to the Consensual Dual Agency Agreement, Lori Kiter's role was to facilitate the transaction between the Holidays and Rick Shyster. As a Dual Agent, Ms. Kiter was not liable to either party for refusing or failing to disclose confidential or privileged information, **except** for known material defects or other specific disclosures required by law. *See* Exhibit A, page 2.	8. Objection, this allegation calls for a legal conclusion. To the extent that this allegation is factual, the contract speaks for itself.
9. The Purchase Agreement stated: "Unless otherwise disclosed in writing, the Seller acknowledges the roof if free of all leaks; there are no problems with the drainage; and the basement is free of water and leaks." *See* Exhibit A, paragraph 12(b).	9. Admit accurate quote. Deny legal import. Allege that one or more answering defendants disclosed to plaintiffs.
10. Mr. and Mrs. Holiday closed on the purchase of the home on June 23, 2014.	10. Deny for lack of information and belief.
11. The State of Utopia requires property disclosures in accordance with statute X be provided to the buyer before a property transaction is completed by the parties.	11. Objection, the document speaks for itself. Calls for a legal conclusion and is therefore denied. Affirmatively allege that mrs. Holiday had lived in the house for months, was working for one or more of the defendants at the time the sale closed and knew full well every aspect of this home.
12. There is no indication in the agreement that any disclosures were provided or made a part of the transaction. *See* Exhibit A, paragraph 15.	12. Deny
13. The Holidays did not receive any disclosures from Rick Shyster, Lori Kiter, or from anyone else at Dream On.	13. Deny
14. The Holidays did not agree to nor execute a waiver of their right to receive the mandatory disclosures.	14. Deny for lack of information and belief.
15. In March 2015, the Holidays had trouble with water coming through their living room ceiling and front entryway. Not knowing what was causing the problem, they contacted their homeowner's insurer.	15. Deny for lack of information and belief. It is affirmatively alleged that the holidays made substantial structural changes to the house before having any alleged problems.
16. In June 2015, State Ranch Insurance sent an engineer to inspect the home and comment on the problems. The inspecting engineer suggested strategies to resolve the problems which the Holidays tried to implement. Despite their best efforts, the problems continued and became worse over time.	16. Deny for lack of information and belief.

(*Continued*)

Complaint	Defendant's Answer to Complaint
17. By chance, Mrs. Holiday encountered Bart McGillicuddy, owner of Tudor Construction, LLC, in December 2015. When the discussion of problems with the Holidays' home came up, Mr. McGillicuddy indicated that he had actual contact with Mr. Shyster regarding the problems at the home and Rick Shyster had given him documents related to problems with the home prior to the Holidays' purchase of the home. *See* Exhibit B.	17. Deny for lack of information and belief.
18. Mr. McGillicuddy provided a copy of the documents that Mr. Shyster had given him regarding the house. The documents explicitly addressed the problems the Holidays were experiencing and stated that much of the ceiling on the inside of the house had signs of water damage due to a defect in the insulation, vapor barriers, or design. *See* Exhibit C.	18. Objection, Exhibit "C" speaks for itself. Alleged same documents previously supplied by Mr. Shyster to plaintiffs. Deny for legal conclusion.
19. Prior to the sale of the house to the Holidays, Mr. McGillicuddy had completed some repairs but was explicitly instructed by Mr. Shyster not to worry about the moisture problems. *See* Exhibit B, page 2.	19. Deny
20. Mr. Shyster, as Broker of Dream On, had a duty to properly supervise his agent, Lori Kiter.	20. Admit
21. Even acting as a Dual Agent to facilitate the real estate transaction, Lori Kiter owed the Holidays a duty to inform them that by law the seller must provide a Property Disclosure Statement unless the parties execute a waiver.	21. Calls for a legal conclusion and is therefore denied. Alleged full disclosure.
22. Mr. Shyster, a licensed Real Estate Broker in the State of Utopia, was aware that he was required by Utopia state law to provide a completed written Property Disclosure Statement, or an agreement in writing exempting the transaction from the statutory disclosure requirement, and yet he provided no written disclosures.	22. Deny. Affirmatively allege that plaintiffs lived in the house, knew of issues with it, and defendants provided full disclosure, but not on the forms required. Calls for legal conclusion and is therefore denied.
23. Lori Kiter owed the Holidays a duty to facilitate the real estate transaction in an expert and professional manner, including informing them that mandatory disclosures, or a written waiver, were a statutory requirement for the sale of the residential real property and ensuring that disclosures of a waiver were incorporated into the real estate transaction.	23. Admit Lori Kiter had a duty to handle the transaction competently. Alleged information withheld by plaintiffs. Allege full disclosure to plaintiffs, though not on statutory forms. Calls for a legal conclusion and is therefore denied.
24. Lori Kiter failed to live up to the terms of the Dual Agency Agreement and Utopia statutes regarding this real estate transaction.	24. Calls for a legal conclusion and is therefore denied.

(*Continued*)

Complaint	Defendant's Answer to Complaint
25. Lori Kiter failed to use such skill, prudence, and diligence as other members of the real estate profession commonly use in facilitating the real estate transaction.	25. Deny
26. Lori Kiter's violation of her duty as Dual Agent and failure to use such skill, prudence, and diligence as other members of the real estate profession commonly use is a proximate and legal cause of damages to the Holidays.	26. Deny
27. Mr. Shyster, as the Broker of Dream On, had a duty to the Holidays to properly supervise his Agent Lori Kiter. Mr. Shyster failed in that duty.	27. Admit duty. Deny that there was a breach of the duty.
28. Mr. Shyster's failure to properly supervise his Agent, Lori Kiter, is a legal and proximate cause of damages to the Holidays.	28. Deny
29. Mr. Shyster's failure to provide the mandatory disclosure statement to the Holidays was a violation of state statute X.	29. Calls for a legal conclusion and is therefore denied. Alleged full disclosure.
30. Mr. Shyster's failure to provide the mandatory disclosures required by statute X when he had explicit knowledge of the ongoing and extensive problems of the house constituted bad faith and is in direct violation of statute Y.	30. Deny
31. Mr. Shyster's failure to provide the mandatory disclosures is a legal and proximate cause of damages to the Holidays.	31. Deny
32. Mr. Shyster and Dream On entered a contract with Millenium 31.	32. Admit
33. Pursuant to that contract, Dream On was provided with information and advertising to further its real estate business. Namely, the Millenium 31 moniker and Millenium 31 system for promotion and operation of a real estate office.	33. If the allegation is directed toward Millenium 31, the contract speaks for itself. Otherwise, it is denied for lack of information and belief at this point.
34. In consideration, Dream On was required to pay a portion of its profits to Millenium 31. Mr. Shyster and Dream On were also obligated to live up to other requirements to assure the Millenium 31 standard was upheld at the Dodge office.	34. If the allegation is directed toward Millenium 31, the contract speaks for itself. Otherwise, it is denied for lack of information and belief at this point.
35. One of the requirements was that Mr. Shyster and Dream On obtain insurance, and that insurance separate endorsements naming Millenium 31, Artimus and Artimus Finance Holding Corporation as additional insureds. Dream On was required to provide Millenium 31 a certificate of insurance proving the existence of said endorsements annually.	35. If the allegation is directed toward Millenium 31, the contract speaks for itself. Otherwise, it is denied for lack of information and belief at this point.
36. Neither Mr. Shyster nor Dream On has been insured since February 2011.	36. Deny

(Continued)

Complaint	Defendant's Answer to Complaint
37. The Holidays contacted a Millenium 31 office. They had full and reasonable expectations that the Millenium 31 office in Dodge met the same high standards that Millenium 31 advertises regionally.	37. Admit
38. Millenium 31 was negligent in its supervision of the local Dodge office in that among other things, its failure to recognize that Mr. Shyster and Dream On had not sent certificates of insurance for three years.	38. Calls for a legal conclusion and it therefore denied. Affirmatively allege that Mrs. Holiday was working for one or more of the defendants during this time and was well aware of the related facts.
39. The Holidays are foreseeable third-party beneficiaries of the contract between Millenium 31 and Mr. Shyster and Dream On requiring insurance.	39. Allegation directed to a different defendant than answering defendants. To extent response is required, this allegation is denied.
40. Mr. Shyster and Dream ON acted in a manner that caused the Holidays to reasonably believe that it was authorized to act as a Millenium 31 office.	40. Calls for a legal conclusion and is therefore denied.
41. Millenium 31 did nothing to let the Holidays of others know that Mr. Shyster and Dream On were not acting with their authority or living up to Millenium 31's standards and requirements of the contract.	41. Calls for a legal conclusion and is therefore denied.
42. Millenium 31 is legally responsible for the actions of Mr. Shyster and Dream On.	42. Admit as to Mr. Shyster; deny balance.
43. Millenium 31's acts and failure to act are a legal and proximate cause of the damages suffered to the Holidays.	43. Deny All allegations not specifically admitted are hereby denied.
WHEREFORE, Mr. and Mrs. Holiday request the following relief: A. That judgment entered in their favor and against Rick Shyster, Lori Kiter, Dream On, and Millenium 31. B. That they be awarded damages for past and future economic loss. C. That they be awarded damages to compensate them for past and future noneconomic damages. D. That they be awarded damages to compensate them for past and future emotional distress. E. That they be awarded treble damages as permitted by statute. F. That they be awarded punitive damages. G. That they be awarded full attorney fees as permitted by rule. H. For such other relief as the Court deems just and equitable. EXHIBIT A: Earnest money receipt and 10-page Purchase Agreement. EXHIBIT B: Todor Construction 1-page letter of repair work for Mr. Shyster. EXHIBIT C: SOS Maintenance emails and notes documenting 2012-13 presale repairs to the subject home.	**AFFIRMATIVE DEFENSES** 1. The complaint fails to state a claim upon which relief may be granted. 2. Statute of limitations. 3. Unclean hands/laches. 4. Failure of third-party beneficiary status. 5. Comparative fault/contributory fault. 6. Statutory avoidance/effective compliance. 7. Failure of one or more conditions precedent. 8. Hold harmless agreement in Consensual Dual Agency Agreement precludes recovery. Defendants reserve the right to assert such other and further affirmative defenses as arise during the course of discovery. **WHEREFORE** defendants request that the complaint be dismissed with prejudice, that defendants be awarded full and reasonable costs and fees for defense of this suit, and such other and further relief as this Court deems just.

The complaint comprised 43 *counts* against multiple defendants: Mr. Shyster, Dream On, Millenium 31, and realtor Ms. Kiter. This is because all the parties sued were linked to the real estate transaction in question, a common practice with unknown insurance coverages and recovery may require drawing from multiple assets. The court decides to indemnify or rule out any defendant based on findings and legal precedent.

Discovery: Investigation, Findings, and Mitigation

Investigation revealed three repair events relevant to this case. The first event, which took place between July and August 2013, was repairs commissioned by the original owner through their property manager to ready the home for sale to realtor Mr. Shyster. Figure CS3.1 is a timeline of repairs

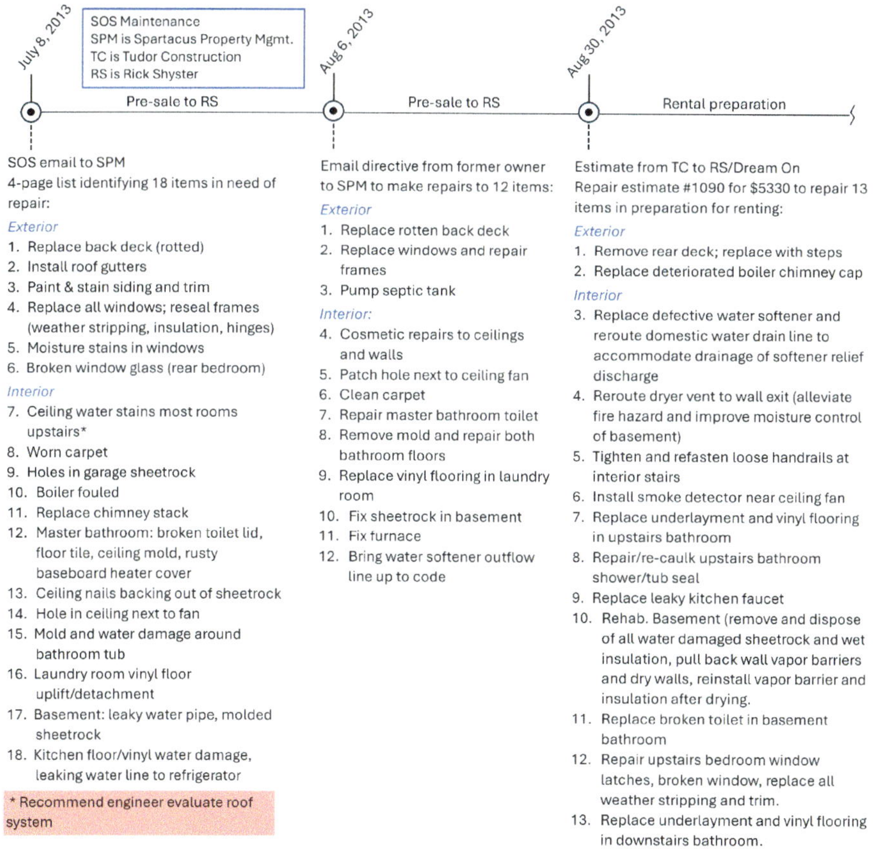

FIGURE CS3.1
Timeline of home repairs in preparation for sale to Mr. Shyster and rental to the Holidays.

related to that event. The second event related to repairs made by Tudor Construction (TC) for Mr. Shyster between August and December 2013 (Fig. CS3.2) in preparation for renting the home to the Holidays. Note that Figures CS3.1 and CS3.2 are linked by two repair lists – a short list and a long list – made by TC. Figure CS3.3 represents a 2-month period of pre-sale repairs and engineering home inspections related to the property transfer to the Holidays in spring 2014. Subsequent to a problematic winter, the Holidays filed an insurance claim, and the insurer commissioned

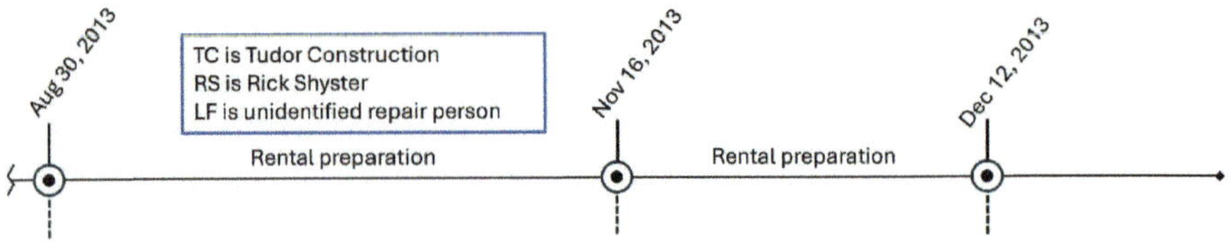

Estimate from TC to RS/Dream On
Repair Estimate #1091 for $37,881 to repair 24 items in preparation for renting:

Exterior

1. Pressure wash/stain exterior siding, entire structure
2. Replace exterior outlets with GFI
3. Build new rear deck
4. Replace broken exterior light fixture
5. Trim trees
6. Remove smaller of two TV antennas
7. Install roof gutters all eaves
8. Replace exterior door at dining room, including jamb and rotted threshold
9. Replace sticking front door lock
10. Replace/repair 18 openings/windows in home

Interior

11. Remove junk from boiler room
12. Patch and repair all cracked/damaged drywall, mask and tape, prime and repaint to match existing texture/color
13. Install GFI outlets near kitchen sink per code
14. Replace all missing/mismatched light bulbs throughout home
15. Replace wobbling upstairs ceiling fan
16. Replace all missing/damaged cable and phone jacks throughout home
17. Remove all damaged/outdated mini-blinds in home
18. Replace substandard lights in dining room and kitchen per code
19. Replace vinyl flooring and trim at garage entry and laundry room
20. Replace upstairs bath/shower damaged sheetrock and re-caulk tub
21. Replace missing trim above entry door
22. Replace rusting electrical outlet in den
23. Replace rusting water tank in basement
24. Install boiler fresh air intake

LF to RS/Dream On
Repair invoice #685 for $6000 (balance due $2000)

Exterior

1. Demo & remove rear deck

Interior

2. Reroute two water pipes
3. Demo all water damaged sheetrock and replace with new in basement
4. Reseal vapor barrier as necessary in walls
5. Cover and mask all areas that need painting
6. Patch all damaged walls and ceilings to match existing texture
7. Prime and paint all walls and ceilings
8. Sand and paint baseboard heaters

TC invoice to RS/Dream On

Inv. #1009 for $23,945 (discounted $1243*, $23,200 prepaid, balance due $745)

*RS identified as loyal customer

FIGURE CS3.2
Tudor Construction's long list of recommended repairs.

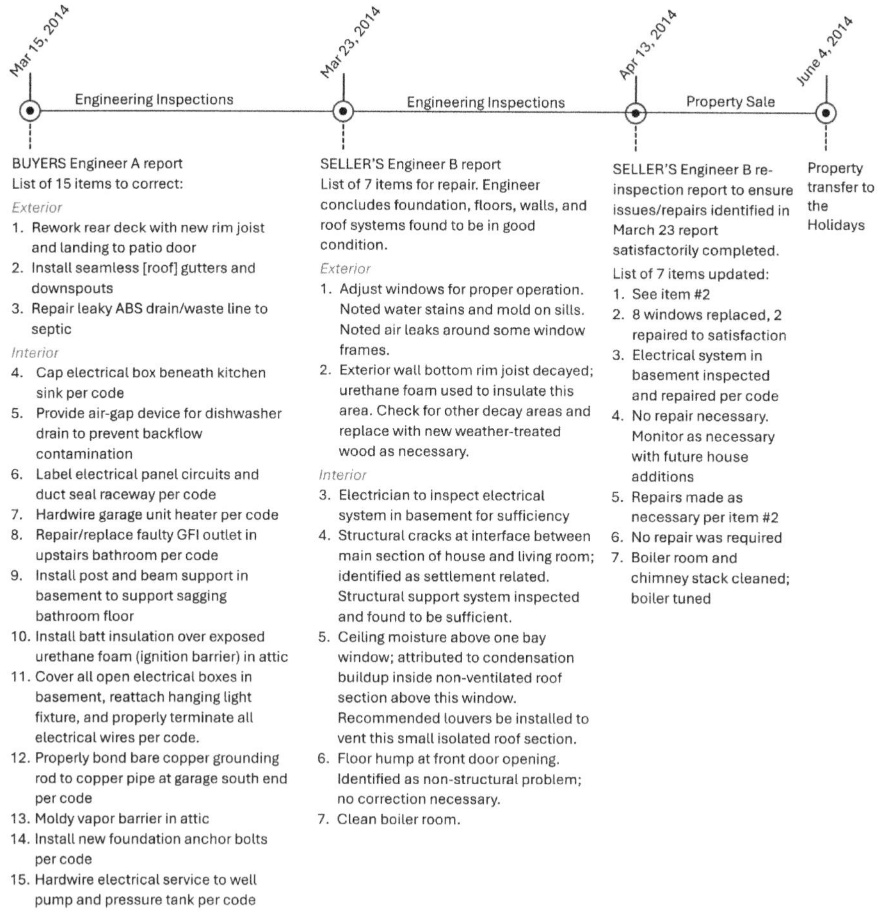

FIGURE CS3.3
Timeline of presale engineering home inspections and property transfer period.

an engineering investigation to figure out what was wrong with the house. Figure CS3.4 represents the postsale insurance claim and engineering investigation period from March to May 2015.

In summary, SOS identified 18 items of concern, the most serious of which were the ceiling water stains and a recommendation to have an engineer inspect the roof system. The former owner authorized repairs to 11 items, notably 'cosmetic repairs' (painting and patching) of water-damaged ceilings and walls. The assessed damages were consistent with a house that had frozen up. The recommendation to have the roof system evaluated by an engineer (red highlight in Fig. CS3.1) was ignored. (Cosmetic painting and patching water-damaged ceilings and walls attests to ignoring the *source* of the water intrusion problem.) Mr. Shyster also approved Tudor

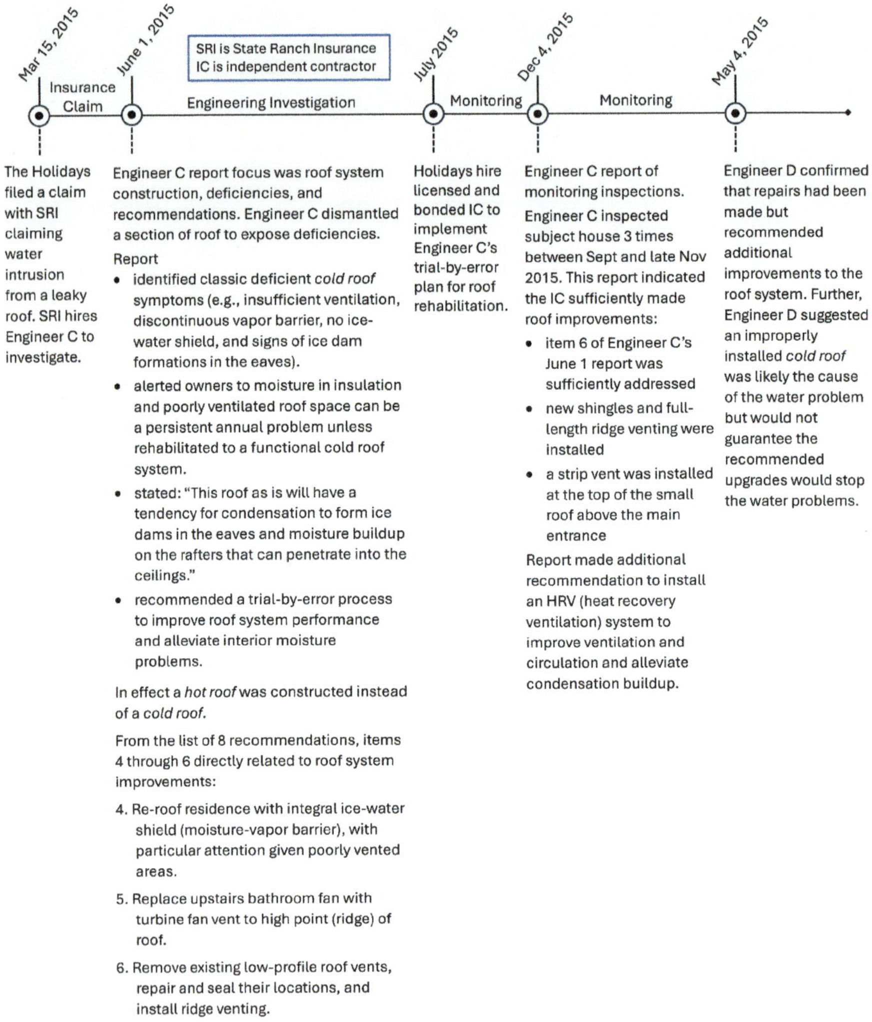

FIGURE CS3.4
Postsale insurance claim and engineering investigation period.

Construction's 'short list' (Estimate #1090) of repairs in preparation for renting the house to the Holidays. But there was also a longer list of recommended repairs, and repairs performed by another contractor identified only as LF.

TCs December 12 invoice amount suggested a level of repairs between the short and long lists were completed. Additionally, a repair person identified only as LF on an invoice performed *water damage* repairs and invoiced

Rick Shyster in November of 2013. Recall Mr. Shyster had told TC that he would take care of the water damage issues. The LF repairs are consistent with what Mr. Shyster communicated to TC about repairs back in August. Note that again there is no mention of *roof system inspection and evaluation* in the LF invoice. This completed the repairs in preparation for renting to the Holidays.

Next, early 2014 presale engineering reports were evaluated (Fig. CS3.3). It is important to note that the engineers were not privy to the specifics of the repairs that were made in 2013. Realtors use preferred home-inspection engineers for this low-budget component of a real estate transaction.

In summary, the home-inspection engineers identified a few electrical and cosmetic interior, and roof drainage issues. Upon completion of the repairs and the electrician's inspection, the engineers approved the house for sale. Neither engineer observed a problem with the roof system, despite attic soffits covered with blown-in cellulose insulation (see Fig. CS3.5). The reports

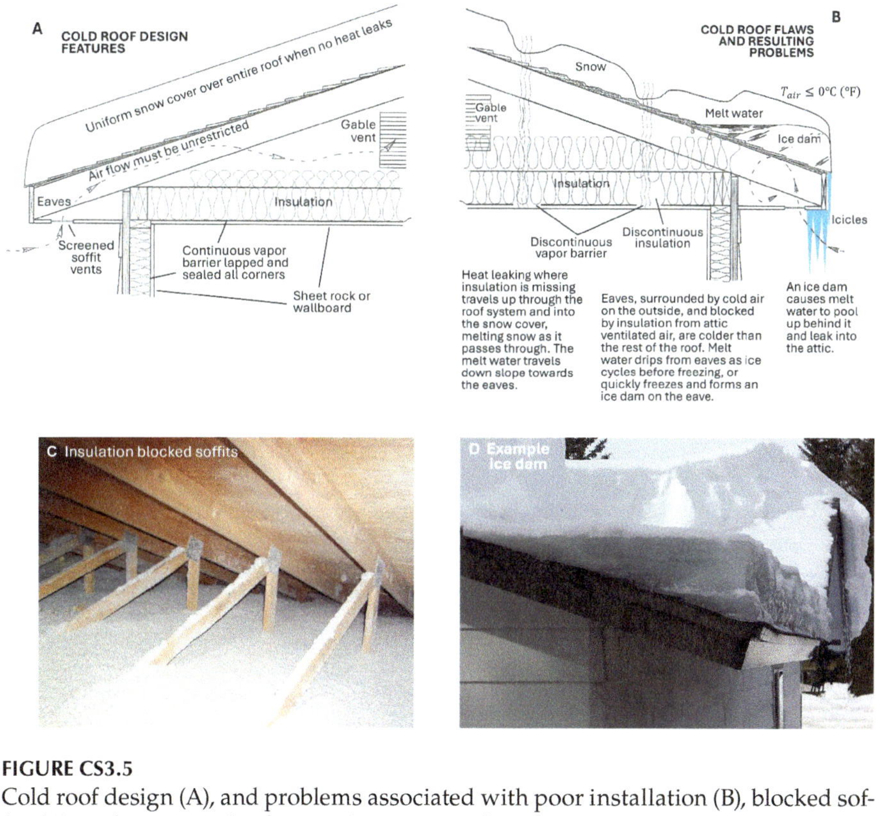

FIGURE CS3.5
Cold roof design (A), and problems associated with poor installation (B), blocked soffits (C), and an example of an ice dam on a roof eave (D).

did not mention if the engineers accessed and inspected the attic spaces. The parties were satisfied and property ownership was transferred to the Holidays on June 4, 2014.

The Holidays then retained an engineer to investigate the roof system. Engineer C hypothesized an improperly installed *cold roof* and dismantled a section of the roof to expose what was effectively a *hot roof* situation. He then recommended upgrades that would rehabilitate to a cold roof system (Fig. CS3.4). Following a series of monitoring events, a fourth Engineer D (recall State Ranch's requirement that an independent engineering validate and sign off on the improved roof system) validated the new cold roof and HRV systems had effectively mitigated water problems through winter 2014-15. However, Engineer D would not ensure (sign off) that the upgraded roof would mitigate water problems moving forward without at least two years of additional annual inspections.

By May 2015, the Holidays had spent tens of thousands of dollars fixing a house that had a bad roof system and experienced two freeze-up events prior to their renting. They claimed that none of this history was disclosed to them prior to their purchase of the home.

Roof Systems and Building Freeze-up

Residential structures have cold, warm, and hot roofs. A *cold roof system* (Fig. CS3.5) for homes in northern states (like this case study) is passively ventilated with soffit and ridge vents and insulation is placed between the rafters, which act to keep the roof cavity (attic) cold compared to the living space below. This conventional design allows air circulation and helps prevent condensation and ice dams from forming along the roof eaves (Fig. CS3.5, frame D). Key features are ventilation and an integral vapor barrier to prevent condensation and moisture buildup in the attic. Advantages are a cooler home in summer, prevention of ice dams in winter, and energy efficiency. However, crucial moisture control is challenged with inadequate ventilation and a discontinuous vapor barrier, and water damage, mold, and wood rot can result.

A *warm roof system* places rigid board insulation directly over the roof deck, making the attic part of the heated space of the building, and the insulation acts as a barrier to cold air penetration. Advantages are that no ventilation is required, and minimal heat loss from the building results in improved thermal control inside. However, this roof system has the same moisture control requirements as with cold roofs, and costs more to install.

Hot roofs are for warm to moderate climates with mild or short winters like Florida. With minimal insulation and ventilation (soffits only), no need for a vapor barrier, and little chance of eaves icing up, hot roofs are the most energy efficient and least costly to install. However, hot roofs are not for cold-climate structures.

Building freeze-ups are a northern winter phenomenon to avoid at all cost. A freeze-up is often the consequence of nobody being home when the heating system fails (e.g., power outage, bad thermostat, clogged fuel injector with a boiler). Left unchecked, the air temperature in the house eventually drops to below freezing, causing water to freeze, expand, and split pipes. One might imagine that with pressurized water in the pipes, bursting discharges more water at multiple locations throughout a building. If a burst pipe is an intake from a well, the warmer water supply might discharge for a long time, causing massive icing inside and total loss as shown in Figure CS3.6. For this to happen requires the three W's: winter, water, and well. It happens more often than you hear about in the news.

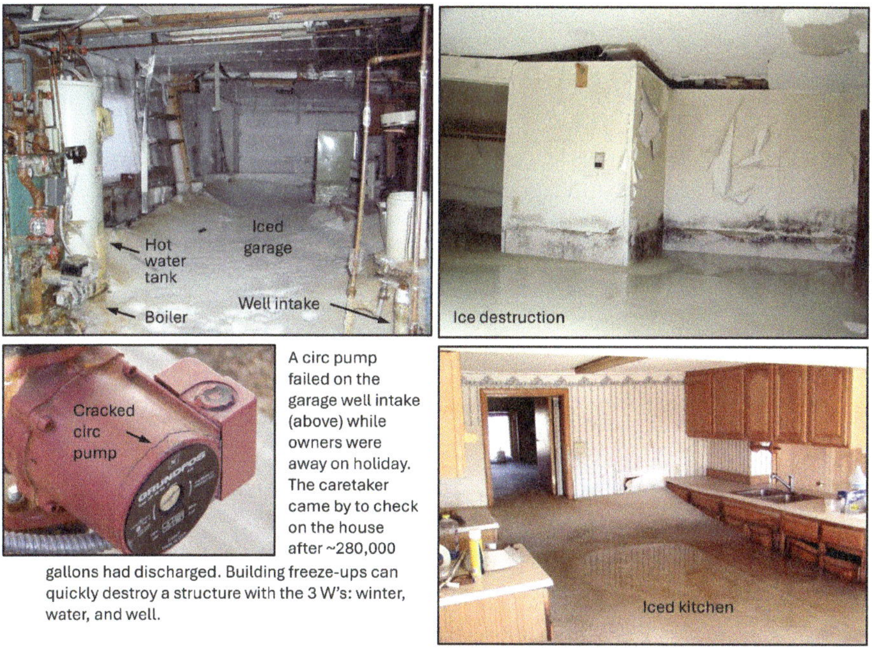

FIGURE CS3.6
Structural freeze-ups in winter.

Findings

So, who is responsible for the dream house catastrophe in this case? We leave this for students to consider with Chapter 3 exercises.

Note

1. A "cold roof" is a type of roof system where the attic space is ventilated with soffit and ridge vents and insulation is placed between the floor joists, keeping the uninsulated roof cavity cold compared to the living space below. This conventional design allows air circulation and helps prevent condensation and ice dams from forming along the roof eaves.

4

Jurisprudence

CONTENTS

CONCEPTS	KEY TERMS	ADDITIONAL READING
Federalism	US Constitution	Exercise 7
Legal System	Jurisprudence	Exercise 8
Federal Courts	Roman and canon law	
State Courts	Justinian Code	Case Study 4
Civil Law	District, circuit, and county courts	
Criminal Law	Common law	
Negligence, Standard of Care, and Performance Evaluation	Statutory law	
	Constitutional law	
	Administrative law	
Evidence Standards	Frye test and Daubert standard	
Interrogation	Predictive similitude	
Cross-examination	Preponderance of evidence	
	Reasonable doubt	

The United States legal system is founded on powers and structure in accordance with the ***US Constitution***. In our engineering capacity to help the courts, it is important that we understand how our system of governance relates to the laws, regulations, and courts that define American ***jurisprudence***.

4.1 Federalism and Our Legal System

The United States of America is a *federated nation*. ***Federalism*** means dividing powers between a central (federal) government and regional (state, provincial,

DOI: 10.1201/9781003623311-8

territorial, autonomous regions) governments. Canada, Brazil, Argentina, Germany, Ethiopia, India, Australia, and Russia are other examples of federated nations that represent approximately 40% of the world's population.

Key features of modern federalism are:

1. Division of Powers
2. Supremacy of a Constitution
3. Supremacy Clause (federal laws, regulations, and treaties are superior to regional laws)
4. General Clause (powers not given to the federal government reside with the states or regional governments)
5. ***Courts*** (play an important role in resolving disputes)
6. New Federalism (devolution of power back to the states or regions)

Figure 4.1 highlights important aspects of US federalism. Our legal system with its various courts of law is where the expert witness plays a role in supporting *due process* and helping to resolve disputes.

USA Federalism

1. Supremacy Clause
   Article VI, Clause 2 of the U.S. Constitution, limitations explained in The Federalist Papers
2. States Rights
   10th Amendment of the U.S. Constitution enumerates states rights
3. Commerce Clause
   Article I, Section 8, Clause 3 of the U.S. Constitution gives Congress the power to regulate commerce
4. Enumerated Powers
   Federal (Congress) and states
5. Federal Right to Declare War
   Article I, Section 8 of the U.S. Constitution
6. Legal System
   14th Amendment of the U.S. Constitution defines legal rights such as citizenship, equal protection, and due process (**courts**)

FIGURE 4.1
Key features of US federalism.

4.2 System of Courts

The US legal system operates courts at the federal and state levels. Figure 4.2 provides a breakdown of the courts that give structure to the US legal system. Emphasis is given to state courts wherein forensic engineers most often serve. More specifically, the bolded text identifies civil, criminal, and special cases that involve engineers as expert witnesses. In rare instances, expert engineers find themselves involved in a federal case that winds its way through the legal system all the way up to the US Supreme Court.

Federal Courts

Article III of the US Constitution establishes the judicial branch of the federal government, including the Supreme Court and the structure of lower courts. The federal court system consists of:

1. The Supreme Court (the highest court in the United States)
2. 13 Courts of Appeals (Circuit Courts)
3. 94 US District Courts
4. Specialized courts (e.g., Court of International Trade, US Tax Court, Court of Appeals for Veterans Claims)

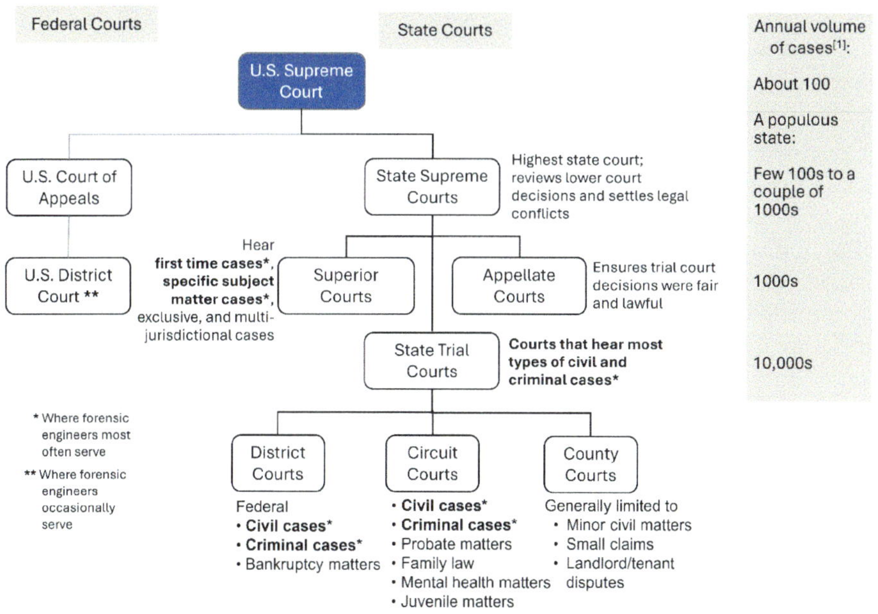

FIGURE 4.2
General breakdown of the US system of courts with emphasis on state courts where forensic engineers most often serve.

Federal cases in which engineers are called upon to testify usually occur in district courts that involve violations of federal law and adjudications (judgments) that sometimes establish precedent.

State Courts

Each state has its own court system, which reflects the federal structure:

- State Supreme Court
- Intermediate Appellate Courts (in 42 of 50 states)
- Superior Courts (at least one in every state)
- State Trial Courts (district, circuit, and county predominately)

State trial courts, usually classified as district, circuit, and county courts, manage most legal cases in the United States. Some states have other more specific state courts (e.g., Family Court, Surrogate Court). Four states – Kentucky, Massachusetts, Pennsylvania, and Virginia – use the term "*Commonwealth*" in describing their judicial structure, and Pennsylvania has a Commonwealth Court, a kind of intermediate appellate court with jurisdiction over administrative and civil public law.

State district and circuit courts differ in the types of cases they oversee and the complexity of those cases. District courts usually manage elementary cases within their jurisdiction (city, county, or group of counties); circuit courts typically oversee more complex and serious cases. Civil cases with high or no monetary limit and the Case Study 3 real-estate predicament are two circuit court examples.

Engineers most often find themselves supporting civil law cases in state district or circuit courts. Should a forensic engineer find themselves testifying in Superior Court, the case involves special subject matter or is a first-of-a-kind, and the experience can be quite interesting. For example, little is known about the subterranean use of proprietary fluids used with *fracking*.[1] Obtaining the recipe for a unique chemical mixture is crucial to spill response and may involve an order from a state's attorney general. Investigating spill impacts on the ecosystem is where the science makes it interesting.

FLORIDA DISASTER RESULTS IN MULTIPLE FEDERAL AND STATE LAWSUITS

Lawsuits from the FIU Pedestrian Bridge Collapse (Case Study 2) were tried in federal and state courts in Florida. Florida International University (FIU) sought damages from the parties (insurance carriers and companies) through district and circuit courts. One lawsuit sought $5M from an insurance policy held by the builder, with FIU offering to relinquish that money to the victims. Another lawsuit filed in a Tallahassee federal court argued that a federal agency violated due process rights in unjustly disqualifying companies from future contracts based on affiliation alone. Failure of infrastructure projects that result in injury and/or death and involve federal funds will lead to complex cases in federal and state courts that can take years to resolve.

4.3 Legal Sources and Types of Cases

The US legal system is a collection of various sources of law[3]:

- *Common law* (based on judicial precedents and inherited from British legal tradition)
- *Statutory law* (enacted by legislatures)
- *Constitutional law* (derived from the United States and state constitutions)
- *Administrative law* (derived from rules and regulations established by government agencies)

Both federal and state court systems oversee criminal and civil cases. Criminal cases involve prosecutions for violations of criminal law. Civil cases resolve disputes between private parties, organizations, companies, or combinations of these. Federal court jurisdiction is limited to cases involving federal laws, constitutional issues, or disputes between parties from different states. State courts manage everything else, and jurisdiction is based on where the dispute occurred.

The legal system operates on the principle of *checks and balances*, ensuring that no single branch of government is too powerful, and providing an opportunity to challenge lower court decisions on appeal. Appeals usually arise from judgments based on alleged legal errors made during trial, and a panel of judges on a higher appellate court reviews cases to determine if applicable law was applied correctly.

In summary, despite the complexity of our federal-state legal system drawing from various sources of law, it provides a balanced approach to justice. With its multiple levels of courts, and checks and balances, the system strives to apply law fairly and equitably to all.

4.4 Civil Law

Civil law governs private or civil rights, duties of individuals and organizations involved in manufacturing things, and engineering failures. It provides restitution for wrongs through compensation rather than punishing the wrongdoer with incarceration. Disputes involving contracts, property, personal injury, consumer problems, inheritance, civil rights, and *torts* are examples of civil issues.

Civil law is based on Roman law, in particular the *Justinian Code*,[2] with influence from *canon law*. Canon law refers to ecclesiastical law developed within Roman Catholicism and Christianity that draws from scripture, papal decretals, decrees of ecclesiastical councils, and secular Roman law.[2] Characterized by broad codes and statutes, civil laws are the rules that civilized societies abide by, and judges play a key role in applying the laws.

Within civil law, professional liability claims against designers, builders, manufacturers, and fabricators arise from ***breach of contract, tort***, or ***negligence***. A breach of contract is when a contractual party does not do what it agreed to do. A violation of any part of a contract can result in a breach of contract. A tort is an act or omission that injures or harms another as a civil wrong for which courts impose restitution. In the context of torts, *injury* means the violation of a legal right and *harm* means a loss or detriment suffered[4].

Negligence is an *unintentional tort* of the severest kind. Requirements for a tort to be negligence are duty, breach of duty, causation, and damages. Duty implies responsibility to act in a reasonable manner; failure to do so is a breach of this duty, and if the breach caused loss or injury that resulted in damages (monetary, property, or other loss), then negligence can be proved. Further, a ***strict liability*** tort does not require negligence or intent to injure. Based on an absolute or *strict* duty to ensure something is safe, strict liability is often associated with injuries from hazardous activities (e.g., machine operation, constructor falls during elevated work on a new building).

Lawyers sometimes involve forensic engineers with ***document review***, in particular, design plans and specifications, *variance requests*, and *change orders*.[3] Expert engineers might also be asked to evaluate design and construction practices, *standards of care*, and *performance evaluation*.[4]

Also with civil law, expert witnesses have the opportunity to review and correct deposition transcripts. Engineers are there to *help* the courts, and the courts want to ensure the facts are accurate. Further, deposition question-answer exchanges are sometimes stressful and exhausting, and transcribers are human too. There is every reason to encourage an *errata process*.[5]

Finally, an *intentional tort* is a deliberate act intending to cause harm and injury to another. This crime falls under the jurisdiction of c*riminal law*.

4.5 Criminal Law

Criminal law involves alleged violations of criminal laws. The type of infraction that most commonly involves forensic engineers is motor vehicle accidents (MVAs) that kill people. Cases are usually filed with the appropriate criminal court by a state attorney's office (SAO), and experts are not involved until after police investigations conclude and the SAO evidence review. Sadly, 25 years of experience in several states suggests high variability in the quality of police reporting, investigating, and evidence security. Often is the case where the expert engineer must rely on years-old police collision scene photographs, limited or no collision scene survey, and no access to physical evidence (collision vehicles left the impound yard long ago). One might imagine the challenges in supporting such cases.

DEFENSE ATTORNEY QUESTIONING THE STATE'S EXPERT WITNESS ON THE STAND:

Attorney: *Mr. Mental Giant, did you inspect the motor vehicles and obtain black box data for this accident?*

Mr. Mental Giant: *No, the vehicles were not available for inspection and crash data retrieval.*

Attorney: *You would agree with me that black box data can provide valuable information about driver actions prior to a collision, and that without this information, your opinions about speed and braking are speculation at best?*

Also, the rules are different than for civil law. In civil cases, there must be a *preponderance of evidence* – the quality and persuasiveness of which is compelling – against defendants. However, in criminal cases, lawyers must prove guilt beyond a *reasonable doubt.* Therein lies a predominant defense strategy to cast doubt on evidence and in the minds of jurors. Doubt about any one particular charge can lessen a sentence. And some states have *no-fault insurance* laws and their courts practice *comparative negligence.*[6]

Experience suggests eyewitness statements are more respected by attorneys on both sides in civil cases than they are by defense attorneys in criminal cases. Imagine five eyewitnesses attesting to a drunk, vile, and belligerent defendant at the bar a half hour before the multivehicle collision that killed two people, and a defense attorney questioning an eyewitness driver about what the defendant was wearing after witnessing the carnage and ensuing emergency response trauma.

Other differences between civil and criminal cases relate to appeals, remedies, and defense tactics. In civil cases, either party can appeal a ruling, but only the defendant can appeal in a criminal case. Also, the remedy for civil cases is payment of monetary damages. A driving conviction in a criminal case can result in fines, community service, loss of driving privileges, and in worst cases manslaughter jail time. The stakes are higher with criminal convictions.

The rules favor the defendant in criminal cases. For example, expert disclosure can be one-sided. The state's expert can investigate, report, and depose before any disclosure by the defense expert. This opens the door to advocacy; the defense strategizes how to poke holes in or refute the work of the state's expert, while the state knows nothing about what the defense expert will testify to. How is this fair and equitable?

EYEWITNESS STATEMENT GIVEN TO POLICE ONE HALF HOUR AFTER A HORRIFIC COLLISION BETWEEN A DARK BLUE SEDAN AND A WHITE SPORTS CAR WITH SERIOUS INJURIES:

"As I was approaching the intersection, the light was green, and the white car was heading towards me when all of a sudden the red car came speeding up from the left and hit the white car so hard that it flew into the air and spun around before it landed up there by the house."

One word in this statement can be used to cast doubt. What is that word?

Hint: The eyewitness, while waiting for police to give her statement, observed several emergency vehicles – fire trucks and ambulances – arrive to triage the scene for over 30 minutes.

4.6 Evidence Standards

Two accepted legal standards used to determine the admissibility of expert testimony and scientific evidence in court are the ***Frye test*** and the ***Daubert standard***. The Frye test, established in *Frye v. United States* (1923), is the general acceptance test of scientific evidence. The protocol requires that scientific evidence and the methodologies used to establish it be *"generally accepted'* within the relevant scientific community and/or peers in the field to be admissible.

The Daubert standard (or Daubert) has replaced the Frye test in federal courts and some states. Established in Daubert v. Merrell Dow Pharmaceuticals (1993), Daubert provides judges with a more comprehensive framework to assess the relevance and reliability of expert testimony. Specifically, Daubert considers several factors with acceptance of a theory or technique used:

1. Can the theory or technique be tested, and if so, has it been?
2. Has the theory or technique undergone peer review and publication?
3. The known or potential error rate.
4. Existence and maintenance of standards controlling its application.
5. Degree of acceptance within the relevant scientific or peer community.

Daubert is more stringent than *"general acceptance"* in that it requires judges to evaluate the methodology and reasoning behind expert opinions. One does not necessarily have to meet all five criteria of this standard, as it provides judges with more discretion as *gatekeepers* of scientific evidence.

However, a problem with these evidence standards relates to the acceptance of novel scientific methods or approaches used to establish evidence. It's called *progress*. New technology often transitions into common use and acceptance over time; EVs (electric vehicles) for example. Further, research develops concepts for wider applications over years or decades of studies. Two examples:

1. Crash data retrieval (CDR) from event data recorders (EDR) and analysis to infer vehicle performance and driver actions. The technology has existed for nearly 40 years, has been widely integrated into US vehicles since the early 2000s, and now has a certification program developed by the Collision Safety Institute for accident reconstructionists. Courts of some states (Pennsylvania, Illinois, Florida, and New Jersey among others) acknowledge EDR crash data as generally accepted within the relevant scientific community and meeting the Frye standard. Some states have enacted statutes that protect driver/owner privacy, requiring consent for downloading EDR data. No court has excluded data downloaded from an EDR to the author's knowledge.
2. Similitude is the concept of geometrically, kinematically, and dynamically relating models to prototypes to predict outcomes. Used by designers of airplanes and dams for nearly a century, it is

cheaper to build and test replicas (models) than to determine design flaws from having the real thing (a prototype) fail. A 2017 PhD thesis applied similitude in a novel way to predict structural behavior due to applied dynamic forces. Realizing that much infrastructure is built as common 3D geometric shapes (e.g., cubes, rectangular prisms, and cylinders) – Can we predict their behavior from testing simple wooden block analogs of similar shapes? – was the question to answer. The answer was yes, some behavior. How long will it take the scientific community to accept *predictive similitude* for structures and wider applications?

The point is it takes time for anything novel to meet all five Daubert criteria. And *preponderance of evidence* and *reasonable doubt* are different legal standards for burden of proof in different courts. Discretion is the better part of valor. Judicial discretion is better than the rigid exclusion of new principled methods and approaches in establishing evidence.

4.7 Interrogation and Cross-Examination

This content is added to remind the forensic engineer to *hope for the best but prepare for the worst.* A prepared expert recognizes and defuses adversarial questioning techniques with confidence.

It is an enjoyable experience when there is mutual respect among professionals in a legal setting. However, this is not always the case, and adversarial tactics are sometimes used to establish or undermine the credibility of experts. Expect this to happen early in one's inexperienced career; every forensic engineer has a first time in the legal system.

Table 4.1 identifies some common *interrogation techniques* experienced by expert witnesses. Figure 4.3 illustrates the role of *cross-examination* and identifies some techniques used to control witnesses. ***Interrogation*** is a questioning that usually occurs during the investigation (or discovery) phase of a case. A variety of tactics (some are described in Table 4.1) are used to get information, assess credibility, control, or trick. ***Cross-examination*** is a legal procedure in which an attorney questions an expert witness for the opposing party in court. Cross-examination is designed to test witness credibility, uncover inconsistencies or weaknesses in their story, and assess bias in testimony. The more credible and experienced an expert, the more the plaintiff's attorney wants them to testify, and the less defense attorneys do.

It is important to say that honorable attorneys apply these techniques for good and principled reasons. However, there are those who will attempt to

TABLE 4.1
Some Interrogation Techniques[4]

Interrogation Technique	Description
Leading Questions	The questions suggest the desired answer, guiding the witness toward a specific response. Beware of "You would agree with me" speak. Depending on the situation, it could be used to build consensus, subtly assert an opinion, or skillfully lead a witness.
Loaded Questions	1. Questions include assumptions or biases within the wording, attempting to influence the witness's answers. 2. Baited questions intentionally framed to paint a witness in a bad light in front of a jury.
Mutt & Jeff Tactic (aka good cop, bad cop)	One lawyer's demeanor is friendly and understanding while another is confrontational during questioning. The approach is designed to create pressure and elicit desired responses.
Establish Credentials	Questions such as • What is your forensic specialty? • Are you <u>specific acronym</u> certified? • How many papers have you published on <u>specific subject</u>? are intended to narrow the witness's expertise or discredit.
Repetition and Rephrase	Intentional repetition of key words or phrases during a deposition to emphasize or reinforce a particular point or central theme. Rephrasing is reiterating in different ways to see if the witness answers consistently.

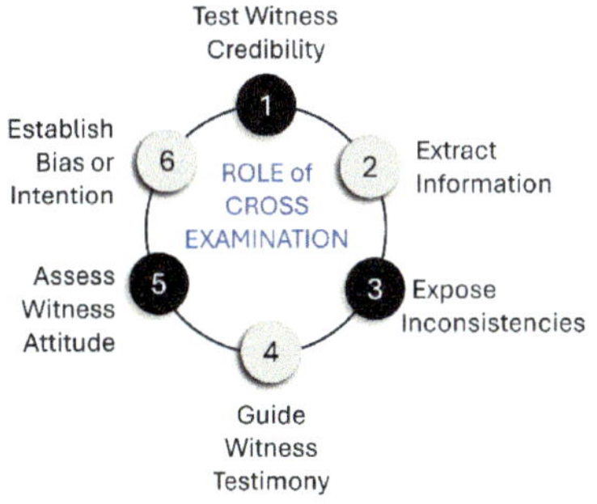

Repeat the question and rephrase
Was the car red?
Was the car red?
Was it the red car that struck the white one?

Repeat twice, then reverse
Did the red car hit the white one?
Did the red car hit the white one?
So, the white car hit the red car?

Make the question Yes or No
Q: The red car struck the white one?
A: No response
Q: Yes or no. The red car struck the white one?

Are you finished yet?
When the witness rambles on when answering, ask, are you finished yet? Then repeat the original question and say: "Please be brief and specific."

FIGURE 4.3
The role of cross-examination (L) and some witness control techniques (R).

discredit for the sake of winning. The expert witness has every right to ask about opposing counsel's credibility, case strategy, and the scope and limitations of a pending deposition. One needs to prepare when heading into a storm.

Another thing that can happen to the state's expert in criminal proceedings is relentless attacks on character, irrespective of overwhelming evidence against the defendant. Beware of the *Mutt and Jeff* routine that some defense attorneys try in depositions. A psychological tactic used to goad a witness into cooperating with a position, or to antagonize, one attorney asks questions with a nice demeanor and the other suddenly chimes in with a hostile tone. An expert witness can refuse to participate in disrespectful proceedings and ask their attorney(s) to warn the other side for the record.

How can such tactics occur within a just legal system? Because they can. This is why many forensic engineers limit their practice to civil law – it is a more respectful forum to work in.

Notes

1. Fracking is a method for extracting oil and natural gas from deep rock formations. Proprietary mixtures of water, sand, and chemicals are injected through high-pressure wells to scour cracks or fracture rock, thereby releasing oil and gas from a subterranean reservoir.

2. The *Justinian Code*, a part of the *Corpus Juris Civilis* (or *Body of Civil Law*), is a collection of laws and legal interpretations codified in the fifth century AD under Byzantine emperor Justinian I. It is the fundamental basis for modern civil law systems around the world.

3. In the context of design, change orders and variance requests are *change events*. A change order modifies a design; a variance request seeks an exception or alternative to the original design. Both require approval from the designer (Architect or Engineer of Record) to implement.

4. Standard of care refers to the degree of care a prudent and reasonable person (e.g., designer or builder) would exercise under the circumstances. Performance evaluation is consideration for the operating environment of things in the built world (e.g., buildings, machines in/on infrastructure, utilities in the ground). For example, a cladding system on a building in Texas may not work on a similar building in North Dakota. Also, a cool-dry mechanical room is easier on pumps and water heaters than a hot humid room.

5. An errata process is a procedure wherein an expert witness in a legal deposition reviews a transcript of their testimony to identify and correct errors or misstatements made during the recording. They detail changes and reasons for them with an *errata sheet*, an official document that allows for minor corrections to the official record while maintaining the integrity of original testimony.

6. A *no-fault accident* means fault is not considered when determining who pays for medical costs resulting from the collision, regardless of who caused the accident. Both parties file claims with their insurers to deal with damages and medical expenses. Should the accident become a *comparative negligence* case, the plaintiff can recover damages up to their percentage of fault as assigned by the court.

References

1. Noah A. Rosenblum (December 13, 2023). *An Overview of the U.S. Courts System,* Foreign Press Center briefing. Source: https://www.uscourts.gov/data-news/reports/statistical-reports/federal-judicial-caseload-statistics/federal-judicial-caseload-statistics-2022; retrieved 27 Dec 2024.
2. Scott N. Carlson (May 2009). *Intro to Civil Law Legal Systems, INPR OL Consolidated Response (09-002).* Int'l. Network to Promote the Rule of Law, pp. 18. Retrieved 22 Nov 2024 from http://www.inprol.org/node/4126.
3. James Apple and Robert Deyling (2024). A Primer on the Civil-Law System. The Federal Judicial Center, pp. 74.
4. Gary N. Holton and Melvin E. Jones (1978). The System of Criminal Justice. Little, Brown & Company, Inc., Boston, USA, pp. 6–18.

Case Study 4: Boy Scout and the Hired Gun

RECKLESS DRIVER ACCOUNTABILITY

Forensic Investigation Rejuvenates Old Case

- Accident Reconstruction
- Car-Bicycle Collision
- Criminal Case
- Small Details, Right Results

Message to students:

> Attention to detail is an essential skill of the forensic engineer

Sometimes cases grow old and go cold with time. Large volumes of cases and state attorneys moving on, especially once they make a name for themselves and the private sector hires them away, are reasons in populous states. And then there was a pandemic, which added two years to case backlogs.

This case study is about a reckless driver who drove his car *through* a bicyclist in 2016. Under the influence of alcohol and other illicit substances, the

 DOI: 10.1201/9781003623311-9

driver had no recollection of what he had done at the time of his field sobriety tests and police interviews. There were many eyewitnesses, an excellent police investigation, blood test results, a medical examiner's report, and suspect self-incrimination. An open-and-shut case, one would think. But legal delays are strategy, evidence can disappear, and a pandemic can halt the courts. An independent forensic investigation was needed to reinvigorate the case years later.

Background

It was tourist season, and the beaches were busy by day and boardwalks and restaurants and bars by night. A man and his son were anxious to see the sun set when another man approached. He was not a happy drunk, and apparently did not like Spanish-speaking people much. His racist, profane language caused the man to pick up his young son and walk away.

Thirty minutes later, a bicyclist lay dead on a road, his bike smashed and belongings strewn along the middle lane. As fate would have it, the nice man and his son and numerous others witnessed this tragedy. And the impaired vile driver was oblivious to what had just happened. He had driven a reckless mile to this place on a five-lane road, using the continuous left-turn middle lane as his personal speed lane. Something had hit his car, but he didn't know what. It all happened so quickly. But drivers and pedestrians saw, and the damage spoke for itself.

The driver was criminally charged with manslaughter. Car repaired, five years and two investigations later, he remained free to drive again. The case lingered, as did his fate.

2016 Police Investigation

Back then, the emergency responders had done their jobs well. Paramedics triaged and tried to save a life, and eyewitness motorists and pedestrians stayed and blocked the suspect's car from leaving the scene and assisted the traumatized, even the suspect. People did what we all should do in similar circumstances.

Upon arrival, police quickly detoured traffic, cordoned off the area (Fig. CS4.1), and secured eyewitnesses for statements. Only emergency responders and the Accident Scene Investigation (ASI) team were allowed in to do their work.

FIGURE CS4.1
North-to-south view of accident scene showing where the bicyclist was triaged and the final resting position of his bicycle within the debris field. Note that the bulk of the debris field is within the middle continuous turn lane.

Review of the ASI team's work indicated experienced FARO surveyors (Fig. CS4.2) with a good eye for photo-documentation of the scene from different angles. The same can be said for the officers who examined the strike vehicle and bicycle in the impound garage, and the deputies who recorded seven eyewitness statements and three interviews with the suspect. Professionals all.

Eyewitness Accounts

There were seven eyewitnesses, five motorists and two pedestrians, who observed the suspect driving recklessly and/or the collision with the bicyclist. Each provided a recorded audio statement to police within an hour or so of the accident. Their accounts are summarized in Table CS4.1. Table CS4.2 is a compilation of excerpts from interviews with the suspect. Table CS4.3 is an excerpt from a police discussion with the General Manager of the collision repair shop that fixed the suspect's car three months after the accident.

> "In much of your talking, thinking is half murdered."
>
> **Kahil Gibran**
> *From his 1923 poem On Talking*

Table CS4.3 demonstrates one way evidence can go away. After reviewing police logs and other case-related communications, the following was surmised. Police responders had placed the bicycle in the trunk of vehicle V1 for

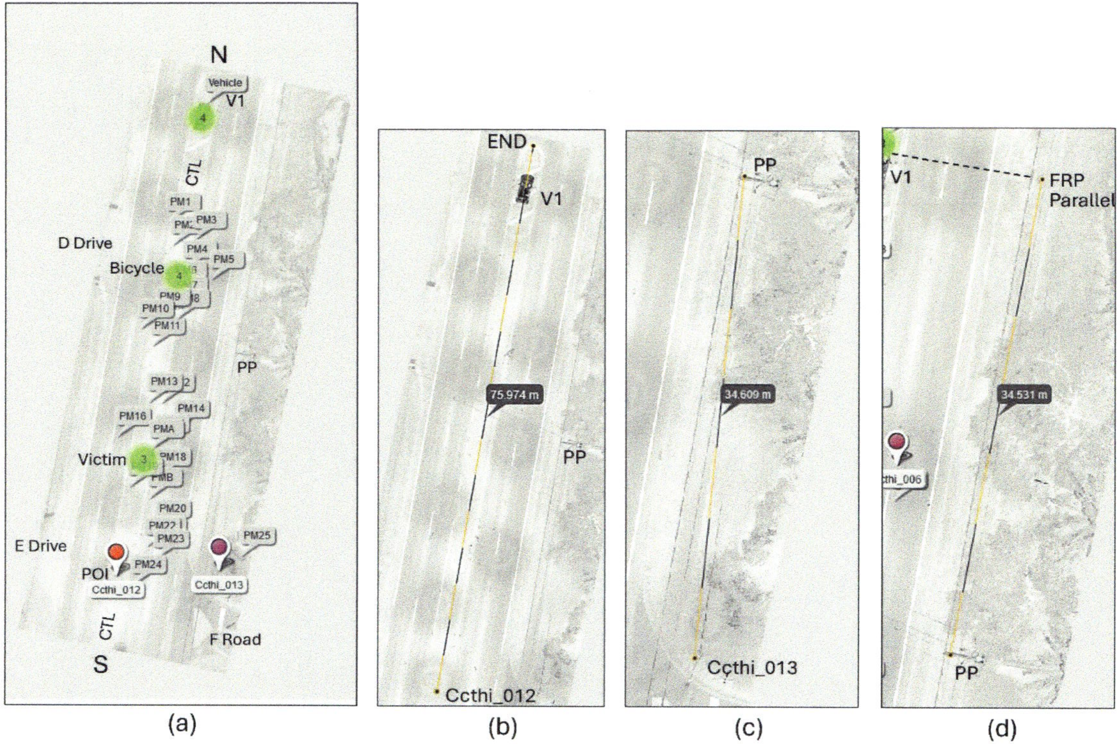

FIGURE CS4.2
FARO survey excerpts demonstrating thorough documentation of the accident scene and debris field. Ccthi_012 and 013 are FARO scanner positions; CTL is continuous turn lane; FRP Parallel is sidewalk position parallel to front end of strike vehicle V1; N is north; PM1 to PM25 are numbered debris items; POI is point of impact; PP is power pole; S is south; and green circles with numbers denote scan clusters and the number of scans taken in each vicinity.

TABLE CS4.1

Summaries of Eyewitness Statements

Eyewitness	Credibility	Interview	Audio Interview Summary
EW1	Male motorist driving truck northbound, inner through lane Credible	At scene, 18 min after accident	Witnessed V1 reckless driving EW1 in truck stopped on red light at Intersection X. Witnessed V1 approached from behind at high rate of speed, then swerved into left turn lane just in time to avoid collision with sedan behind. On green light, V1 sped out in front of the two through lanes. EW1 next came upon smashed bicycle and bicyclist on the road. EW1 did not witness the collision. EW1 provided police with V1 license plate number.
EW2	Female motorist driving sedan northbound behind EW1, inner through lane Credible	At scene, 28 min after accident	Witnessed V1 reckless driving EW2 in sedan in inner through lane behind EW1 truck, stopped on red light at Intersection X. EW2 was passed by V1 traveling fast in the left turn lane. On green light, V1 sped out in front of through lane traffic. EW2 next came upon debris and a body on the road; swerved to miss the body. She immediately thought that V1 had hit the bicyclist. EW2 did not witness the collision.
EW3	Male motorist driving SUV northbound, outer through lane Credible	At scene, 40 min after accident	Witnessed V1 reckless driving EW3 in SUV in outer through lane, stopped on red light at Intersection X. Observed V1 fast approach from rear and swerve hard into left turn lane to avoid hitting EW2 car. V1 then sped off through intersection in continuous left turn lane. EW3 lost sight of V1 as traffic moved on ahead. EW3 next saw traffic ahead brake and slow. EW3 saw debris and a body on the road. EW3 did not witness the collision.
EW4	Male tourist motorist driving northbound in rental van, behind EW3 in outer through lane Witnessed V1 driver behavior in restaurant and bar prior to accident Credible	At scene, 56 min after accident	Witnessed suspect driver drinking, stumbling, using profanity at restaurant/bar Witnessed V1 reckless driving Witnessed suspect driver shaking victim in road First contacted suspect drinking a beer, stumbling, using profane language at them (EW4 and son were Hispanic) in restaurant bathroom. EW4 and son quickly left bathroom, went to take sunset photos, and then left to avoid confrontation with suspect. When trying to back up in parking lot, suspect in V1 pulled behind their van, stopped, and blocked their exit for about a minute. V1 then left. EW4 watched V1 go northbound, and then detoured through neighborhood to avoid V1 before turning back to main road several blocks later.

(Continued)

TABLE CS4.1 (*Continued*)

Summaries of Eyewitness Statements

Eyewitness	Credibility	Interview	Audio Interview Summary
			A moment later, while driving in the right through lane, V1 nearly swiped their van before veering left across lanes as they approached Intersection X. "V1 passed us going about 80 mph and then hit the brakes hard. When the light turned green, he *burned out fast,* still driving in the middle turn lane." EW4 observed "...things flying up into the air from the front of V1 car. We didn't know what it was. Then we saw a man on the road, and noticed several other cars stopped, and one lady blocked V1 car in the median with her car. V1 drove forward a little bit, so she blocked him in so he wouldn't take off. Then the suspect and the lady ran back to man on the road. I saw the suspect kneel and he started crying and shaking the guy." EW4 did not witness the impact due to other traffic in his line of sight but did see debris fly up above the traffic.
EW5	Male tourist pedestrian walking southbound on east sidewalk at time of accident Credible	At scene, 32 min after accident	Witnessed V1 reckless driving Witnessed collision Was walking southbound on sidewalk, approaching Intersection X. Witnessed V1 speed out from left turn lane in front of through traffic, wheels screeching. EW5 turned to watch V1 drive on and hit bicyclist who was in the middle continuous turn lane. EW5 witnessed bicyclist go up and over V1, debris flying into the air, and victim land on the road. EW5 witnessed entire accident sequence.
EW6	Male motorist driving sedan southbound, approaching Intersection Y (intersection closest to collision) Credible	At scene, 46 min after accident	Witnessed suspect actions/statements immediately after collision and during victim assist EW6 was traveling southbound in left turn lane ready to make left turn at Intersection Y. Observed accident scene, debris in the road in northbound inner through and left turn lanes and saw bicyclist body on road on double yellow line that separates southbound lanes from middle left turn lane. EW6 did not witness collision. EW6 used his vehicle to help block off traffic, got out, and went to assist victim. Checked victim's pulse. Observed suspect *"freaking out,"* heard suspect say: *"I was the one that hit him,"* observed suspect crouch down and talk to the victim, and heard suspect say: *"Come back, please come back."* EW6 then "...tried to get the suspect off of the victim because he was putting too much weight on him."

(*Continued*)

TABLE CS4.1 (*Continued*)

Summaries of Eyewitness Statements

Eyewitness	Credibility	Interview	Audio Interview Summary
EW7	Male pedestrian walking southbound on east sidewalk. Was last to talk with bicyclist moments before accident Credible	At scene, 64 min after accident	Witnessed suspect actions immediately after collision and during victim assist Was coming from a beach festival, had talked with victim (bicyclist) at a sidewalk bench moments before the collision. Victim was securing groceries on bike as they talked. EW7 did not witness bicyclist cross the road. EW7 heard impact, turned to look, and saw victim land on the road. EW7 witnessed suspect and a woman assist victim. Observed suspect lay his head on victim's chest in grief. Observed EW6 comfort suspect and move him away from victim.

TABLE CS4.2

Excerpts from Police Interviews with the Suspect, Male Driver of Vehicle V1

Interview	Audio Interview Excerpts
Interview 1 At scene, 16 min after collision, shortly after police arrived	Suspect on approach to Deputy 1: "*Hey, how's it going Sheriff?*" Deputy 1: Good, how are you sir? Suspect: "*Pretty good.*" Deputy 1: "I'm here to investigate a traffic crash…." Suspect: "*Could be worse.*" Deputy 1: "Are you involved with the traffic accident?" Suspect: "*I'm very involved! I was coming down the road, and there was this accident, and I pulled off the road, and I met this guy.*" Suspect: "*I was in the fast lane. Some people don't know what the* -obscenity- *the law is.*" Deputy 1: About how fast were you going? Suspect: "*35-40…I think. I'm a pretty savvy driver. I'm a professional driver. I pay attention to* -obscenity-" Deputy 1: So, what happened just before the collision? Suspect: (flapping lips with fingers) – "*I don't remember a collision.*" Deputy 1: What happened just after the collision? Suspect: (gets emotional and starts to cry) "*Is he OK? What's his name? I will go to the hospital and hold his hand.*" Deputy 1: What caused you to stop? Suspect: "*I never saw him!*" Deputy 1: What caused you to stop? Suspect: "*I guess an accident.*" Deputy 1: You just knew that you were in an accident, a traffic crash? Suspect: "*Yes. It didn't sound like a wa-baam, where everything crashes and breaks. I've been in accidents before, you know, I'm not a great driver. It wasn't that popping sound; it was more like just kind of like a flow through….you wake up and realize like what the* -obscenity- *happened.*" Deputy 1: What did you do after you stopped?

(*Continued*)

TABLE CS4.2 (*Continued*)

Excerpts from Police Interviews with the Suspect, Male Driver of Vehicle V1

Interview	Audio Interview Excerpts
	Suspect: *"I got out, looked back, saw him in the road and went to help him. I need to see movement, eye movement, toe movement. I need to see he's* -obscenity- *OK."* Suspect: *"All I know is that I hit the brakes, and now I'm in this situation. I have pretty good reflexes. I play video games. I have good reflexes."* **Deputy 1 impressions: Suspect is high, incoherent, flippant, and emotional. Suspect under the influence and implicates himself.**
At scene, Deputy 2 recording 20 minutes after Interview 1	48-minute recording of suspect rambling immediately after throwing up, crying, and sitting down to chat with Deputy 2. Suspect: *"How often does this happen? People crossing the road dude, is a lot, huh? There should be a bridge. Yep, yep, yep. Because this place is so* -obscenity- *popular, so* -obscenity- *awesome.* -Obscenity- *like this happens and we gotta have a spot where people can cross safely."* Suspect: *"This world throws a lot of* -obscenity- *on you.* – sobs – *This world isn't fair. It doesn't give a crap who you are, how you live, or what you are. He could be a drug addict riding his bicycle, doing the only thing he knows how to do, sell drugs, and he's selling his* -obscenity- *on a bicycle, and he gets hit by a car, and everybody blames him. And he was doing all he could do to make people happy. Drugs are bad, sure they are, sure they are. But they give people perspective (slurred). You guys have no idea what life is even about, or else you'd be stoked. Cause if you ain't stoked, you have no* -obscenity- *idea. Stoked is when you drop in on a wave. I should have been institutioned. Come on in boys, the water is fine. I should stand up; I don't want to look tough. I don't want to look wimpy."* Suspect rant about police, sister, and himself: *"You guys are so good and trustworthy, and my sister too. She's my hero. I can trust her. But I'm an abominable person. Even though I'm abominable, I wouldn't hesitate to save someone's life."* Suspect: *"You me, against the government. I know you're not stupid. The government… is a tricky mother* -obscenity-. *Very* -obscenity- *tricky. They make guys like you think that you are doing something….lawful… -inaudible-….heartful, heartful. If I smack you in the face, that would be very* -obscenity- *real. It wouldn't be good because the world would come down on me. I wouldn't smack you in the face because you're on my side."* Suspect implicates self with surfing rant: *"Life throws* -obscenity- *at me constantly; so do the waves. I started drinkin.' I started drinkin' and stuff. Got a good buzz on. Waves were killer. Curls were rad man. I was inside, man. I mean, waves are so much better buzzed."* **Deputy 2 impressions: Does suspect have a history of mental illness? State Attorney should interview his sister to find out.**
At scene, Deputy 3 recorded interview sometime after Deputy 2 recording	Deputy 3 identifies herself, the nature of her criminal DUI case investigation, and read Miranda rights to suspect. Asked if he understood each of his rights; suspect answered yes. Deputy 3: Would you like to talk to me about what happened tonight? Suspect: "*Suuure*" Deputy 3 explained that the person that was hit was deceased. Suspect: – uncontrollable sobbing. " -obscenity- *the law,* -obscenity- *this,* -obscenity- *that,"* angry, yelling at Deputy 3. Suspect: *"I care if it was his fault. I forgive him.* -obscenity- you and your -obscenity- *technicality. I don't really remember what really happened. He fell off his bike. I guess it was my fault…. I love people man. I'm a good person. I had one Corona; it wouldn't affect my judgeability* (slurred)." Deputy 3 discusses blood draw and introduces paramedic.

(*Continued*)

TABLE CS4.2 (*Continued*)

Excerpts from Police Interviews with the Suspect, Male Driver of Vehicle V1

Interview	Audio Interview Excerpts
	Suspect during blood draw: "*I don't do heroin, so I'm not used to this* -obscenity-. *I like to cooperate. I'm a scenic artist. I work in the forest all the time, cuttin' down trees. I got talent. I just had two drinks.*" Deputy 3: How did you know you were in an accident? Suspect: "*I don't know.*" Deputy 3: OK, what made you stop? Suspect: "*I think it was the freakishness of the people around me. I'm not quite sure; people were yelling at me from around me.*" Field sobriety exchange: Deputy 3: Explains field sobriety tests, ask suspect if OK to administer. Suspect: "*No, it would probably be best right?*" Officers explain pros/cons of not doing field sobriety tests. Suspect: "*Sounds like you guys are trying to catch me in a 22. Well, I had a drink, so isn't that gonna make me fail?*" Deputy 3: Partner, it's not a pass or a fail, we use…. Suspect: "*It is a pass or fail….you are* -obscenity- *wrong, or you're a piece of* -obscenity-, *or, or you're somebody of the world whose just….*" Field Sobriety officer asks suspect if he recently smoked pot. Observes telltale signs. Suspect: "*I've been smokin' weed all my life.*" **Deputy 3 impressions: Suspect high, incoherent, flippant, profane, disparaging to law enforcement, and may be mentally ill.**

TABLE CS4.3

Excerpts from Police Discussion with Collision Repair Shop General Manager

Interview	Audio Interview Excerpts
Collision repair shop General Manager (GM) Three months after accident (Disposition of bicycle)	GM explained that V1 and bike were towed to impound yard. Described bike as blue, rusty, and severely damaged from rear. Deputy 3: "Do you recall what was done with the bike after it was delivered with the car?" GM: "Well, the bike was here until the owner came in and signed authorization for us to take it, which was on date. Shortly after we disposed of it with all the scrap metal that we throw away." GM tried to identify who the owner was, but then confused the car owner (suspect) with the [deceased] bike owner. GM could not produce paperwork on who signed for the bike.

towing to the impound yard. Both were inspected and photo-documented (see Fig. CS4.4). How they both ended up at the collision repair shop is a mystery no one could recall. The suspect was allowed to have his repaired car after signing the paperwork. Further, he signed another form releasing the smashed bike to the shop for disposal. The shop apparently thought the man owned both.

2022 Forensic Investigation

The State Attorney's Office (SAO) retained a forensic engineer to review their case file, further investigate as necessary, and provide expert witness support at trial. Physical evidence was gone – no black box crash data retrieval possibility – and the case was on the docket scheduled for trial. The defense strategized *comparative fault*. The bicyclist was crossing the road (not at a crosswalk) and the suspect drove into the middle turn lane to avoid. Success of this strategy hinged upon a single police photograph, the one that is the cover image of this case study. That diagonal black dashed line.

A site visit by the engineer validated the FARO survey data, review of the witness audio statements proved consistency beyond doubt; the suspect incriminated himself in his interviews with police, and the police photographs were thoroughly scrutinized for anything that may have been overlooked. The team scrutinized that one photograph and another from a different angle. Magnification, image lightening, scrutinizing the details… and there it was: Figure CS4.3.

Referring to the detail in Figure CS4.3(c) and Figure CS4.4, upon impact with the front bumper of strike vehicle V1, the tire was pushed forward and downward, causing the initial full-tire-width skid portion. As the tire depressed into the pavement and rim began to deform, it deflated, causing

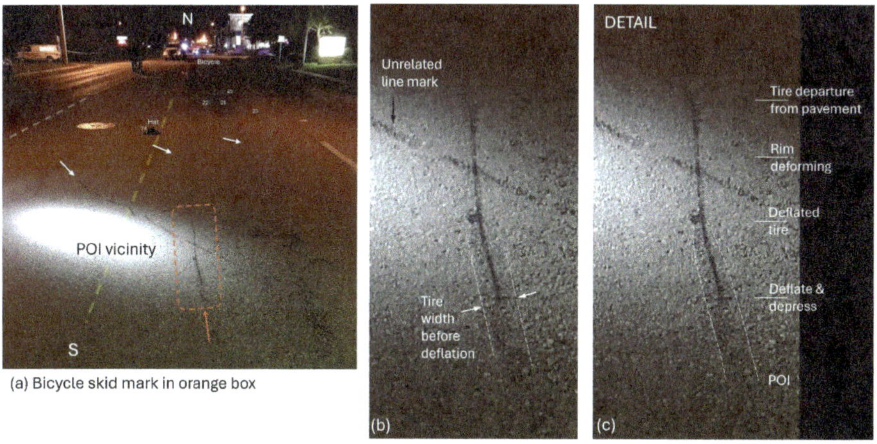

(a) Bicycle skid mark in orange box

Bicycle (a) skid mark, and (b) tire deformation detail.

FIGURE CS4.3
(a) Police photograph of POI vicinity, lightened to distinguish bicycle skid mark (orange arrow/box) from line marks (white arrows) unrelated to this accident. Debris items are numbered in white in the background. The smashed bicycle is lying on the double-yellow line in the far background below the N (north) indicator. (b) and (c) describe the bicycle skid mark in detail.

FIGURE CS4.4
Matching bicycle deformation and rack/basket assembly with hood damage inside police impound garage.

the thinner and darker skid line. The deforming rim crushed inward, taking the shape of the car's front end (Fig. CS4.4R). Ultimately, the tire then departed from the pavement as the bike (and bicyclist) flipped upward, into, and over the car. The police got it right in aligning the bicycle with the car as shown in the upper images of Figure CS4.4.

It is important to recognize the position and alignment of the bicycle skid mark, slightly angled toward and outside of the yellow line denoting the middle continuous turn lane (Fig. CS4.3(a)). This is consistent with the bicyclist having crossed the northbound through lanes and turning into the middle turn lane to wait for southbound traffic to clear before continuing across the road. The suspect vehicle did not hit the bicyclist broadside, but rather ran into the bicyclist from the rear.

The bike rack with affixed basket crunched with the bike seat before being severed as the bike wrapped onto the hood of the car. The rack/basket assembly compressed into the hood of the car, causing the scrapes and gouges visible in Figure CS4.4B, and the bicyclist impacted with the rear of the hood and windshield. Bike and bicyclist then separated, went over the car, and landed at separate locations along the yellow line of the middle turn lane (Fig. CS4.1).

The reckless driving sequence from Intersection X through the accident scene was pieced together using eyewitness accounts, police investigation photographs, and the FARO survey. Distance from Intersection X to the strike vehicle FRP (Fig. CS4.5(a) and (b)) was about 300 meters. Vehicle V1 drove in the continuous turn lane to the bus stop (Fig. CS4.5(b)) before starting to veer right into the inner northbound through lane. Shortly after, he hit the bicyclist in the intersection at F Road (Fig. CS4.5(c)). The red dashed line path of the strike vehicle as shown in Figure CS4.5(b), from POI to the car's final resting position in the continuous turn lane, is consistent with the debris field and final positions of the bike and bicyclist.

Recall the set of black lines on the pavement at the POI vicinity denoted in Figure CS4.3(a). Also depicted in Figure CS4.5(c), these lines are suggestive of motorcycle tire skids with turns to and from F Road.

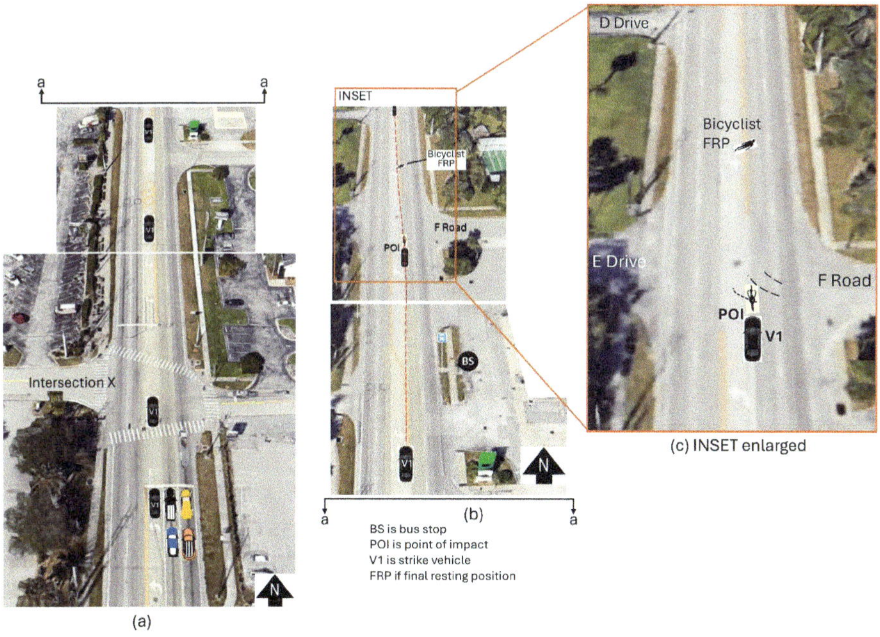

FIGURE CS4.5
Strike vehicle V1 travel path from Intersection X to the accident scene.

Trial Proceedings

A forensic investigation report was filed with the court, and the defense was provided a copy in accordance with law. The defense then tried to depose the state's engineer before their expert could be deposed. In essence, the state (plaintiff) played their hand in good faith, while the defense held theirs until forced to disclose. How can one side investigate, report, and depose before the other side discloses anything? Because the system allows it.

The trial went on without the state's expert witness, who disengaged once the shenanigans went too far. Always provide an out in the contract.

> We reserve the right to disengage from expert witness and trial support for any reason (e.g., disagreement with basis for case, conflict of interest, non-payment).

5

Report or No Report

CONTENTS

CONCEPTS	KEY TERMS	ADDITIONAL READING
Civil Rule 26	Disclosure	Exercise 9
Write to Settle	Discovery	Exercise 10
Cohesive Paragraph	Nuances of law	
Communication Liability	Constellation of conclusions	Case Study 5
Privacy	Grandstand	
Legal Motions	Protected communication	
	Subpoena	
	Motion in limine	
	Translational engineer	

> There is no such thing as the unknown – only things temporarily hidden, temporarily not understood.
>
> **Capt. James Tiberius Kirk**
> *Starship Enterprise*

In our capacity to support the courts as experts, our mission is to promote efficiency, fairness, and cooperation during the legal discovery process. A cornerstone of this mission for attorneys and engineers alike is Rule 26 of the Federal Rules of Civil Procedure (FRCP), or ***Civil Rule 26***, or simple Rule 26. Civil Rule 26 establishes general provisions for discovery in federal legal proceedings and outlines the responsibility to disclose information. Thirty-five of the 50 states have adopted rules for civil law based on the most recent 2006 improvements to FRCP 26, and most other states incorporate parts of FRCP 26 in their civil statutes.[1] Nevertheless, some states have their own rules for civil law discovery, and *nuances of law* – the subtle differences in how laws

DOI: 10.1201/9781003623311-10

are interpreted and applied – complicate things and can challenge fairness in certain situations. Without a uniformly applied discovery rule, it is imperative that forensic engineers understand the complexities of disclosure relative to the states they choose to serve. Civil Rule 26, at the core of *report or no report,* is the focus of this chapter. Other important concepts discussed are *communication liability* and tips for writing expert reports.

5.1 Civil Rule 26

Civil Rule 26 is the fundamental rule governing the discovery process in civil litigation. It strikes a balance (Fig. 5.1) between providing necessary information and safeguarding parties from overly burdensome discovery requests, ensuring that the scope of discovery remains proportionate to the requirements of the case. Professionals serving the legal system have a duty to disclose information pertinent to resolving disputes brought by the courts. Central to Civil Rule 26 is efficiency, fairness, and cooperation between the parties in resolving a dispute.[2]

Key components of Civil Rule 26 are:

FIGURE 5.1
Scales of *Aequitas*.[1]

Initial Disclosures

A requirement of Rule 26 is that the parties provide initial disclosures:

- Names and contact information of individuals likely to have discoverable information (e.g., litigants and anticipated witnesses).
- Copies or descriptions of relevant documents and tangible items (e.g., contracts and physical evidence).
- A calculation of damages claimed.
- Applicable insurance agreements (i.e., insurance policies and coverages).

Scope of Discovery

Scope of discovery means any nonprivileged material relevant to any party's claim or defense, *proportional to the needs of the case.* This last phrase means a level of discovery or investigation appropriate and relevant to the importance and complexity of the dispute. In other words, avoid unnecessarily excessive or burdensome requests that are not needed to resolve issues at stake. Attorneys must be reasonable in considering the importance of issues, their significance, the parties' access to information, and the burden to get it.

Expert Testimony

This part of Civil Rule 26 defines procedures for disclosing expert witnesses, including their identity, scope of their testimony, and reports.

Discovery Limitations

Attorneys should only seek information that is relevant to the case, likely to be admissible, and not overly burdensome to obtain. Discovery cannot be used as a *"fishing expedition."* In other words, discovery cannot be used to demand too much information based on hunches and accusations.

Protective Orders

This part of Rule 26 allows the parties to seek protective orders to limit or forbid certain discovery requests. For example, a motion to prohibit discovery is filed with the court when the information sought is not reasonably accessible.

Timing and Sequence

Civil Rule 26 sets guidelines for the timing and sequence of discovery, including when parties can begin seeking discovery and its completion. States usually provide judges with discretion to amend the deadline for discovery completion as necessary based on caseloads and special circumstances.

Conference of Parties

This part of the rule requires the parties to meet early in the litigation process to discuss discovery plans and potential settlement. Consistent with promoting efficiency, fairness, and cooperation, the conference of parties is designed to ensure that all parties have access to relevant information for their case.

Supplementation of Responses

Rule 26 requires the parties to periodically update their disclosures and discovery responses with new or revised information that becomes available.

Electronically Stored Information

A new addition to Civil Rule 26, modern litigation recognizes the various forms of electronically stored information (ESI). Requesting parties must specify the form in which ESI should be produced, and respondents must identify any issues with producing the requested information in the requested form.

Expert engineers will want to pay particular attention to this aspect of disclosure. Defense attorneys have been known to request a copy of the plaintiff expert's complete case-related ESI. Imagine how draft reports, emails, notes, and other information might be misconstrued and used against the expert.

Claiming Privilege or Protection

A party can withhold information otherwise discoverable by claiming the information is privileged or protected as pretrial or trial preparation material. For example, the requesting party does not wish to disclose to the defense a novel energy method used to evaluate crush damage because their expert is a competitor. Trial exhibits are another example. When doing so, the requesting party must expressly make the claim, describe the general nature of the material (documents, communications, or physical things), and provide the judge with good reason for nondisclosure privilege or protection. If granted, the judge may order the information sealed – the energy method example – until such time as necessary to unseal for use at trial.

Tables 5.1 and 5.2 provide examples of how two states regulate discovery differently. With the Florida example, an expert retained as a *consultant* for hundreds of cases without ever having submitted a report may be hired to specifically refute points and cast doubt, telltale signs of the *hired gun* advocate. In contrast, under Washington state civil law, expert work product drafts are protected from disclosure, and neither party is required to submit expert reports. Also, protection from disclosure of consulting experts is waived if they provide facts or opinions that are otherwise impracticable to obtain by the opposing party. Further, the Washington statute allows for judicial discretion with inclusion of *"and related court decisions."* The Washington example treats information disclosure and experts more equally and fairly, thereby reducing the potential for advocacy that may taint juries.

TABLE 5.1
Florida Disclosure Example

Florida Rule of Civil Procedure 1.280(b)(5) Governs Disclosure of Expert Reports. Key Points Are[3] :

1. Experts are divided into two categories:
    - Testifying experts
    - Consulting experts
2. For testifying experts, the retaining party must:
    - Disclose their expert's identity, scope of employment, compensation, litigation experience, and list of recent cases in which they have testified.
    - Disclose what their expert will testify about, facts and opinions to be presented, and the bases for those opinions.
3. Expert reports:
    - Florida does not explicitly require written expert reports for all experts.
    - If a report is prepared, it may be discoverable along with the expert's work products and notes.
4. Financial records: An expert's financial records are only discoverable under "unusual or compelling circumstances."
5. Timing: Expert disclosures are typically required 90 days before trial.
6. Sanctions: Should a party unreasonably delay disclosing their expert's identity or required information, the courts have authority to exclude the expert witness.

TABLE 5.2
Washington State Disclosure Example

Civil Rule 26(b)(5) ***and Related Court Decisions*** **Govern Expert Disclosure in Washington State. Key Aspects Are:**

1. Experts are divided into two categories:
    - Testifying experts
    - Consulting experts
2. Testifying experts:
    - Parties must identify experts expected to testify at trial.
    - Parties must disclose the subject matter of expert testimony, the substance of facts and opinions, and the basis for these opinions.
3. Consulting experts: Information about experts retained solely for consulting purposes is protected from discovery, except under exceptional circumstances.
4. Expert reports: Expert reports are not automatically required but can be obtained through a formal request for production of documents.
5. Draft opinions: Draft opinions are "preliminary or unfinished reports," protected from disclosure during legal proceedings under the work product doctrine and are therefore undiscoverable to the opposing party.
6. Timing: Expert disclosures must meet the deadline as specified in the Case Schedule.

(*Continued*)

TABLE 5.2 (*Continued*)

Washington State Disclosure Example

Civil Rule 26(b)(5) ***and Related Court Decisions*** **Govern Expert Disclosure in Washington State. Key Aspects Are:**
7. Deposition of experts: Parties are allowed to depose all experts their opponent expects to call at trial.
8. Attorney-Expert Communications: Attorney work product shared with their testifying expert(s) is protected, except for facts given for consideration by the expert. The work product doctrine stipulates that facts once known shall be disclosed to both parties.
9. Cross-examination: Experts may be required to disclose underlying facts or data during cross-examination, even if not previously disclosed.
10. Sanctions: Failure to comply with any of the disclosure rules may result in sanctions, including potential exclusion of expert witnesses.

5.2 Write to Settle

Write to settle with regard to expert reports means crafting a report with the goal of helping settle the dispute without the need for alternative (usually more expensive) resolutions. When an expert engineer adopts the write-to-settle approach to reporting, he or she is highly valued toward potentially reducing the need for lengthy and costly trials. Remember, our goal as expert witnesses is to support the courts with their endeavors to resolve disputes efficiently and equitably.

The write-to-settle approach involves[4]:

1. *Clarity and Accessibility.* Effective reporting prioritizes clarity and simplicity of subject matter, enabling readers from various backgrounds and expertise levels to grasp key insights and conclusions effortlessly.
2. *Balanced Perspective.* The report presents a fair and objective analysis of the facts and issues, acknowledging strengths and weaknesses as necessary for truth in disclosure. If a particular line of evidence is weak, state so, and likewise for strong evidence.
3. *Sound Science.* With investigation, testing, and analysis, apply scientific methods, principles, and industry standards with impartiality and without *a priori* bias.
4. *Facts and Key Issues Focus.* Report only facts in addressing and prioritizing the most critical aspects of a dispute, as they are likely to influence settlement. There is absolutely no place for conjecture and speculation in expert reports.
5. *Consideration of Alternatives.* Sometimes evidence leads to more than one possibility. If an expert report explores various scenarios with

potential outcomes, prioritize them based on the strength of evidence, and quantify probability or rank them based on consilience of evidence when possible.

A case leads to mediation, and an architect hires a lawyer with expertise in mediation to help defend. Significant time is spent with document review, no reports were generated, and $40K later, the mediator arrived at unresolved. Might a report of facts substantiated by applicable codes and standards have helped the mediator arrive at a definitive solution? A real example of a lot of money spent for nothing.

5.3 Expert Report Writing Tips

An outline for an expert investigation report was introduced in Chapter 3. Reiterated here as Figure 5.2, this outline is useful for writing narrative reports about cause-and-origin and failure analysis type investigations. Following the outline, Table 5.3 offers some useful tips for writing investigation reports.

Title and Case/Project #

1. Introduction
    - Define case
    - Identify investigators
2. Information Reviewed
    - List all relevant information sources
    - Identify all interviewees
3. Investigation
    - Scope
    - Methods and techniques used
    - Who did what work
4. Analysis
    - Data evaluation
    - Results
5. Findings
    - Numbered list
6. Informed Opinions
    - Alpha list, prioritized
7. References used
8. Limitations (liability statement)

Appendix (Figures, photographs, data tables, etc.)

FIGURE 5.2
Example outline of a forensic investigation report.

TABLE 5.3

Tips for Writing Expert Reports

1. *Cover*: Identify recipient and issuer, report title, and case name and number as the cover page or first information block of a report.
2. ***Introduction***
   - Define the case with a brief synopsis.
   - Follow with a second short paragraph that identifies the investigators, their roles, and defines the objective of the report.
3. ***Information and Evidence Reviewed***
   - Request and review ALL (emphasis intended) pertinent case file information from the hiring attorney.
   - Review this information for consistency and relevance to causation.
   - Summarize and prioritize salient points as a bullet list or in a table for reference or later use. For example, in Case Study 4, seven eyewitness statements were summarized as Table CS4.1. It is easier to discern consistency (or inconsistency) with a line of evidence when compiled in an ordered format.
   - Inspect and thoroughly photo-document relevant evidence.
   - This initial information-gathering exercise may be the basis for hypothesizing causation. Therefore, it is imperative to pay close attention to details, for this information may guide the investigation scope.
4. ***Investigation***
   Perform tasks as necessary to verify, substantiate, validate, or disprove. Some examples of investigation work are:
   - Site reconnaissance and fieldwork to evaluate, substantiate, or validate conditions and/or evidence
   - Exemplar performance testing, reconstruction, and simulations
   - Similitude modeling and testing to simulate outcomes
   - Materials testing for strength and deformation characteristics, pressure-temperature relationships, performance evaluation, defects, variability, and corrosion
   - Design documents review for consistency and code/standards compliance
   - Construction documents review for conformance with design documents
   - Chemical fingerprinting using analytical techniques to identify and characterize substances
   - Destructive testing for impact analysis
5. ***Analysis***: Review and evaluation of investigation work products (e.g., data, results, and outcomes)
6. ***Findings***
   - Short statements or paragraphs of fact as determined through analysis.
   - Group related findings and prioritize them according to order of importance or significance.
7. ***Informed Opinions***
   - A group of related findings leads to an informed opinion (or conclusion).
   - Opine relevant to factual outcomes or cause.
   - Informed opinions carry less liability than conclusions.

(*Continued*)

TABLE 5.3 (*Continued*)

Tips for Writing Expert Reports

8. ***References***
    - Use references only as necessary to support information in the report body.
    - Cite references used in the report body.
    - The two most common reference styles for engineering documents are APA and IEEE.
    - Associated citation styles are "author-date" (APA) and "numbered" (IEEE) formats. Examples: (Smith & Jones, 2022) and "Samples were tested for ..... and 8 percent on average are found outside of two standard deviations from the mean."[3]
9. ***Limitations***
   A limitations statement in an expert report offers several key benefits:
    - It specifies the intended audience and expresses report focus and limits.
    - It provides context for the findings, guiding readers to interpret informed opinions and/or conclusions within specific conditions.
    - It helps readers better understand investigation conditions, limits, and challenges encountered, thereby reducing the chance of criticism.
    - It expresses no guarantees that others will reach the same outcomes.
    - It leaves the door open to amend content should new information arise.
    - It enhances credibility and transparency.
    - It supports a fair and informed decision-making process in civil litigation.
10. ***Appendix***
    Typically reserved for large-scale investigations and forensic studies.
    Content includes supplemental supporting information of findings too detailed or lengthy for inclusion in the report body. Example contents:
    - Raw data, calculations, and detailed statistical analyses.
    - Detailed descriptions of research, testing, and experiment methods or protocols.
    - Additional tables, figures, maps, photographs, or diagrams that are relevant but not essential to the main text.
    - Audio or video clips related to the investigation.
    - Full interview statements and transcripts.
    - Copies of supporting documents, such as notes, and related published papers or expert reports.
    - Correspondence related to the investigation, such as memos or letters containing important disclosures.

Notes: APA is the American Psychological Association; IEEE is the Institute of Electrical and Electronics Engineers.

Although attorneys have different philosophies about what constitutes an expert report, civil law ensures independence in expert investigation and reporting. Independence means free from influence by others. Active voice – logic flow reports based on scientific methods, limited to less than ten pages, and without an appendix serve well for many cause-and-effect and failure analysis reports. Report only facts with short paragraphs, less is best, and refrain from unnecessary elaboration is a good basis for expert report body text.

Two types of reports to avoid are (1) ***constellation of conclusions,*** and (2) ***grandstand.*** A report with a half-page of introduction and four pages of conclusions is nonsense. Two common attributes of these types of reports are unsubstantiated content and an overabundance of unnecessary references. From introduction straight to conclusion is a *leap of faith,* and *jumps to conclusion* are hasty judgments unsupported by analysis and facts. The author of such a report and those who commissioned it are not interested in settlement. Also, some experts try to impress with a list of unnecessary references. Audiences of expert reports are often not engineers with PhDs but are wise to recognize baseless and irrelevant work. A rule of thumb about references: use only as necessary and cite where used in the report.

5.4 Cohesive Paragraphs

A ***cohesive paragraph*** introduces a single topic with supporting sentences. The first sentence is the topic sentence, the subject of the paragraph, followed by sentences of related statements. These supporting sentences explain, emphasize, illustrate, or otherwise relate to the topic with a level of detail to establish or support a fact. An example cohesive paragraph follows. The sentences are numbered, and each is explained after the text box.

DRAWBACKS OF WATER IN WATER-BASED ELASTOMERIC SEALANTS

1. Water-based elastomeric sealants face several challenges during application due to their water content.
2. Although water serves as a carrier for the solid components during application, it evaporates as the sealant dries and cures.
3. This evaporation leads to some volume loss in the sealant, which results in thinning and the need for additional coats.
4. If not applied thick enough or if the surface is not properly prepared, these sealants may peel soon after application.
5. Also, moisture can interfere with proper adhesion and cause bubbles and blisters, which compromise sealant effectiveness.
6. In cold climates, water-based sealants can become extremely stiff, making them brittle and susceptible to cracking.

Topic sentence 1: Challenges of water-based elastomeric sealants
Support sentence 2: An advantage and a disadvantage of the water component
Support sentence 3: Emphasis on the disadvantages of the water component
Support sentence 4: Potential consequences of the water component
Support sentence 5: Additional potential consequences of the water component
Support sentence 6: Other disadvantages of water with regard to cold-climate applications

The established fact with this paragraph is that water is problematic in elastomeric sealants. Imagine the relevance of such information with a product that is 35% water and sells for US$125 for a 5-gallon container. Consumer products sell on the high notes but rarely disclose limitations or drawbacks.

5.5 Report Challenges

Every project or case has its challenges. Time is a well-recognized constraint. A client who needs a report yesterday does not value the need for research, investigation, and analysis in establishing facts. Professional engineers do not let procrastinators dictate work products. Other common challenges to preparing expert reports are summarized in Table 5.4. By addressing these challenges, forensic engineers and attorneys can work together to produce more effective and defensible expert reports.

TABLE 5.4

Challenges to Preparing Expert Reports

1. ***Insufficient Access to Evidence.*** Independent investigation means direct access to evidence. When circumstances preclude access, the forensic engineer must rely on evidence documented by others. This is considered less than ideal and can limit determinations.
2. ***Insufficient Research and Preparation.*** Without adequate time, experts might quickly review materials and analyze evidence without conducting experiments or validation, which can lead to incomplete or flawed opinions.
3. ***Data Issues.*** Data that is suspect is a common reason for testimony deemed unreliable. Insufficient data collection, selective use of data, and failure to evaluate the accuracy of third-party data are common mistakes with data.

(*Continued*)

TABLE 5.4 (*Continued*)

Challenges to Preparing Expert Reports

4. ***Lack of Clarity and Communication***. Experts often struggle to simplify complex concepts into simple terms without distorting meaning, making it difficult for judges and jurors to grasp technical information. The ***translational engineer*** is highly valued in courts of law.
5. ***Scope and Formatting Issues***. A poorly defined scope and format of a report result in unmet expectations and unnecessary work for the attorney. The forensic engineer must be mindful that their reports are routinely disclosed to other parties, and vigilant to not confound their part in helping the courts resolve disputes.
6. ***Managing Bias and Conflicts of Interest***. These issues can undermine the expert's credibility and the validity of their testimony. Fairness and the Golden Rule are armor for the forensic engineer.
7. ***Staying Within Expertise Boundaries***. Experts may inadvertently offer opinions beyond their field(s) of expertise, which can damage credibility. This can be a problem for the forensic engineer who practices across disciplines and develops broad expertise. Legal professionals often think of engineers as specialists in a particular area. Forensic engineering is a young field, and experts can describe broad-based experience in their resumes and on websites. They can also make this point clear if necessary under oath.
8. ***Inaccuracies in Factual Foundations***. Misstatements or faulty assumptions in expert reports can lead to experts being challenged or barred from testifying. To err is human and the errata process invaluable. However, errata correction may be significantly limited in criminal cases,[2] which can unintentionally influence judicial decisions.
9. ***Maintaining Credibility***. Inconsistencies, errors, or doubts about an expert's testimony can undermine their credibility and weaken the persuasiveness of their opinions. With reports, the *three-pair-of-eyes rule*[3] is indispensable. Preparation for testimony starts with a review of one's expert report and rehearsal, the best ways to ensure consistency with information and facts.

5.6 Communication Liability

Communication liability has a twofold meaning with respect to expert reports: discoverable or ***protected communication***, and grammatical. Under the FRCP, certain protections limit the discoverability of communications between attorneys and experts. Grammatical communication liability refers to the potential legal and professional risks associated with poor grammar in contractual communications.[5] Tables 5.5 and 5.6 summarize key points for each.

To mitigate grammatical communication liability (Table 5.6), experts should prioritize clear, accurate, and grammatically correct communications to maintain professionalism, avoid misunderstandings, and protect themselves from potential legal issues.

TABLE 5.5

Key Points for Protected Communications

Disclosures	Since 2010, communications between testifying expert witnesses and attorneys have received greater protection from disclosure. This includes draft expert reports and most communications between experts and attorneys.
Exceptions	There are three specific exceptions where communications must be disclosed: 1. Those relating to the expert's qualifications, credentials, publications, and compensation. 2. Communications providing facts or data that the attorney supplied, and the expert considered in forming opinions. 3. Assumptions provided by the attorney that the expert relied on in forming opinions.
Expert Reports	Expert written reports must include: • A complete statement of all opinions and their basis • Facts or data considered in forming opinions • All exhibits used to support opinions
Work Product Protections	In most states, federal work product protection applies to report drafts and disclosures for other expert witnesses. However, rules may vary by state. For example, California law allows for broader discovery of expert draft reports and communications compared to federal rules. Attorneys need to make a concerted effort to inform their experts of the applicable rules that apply to attorney-expert communications. When not disclosed up front, the expert should ask: "What are the rules that govern our communications and my investigation, reports, and expert witness testimony?" In doing so, attorneys maintain appropriate protections for their experts while fulfilling disclosure obligations.

TABLE 5.6

Key Issues with Poorly Written Reports

Misinformation and Misunderstandings	Grammatical errors can change the meaning of a message, potentially resulting in misinterpretations and misunderstandings.
Legal Consequences	Grammatically incorrect documents may convey false or misleading information that can result in expert disqualification, financial reimbursement, and damaged reputations.
Loss of Credibility	Poor grammar can damage an engineering firm's reputation and trustworthiness, potentially costing business opportunities.
Reduced Effectiveness	Grammatical mistakes can distract readers from the intended message, making communications less impactful.
Competitive Disadvantage	Poorly written reports negatively reflect the quality of products and professionals who write them.

5.7 Legal Motions

Lawyers often use motions to gain an advantage in legal proceedings. Motions are powerful tools for shaping the course of litigation and influencing the outcome of a case. Table 5.7 summarizes common motions used in

TABLE 5.7

Summary of Common Motions Used in Legal Proceedings

Motion	Description	Court
Motion for Discovery	Requests specific evidence, witness statements, or records from the parties (or a party).	Civil & Criminal
Motion to Compel Discovery	Used to request the court to order a party to respond to discovery requests or provide information they have refused to disclose.	Civil & Criminal
Motion to Quash or Modify a Subpoena	Used to challenge or alter a subpoena that improperly seeks documents or testimony.	Civil & Criminal
Motion for a Protective Order	Seeks to protect sensitive information from disclosure or limit its use.	Civil & Criminal
Motion to Dismiss	Used during the early stages of a lawsuit to dismiss a case or party. May impact the discovery process.	Civil
Motion for Sanctions	Requests penalties for a party's failure to comply with discovery obligations.	Civil & Criminal
Motion for Summary Judgment	Typically, a discovery motion, used to end a case without a trial. Relies on information gathered during discovery.	Civil
Motion to Compel Deposition	Specifically requests the court to order a party or witness to appear for a deposition.	Civil
Motion to Suppress	Seeks to exclude certain evidence from being used at trial, often due to constitutional violations.	Criminal
Motion in Limine[4]	Requests the court to rule on the admissibility of certain evidence before trial.	Criminal

legal proceedings. Note that some are specific to civil or criminal proceedings, and others are common to both.

Some strategic advantages of motions include:

- *Efficiency.* Motions can streamline the legal process by resolving issues before trial.
- *Shaping the Case.* Motions can exclude certain evidence or arguments, thereby limiting the scope of the trial.
- *Psychological Impact.* Successfully arguing motions can create momentum and pressure on the opposing party.
- *Settlement Leverage.* Favorable rulings on motions can strengthen a party's position in settlement negotiations.

It is important to note that filing motions should be done judiciously, and that motion abuse can work against a litigating party. Excessive or frivolous motions and motion manipulation can irritate judges and potentially harm a lawyer's credibility. Motion manipulation can be a particularly irritating strategy when it involves an expert witness.

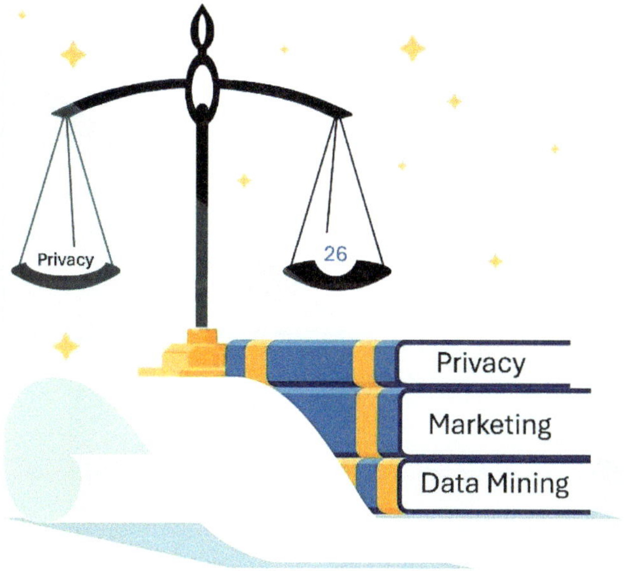

FIGURE 5.3
Civil Rule 26, Privacy, and Rebalancing the Scales of Justice.

5.8 Current Debate about Civil Rule 26

With the proliferation of social media platforms and other new technologies has come a renewed legal focus on privacy. Most of this focus has centered on data collection, storage, sharing, and third-party transactions in which customer information is mined for advertising purposes. But what about other contexts? For instance, could a party decline to produce, review, or collect certain types of data due to privacy concerns? Should privacy be considered a *"burden"* under the proportionality analysis required by FRCP Rule 26(b) (Fig. 5.3)? Privacy protection as it relates to discovery is a hot topic today.

If the title of this chapter were a question, the answer would be Report. *Report to settle* is the quickest and best way forensic engineers can help the courts resolve disputes. It's not about winning; it's about justice.

Notes

1. *Aequitas* is Latin for equity or fairness.
2. The *"I versus we conundrum"* is a classic example. A forensic engineer who works with a team of professionals may author a past-tense report and answer deposition questions in terms of "we." This may confuse the interrogator who

is trying to discern who did what specific investigation work. A "Yes, we did" answer may need to be corrected with errata to "she did," or "I did" later for clarification.

3. The three-pair-of-eyes rule is an editing standard for professional reports. It means three people qualified to review a report for technical and grammatical accuracy, and liability. The report author is the first pair of eyes, a senior expert in technical content and liability the second, and someone well versed in English composition the third is a strong editing team.
4. Motion in limine comes from the Latin *"motion at the start."*

References

1. The National Court Rules Committee (2024). https://www.federalrulesofcivilprocedure.org/frcp/title-v-disclosures-and-discovery/rule-26-duty-to-disclose-general-provisions-governing-discovery/; retrieved 5 Jan 2025.
2. The Committee on the Judiciary House of Representatives (2017). Federal Rules of Civil Procedure, 115th Congress, 1st Session, Committee Report No. 3, pp. 137.
3. Florida Rule Civil Procedure 1.280, General Provisions Governing Discovery (as amended through Jan. 1, 2025.
4. https://www.findlaw.com/litigation/legal-system/what-does-it-mean-to-settle-a-case.html By Lyle Therese A. Hilotin-Lee, J.D. | Legally reviewed by Melisssa Bender, Esq. | Last reviewed May 14, 2024; retrieved 12 Jan 2025.
5. Valerie C. Brannon (2019). Liability for Content Hosts: An Overview of the Communication Decency Act's Section 230, Congressional Research Service Report LSB10306 (Ver 1).

Case Study 5: Tactics & Maneuvers

Accident Reconstruction, Criminal Case

- Plaintiff's Expert Report
- Defense Consultant Expert
- Motion Manipulation
- Protective Order

Message to students:

> Calm confidence in the face of adversity; let truth speak loudest.

Background

A motor vehicle collision occurred during the early evening on a day in 2018 at the intersection shown in Figures CS5.1 and CS5.2. The driver of a dark sedan driving westbound on Road X collided with a white sedan traveling north on Road Y. The collision resulted in serious bodily injuries to both drivers.

DOI: 10.1201/9781003623311-11

FIGURE CS5.1
Collision scene.

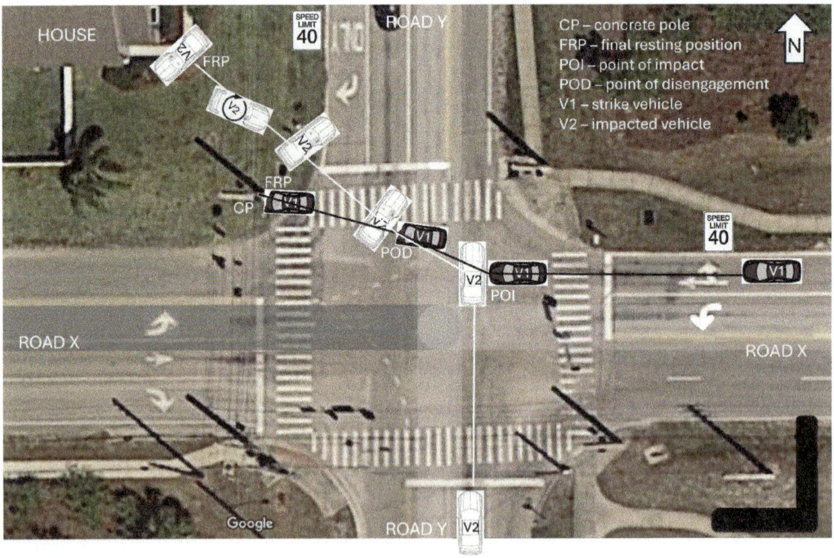

FIGURE CS5.2
Vehicle travel paths from intersection approaches to final resting positions (FRP). (Base image courtesy of Google Maps.)

An eyewitness driving south on Road Y stated that she had a green light as she approached the intersection, and that

> "….the *red car* was racing toward the intersection at high speed when it hit the white car so hard I could hear the crash even with my windows up.

The white car flew into the air and was spinning before it landed. The fast car hit the pole and stopped. I pulled over and called 911 and told them to send an ambulance because I knew people were hurt. Then I saw the driver of the *red car* get out and stumble across the grass before falling; he was bleeding bad. No one got out of the white car. A fire truck and ambulance arrived a couple of minutes later along with the police. An officer could tell I was panicked, asked me if I was OK, told me to please try being calm and relaxed, and said he would come back soon to get my statement. There was so much blood on the grass....and a broken fire hydrant."

Based on the eyewitness account, a toxicology report indicating narcotics influence, and a police crash report that stated failure to yield the right of way in running a red light, the driver of the dark sedan was criminally charged with causing an accident that resulted in serious bodily injury.

Police surveyed and photographed the collision scene, obtained a detailed statement from the eyewitness, and thoroughly photo-documented damages to both vehicles at their final resting positions and the next day in daylight at an impound yard. The defendant lawyered up, and the defense and the State Attorney's Office (SAO) took sides, each accusing the other driver of speeding and running a red light. The case moved slowly along in the court system for two years before the COVID pandemic[1] shut everything down and caseloads backed up.

The hiatus provided an opportunity for police and departments within the legal system to *house clean*. Impound yards were cleared and stored evidence reorganized. The vehicles in this case went to the crusher before crash data retrieval (CDR)[2] was performed. In fall 2021, as the court system transitioned back to normal operating procedures, the SAO realized important evidence was lost. A search for a forensic expert was initiated in hopes their case could be sufficiently supported with the photo-documentation and remaining case file evidence.

Investigation

A 2021 forensic investigation comprised a collision scene survey that established vehicle trajectories and dimensioned travel paths, and locations of ground features (e.g., pavement and grass skids, fire hydrant, concrete pole, and traffic signal control box). Also, posted speed limits, lighting, and other landmarks surrounding the intersection were confirmed. Traffic signal sequencing records for the day of the collision were also obtained.

The new information together with the SAO case file substantiated the state's case; the defendant was the responsible party. The state's expert submitted a report, and the defense hired their own expert as a consultant. That expert did not visit the accident scene, perform a physical investigation, nor submit a report.

Pre-Trial Motions

In criminal proceedings, *discovery motions* can be filed to obtain or protect information relevant to a case. Motion tactics are crucial tools used by lawyers to shape the course of a case. These tactics involve filing various types of motions to address legal issues, challenge evidence, or seek favorable rulings from the court.

For this case, the SAO expert's report was submitted into evidence and the defense attorney was copied. No report from the defense expert was entered into evidence. The SAO's expert was then subpoenaed to testify at pre-trial hearings.

The defense attorney first attempted to disqualify the SAO expert and then obtain an explanation of the expert's investigation methods and means (M&M). The plaintiff's expert cited a conflict of interest in disclosing M&M to a competitor, and subsequently requested protection of investigation M&M, citing proprietary information. Protection was granted by the court, pending motions. Table CS5.1 summarizes the motion tactics that ensued.

At the first pre-trial hearing, each of three motions to first obtain M&M and then suppress the SAO expert's report was denied by the court. The judge granted a second pre-trial hearing, the outcome of which sealed the SAO expert report for trial. This meant the defense had to proceed without knowing the SAO expert's investigation findings. This court decision leveled the playing field so to speak. The defense tactic of hiring a *consultant expert* would have allowed one side to predict the other's strategy without disclosing anything themselves. Further, the defense could then strategize to refute and/or cast doubt in the eyes of a jury. Looking to poke holes in honest forthright work is not justice.

A last attempt by the defense to coerce information came through pre-trial deposition motions. It is common practice to depose investigation experts in civil and criminal cases. This allows both sides to get an idea about what opposing experts might testify about in a jury trial. However, in this case only one expert investigated and reported. The defense filed a no-holds-barred *motion to compel deposition duces tecum*, an attempt to orally depose the SAO expert and obtain a copy of his entire file. The SAO recognized this last-ditch effort and filed a final *motion to quash* the defense's motion, which the court granted.

TABLE CS5.1

Motions Applied to Coerce Expert Disclosure of Information

First Pre-trial Hearing	
1. Defense **Motion to Compel Discovery**: Used to request the court to order SAO expert to respond to discovery request for information he refused to disclose.	Denied
2. Defense **Motion to Direct an Opposing Party to Divulge More in Discovery**: Used when a party believes the other side has not provided sufficient information.	Denied
3. Defense **Motion to Suppress**: Sought to exclude certain evidence (SAO expert's report) from being used at trial.	Denied
Second Pre-trial Hearing	
4. Defense **Motion in Limine**: Requests the court to rule on the admissibility of certain evidence (SAO expert report) before trial.	Judge called a Chamber Conference with the attorneys.
5. Plaintiff (SAO) **Motion for a Protective Order**: Sought to maintain the confidentiality of sensitive information (proprietary M&M and file information).	Granted. SAO expert report sealed for trial. Trial to proceed based on pre-existing evidence. Attorney's told to provide the Court with trial exhibits.
Pre-trial Depositions	
6. Defense **Motion to Compel Deposition Duces Tecum**[3]: Specifically requests court to subpoena SAO expert to bring his entire investigation file to deposition and provide copy to defense.	
7. Plaintiff **Motion to Quash** Subpoena for Deposition Duces Tecum: Specifically requests court to quash defense motion to compel deposition duces tecum.	Granted. Defense motion to compel denied.

Motions play a crucial role in ensuring a fair trial and facilitating the exchange of information between parties in criminal proceedings. However, a savvy judge can use the opportunity to educate a manipulating attorney.

That *Red Car*

The defense attempted to discredit the sole eyewitness to the collision because her story revolved around a speeding *red* car. There was no red car involved in this collision. All aspects of her account were consistent

FIGURE CS5.3
Emergency response at the intersection collision scene.

with the police and SAO expert investigations and signal timing records, except for that one point....the *red* car. How could the eyewitness mistakenly think the speeding car was red? Thirty minutes observing red ambulances and firetruck flashing red lights waiting to give a statement to the police at a collision scene awash in red. When you look at Figure CS5.3, what draws attention? Might a dark sedan against a black backdrop be inconspicuous?

Traffic Signal Sequencing and Signals at Collision

Historic traffic signal information was obtained from the local public works transportation department. Intersection controller data sheets together with the signalization design plans identified the signal phase sequence (or Signal Movement Chart) and phase timing (or Signal Operation Chart) (Figs. CS5.4(a) and (b)) that governed traffic flow at the time of the crash.

The intersection was equipped with under-pavement quadrupole loops at all approaches. These detectors tell the controller when to integrate turn signal phases (e.g., phases 1-5, 1-6/2-5, 3-7, and 3-8/4-7) between through-traffic phases 2-6 and 4-8. When no turn traffic was detected, Road Y through-traffic phase 2-6 was followed by Road Y 4-8 through-traffic phase. The "ALL RED" and "2.0" values in the Signal Operation Chart indicate that all other traffic flows were stopped on red lights when a 2-6 or 4-8 through-traffic phase was green-lighted, and two seconds was afforded to clear the intersection of traffic before the next phase change. What this means in relation to the subject collision is that with a northbound/southbound 2-6 green-light phase, westbound/eastbound Road X traffic flow would have been stopped on red lights.

The eyewitness had described a 2-6 signal phase for her southbound approach to the intersection in her police statement. Further, she stated there was no other traffic approaching the intersection other than the two vehicles that collided. If she had a green light to continue southbound, then the northbound white car (V2 in Fig. CS5.2) did also, and the westbound dark sedan (V1 in Fig. CS5.2) must have approached the intersection on a red light. Together, the eyewitness account and signal sequence records are two independent lines of converging evidence toward a cause.

> The speeding car was dark blue, but the color of trauma was memorably red.

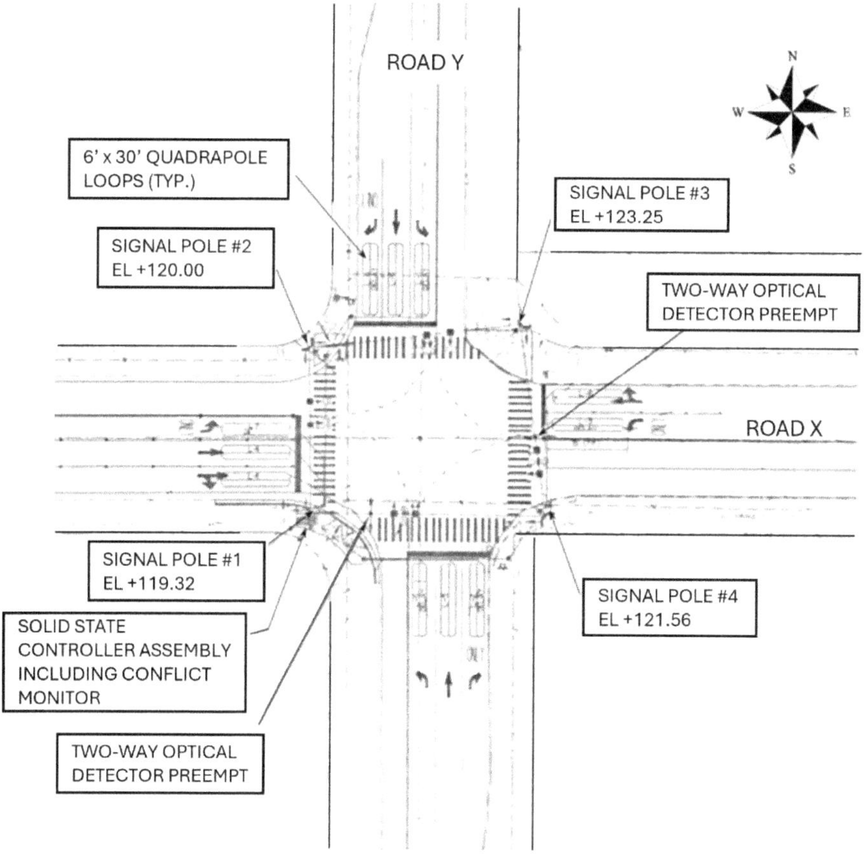

(a) Signalization

FIGURE CS5.4
(a) Signalization equipment and (b) traffic signal movement and operation chart. (*Continued*)

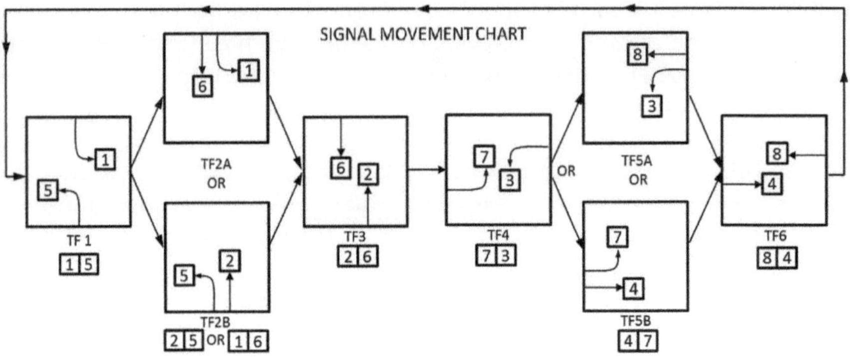

SIGNAL OPERATION CHART									
MOVEMENT	MIN	EXT	MAX	CLEAR	ALL RED	WALK	FL D/W	RECALL	DETECTOR
1 + 5	6	3	10	3.5	N/A	N/A	N/A	N/A	NL
1 + 6 OR 5 + 2	6	3	10	3.5	N/A	N/A	N/A	N/A	NL
2 + 6	10	6	25	3.5	2.0	N/A	N/A	N/A	LOCK
3 + 7	6	3	10	3.5	N/A	N/A	N/A	N/A	NL
7 + 4 OR 8 + 3	6	3	10	3.5	N/A	N/A	N/A	N/A	NL
4 + 8	10	6	25	3.5	2.0	N/A	N/A	N/A	LOCK
EXCLUSIVE PED.	--	--	--	--	--	9	16	N/A	N/A

(b) Traffic signal movement and operation chart

FIGURE CS5.4 *(Continued)*

Notes

1. The COVID-19 pandemic (aka COVID pandemic) began as an outbreak of severe acute respiratory syndrome coronavirus 2 (SARS-CoV-2) in Wuhan, China, in December 2019. It spread through Asia and then worldwide through the first half of 2020. The pandemic led to widespread illness and death. Through March 1, 2025, the World Health Organization reported ~778 million cases and 7.1 million deaths worldwide. Restricted mobility, economic disruption, social isolation, education disruption, and mental health challenges were some of the negative impacts during the first two years of the pandemic. The WHO ended the COVID-19 pandemic on May 5, 2023.

2. Crash data (aka *black box* data) can be retrieved from most 2004 and newer model vehicles for diagnostic evaluation of engine performance, braking, and other systems. The information can be used to determine speeds, acceleration/deceleration, and braking actions surrounding a collision event. Modern investigations routinely rely upon this information to infer driver actions relative to collision events.

3. Subpoena duces tecum – Latin for "under penalty, you shall bring with you."

Part II

Forensic Engineering Practice

6

Accident Reconstruction

CONTENTS

CONCEPTS	KEY TERMS	ADDITIONAL READING
Coefficient of Restitution	Momentum–energy dissipation	Exercise 11
Elastic and Inelastic Collisions	Exemplar	Exercise 12
Momentum Clock	Point of impact	
Energy Dissipation Analysis	Final resting positions	Case Study 6
Crash Data Retrieval	Accident reconstructionist	
Event Data Recorder	On-board diagnostics	
	Keep alive memory	

This chapter introduces students to the fundamentals of collision dynamics. It also serves as an update to Randy Noon's work on vehicular accidents. Advances in automotive performance technology and National Highway Traffic Safety Administration (NHTSA) crash test studies have led to crash data retrieval (CDR) and new momentum-energy analysis opportunities. Discussions of these new concepts with examples are presented here. First, we look at some collision fundamentals.

6.1 Collision Momentum Fundamentals

Many engineering students learn about ***collision momentum*** in a physics class, which teaches that momentum is conserved through a perfectly inelastic collision, how to determine the change in velocities and kinetic energy (KE) of the vehicles (Fig. 6.1), the difference between elastic and inelastic collisions, and other related aspects as collision fundamentals. With the depicted analysis, we see how ***conservation of momentum*** is

DOI: 10.1201/9781003623311-13

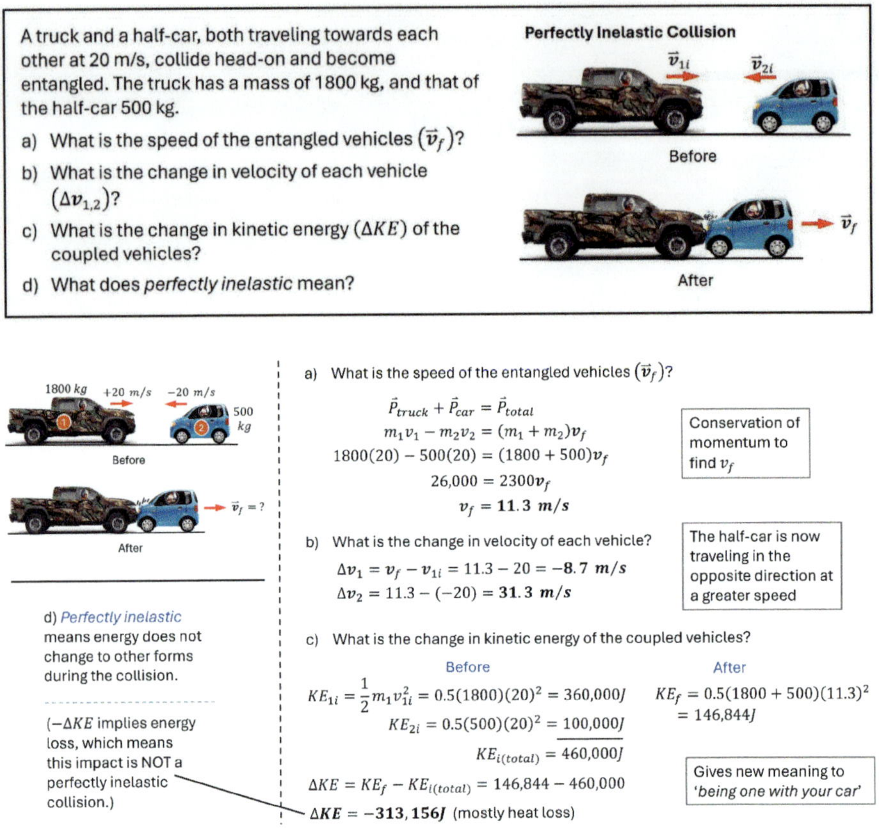

FIGURE 6.1
Example two-vehicle collision momentum analysis problem.

used to find the postcollision speed of the entangled vehicles, a result that is then used to determine the change in speed of each vehicle. Further, calculated kinetic energies before and after the collision are then used to determine the change in the KE of the *coupled* vehicles. Hopefully, students recognize from the outcomes of this example that the half-car doesn't stand a chance of survival in a head-on collision with a much larger vehicle. Imagine the forces on the driver in the mini car that was traveling to the left at 20 m/s (about 45 mph), suddenly one with the truck and moving in the opposite direction, having experienced an instantaneous 31-mph speed transition.

A couple of other momentum concepts are next discussed before going into energy dissipation and crush, a more in-depth method that compares vehicle damage to available crash test information to infer relative impact forces and speeds of the collision vehicles.

Coefficient of Restitution

The ***coefficient of restitution*** (COR), often denoted as "e," is a dimensionless parameter that quantifies the elasticity or *bounciness* of a collision. Defined as the ratio of the relative velocity of separation after a collision to that of the approach velocity, it is quantified as

$$e = \frac{|v_{2f} - v_{1f}|}{|v_{1i} - v_{2i}|}$$

where subscripts i and f denote initial and final speeds of the two objects. The COR essentially measures how much KE is lost during a collision. The range of e is usually a positive value between 0 and 1 for practical collisions:

- $e = 1$: Perfectly elastic collision with no loss of KE.
- $0 < e < 1$: Inelastic collision where some KE is lost to heat, sliding, and deformation (e.g., car crashes).
- $e = 0$: Perfectly inelastic collision where the objects stick together and move as one after the collision, and maximum KE is converted to heat, sliding, deformation, etc.

Values outside of this range are possible but rare. A bullet impacting and passing through a target has a value of $e < 0$. For $e > 1$, a superball's bounce may be the most relatable superelastic example in which an object rebounds with a greater speed than its approach speed. Some additional stored energy is released at the collision to cause an extraordinary bounce.

Although e is independent of mass, the final velocities of collided objects are mass-dependent due to conservation of momentum. For an elastic collision:

$$v_{1f} = \frac{m_1 v_{1i} + m_2 v_{2i} + m_2 e(v_{2i} - v_{1i})}{m_1 + m_2}$$

$$v_{2f} = \frac{m_1 v_{1i} + m_2 v_{2i} + m_1 e(v_{1i} - v_{2i})}{m_1 + m_2}$$

where v_{1i} and v_{2i} are the precollision speeds of objects 1 and 2, m_1 and m_2 are their respective masses, and v_{1f} and v_{2f} their postcollision velocities.

Elastic and Inelastic Collisions

Collisions can be classified as elastic or inelastic. In an ***elastic collision***, the total KE of the system before the collision is equal to the total KE after the collision. No KE is lost; it is simply transferred between the objects involved. The collided objects typically separate after the collision, often bouncing off each other. Kids shooting marbles, billiard ball collisions, and the collision of subatomic particles can be approximated as elastic collisions.

Colliding vehicles, on the other hand, often experience inelastic collisions. An ***inelastic collision*** is one in which there is a loss of KE, yet momentum is conserved. Some of the KE is converted into other forms of energy such as heat or deformation (or *crush*) of the colliding vehicles. In a perfectly inelastic collision, the collided vehicles stick together and move as a single unit after impact. Often collided vehicles travel together inelastically as energy is transferred through crush, before separating. The duration through which the vehicles are coupled can be milliseconds.

Momentum Clock

A basis for understanding collision outcomes is provided with the use of the ***momentum clock*** to the broad-side (or T-bone) collision scenario (Fig. 6.2(a)). White car V1 is traveling west from 3 o'clock and blue car V2 north from 6 o'clock. Momentum (P) of a vehicle is equal to its mass (m) times its velocity (v), or $P = mv$ (in $kg \cdot m / s$ or $N \cdot s$). For a perfectly inelastic collision of the two vehicles with momentum conserved:

$$m_1 v_1 = m_2 v_2$$

With the sedan-on-sedan collision of similar-weight cars traveling at equal speeds, $P_1 = P_2$ and their coupled trajectory will be 45° from the horizontal, toward 10:30 on the clock. If strike vehicle V1 is traveling at a lower speed than that of the blue sedan, its momentum will be less and the trajectory of the coupled vehicles will be at an angle greater than 45°, say 11 o'clock – the red arrow. This means the blue car's momentum dominates because it

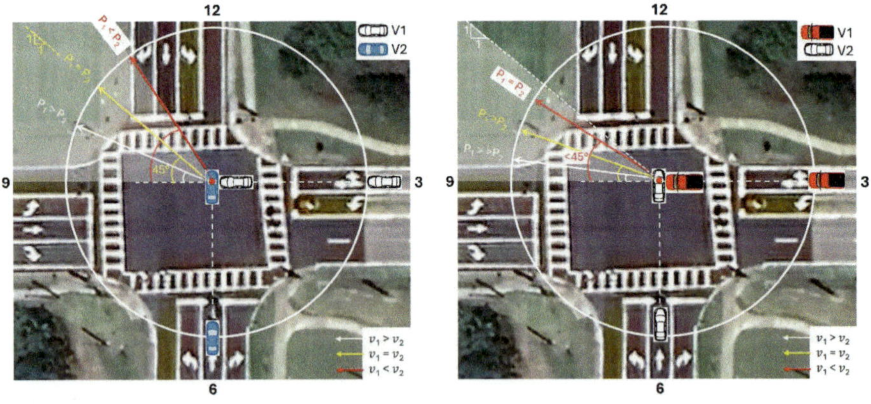

(a) Sedan – sedan collision ($m_1 \approx m_2$) (b) Truck – sedan collision ($m_1 >> m_2$)

FIGURE 6.2
Momentum clock and collision trajectory: (a) collision of similar-mass sedans and (b) a pickup truck colliding with a much lighter sedan.

is traveling faster. Conversely, sedan V1 traveling at a higher speed than V2 will yield a postcollision coupled trajectory at a lesser angle than 45, toward 10 o'clock, for example – the white arrow.

Now consider a half-ton pickup truck-on-sedan T-bone collision, with red truck V1 weighing twice that of the white sedan (Fig. 6.2(b)). For all three speed scenarios, the heavier truck will dominate momentum, and the coupled vehicles will travel along trajectories less than 45° from the horizontal, or between 9 o'clock and 10:30. The higher the truck's speed relative to that of the smaller car, the greater is its momentum.

The white sedan traveling significantly faster than the blue will dominate coupled momentum and lead to a shallower departure angle from 9 o'clock. Likewise, the much heavier truck traveling at the same or lower speed than V2 will tend to push the coupled vehicles more toward the direction the truck was traveling in. The point is: *the vehicle with the most momentum dominates the postcollision trajectory.*

6.2 Momentum–Energy Dissipation and Crush

Energy diminishes from a moving vehicle when it collides with another. *Energy dissipation* is related to *crush,* the damage associated with deformed components of the collided vehicles. Crush measurements can be used to quantify energy dissipation, which can then be used to infer impact speeds.

Energy dissipation analysis (EDA) of an automotive collision is an inverse analytical problem,[1] determining energy requirements to cause observed crush damage. A common assumption with this type of accident reconstruction is that a predominant part of the energy lost during collision is consumed by permanent deformation of vehicle components, and that other parts of the dissipated energy are insignificant or negligible.[1] Also, it is possible to estimate vehicle impact speed from EDA with a limited data set and use of an applicable NHTSA (National Highway Traffic Safety Administration) crash test study.

The discussion that follows makes use of an actual collision between a dark blue 2006 Hyundai Sonata that T-boned a 2012 Ford Mustang at an intersection at night, and an exemplar 2012 Mustang study downloaded from the NHTSA crash test database.[2] An exemplar 2006 Sonata study of frontal impact damage was also reviewed in the database. The primary objective of this EDA approach is to determine all parts of the energy balance dissipated during the collision to obtain a conservative estimate of impact speed for the strike vehicle.

It was determined that the Mustang had the right-of-way, and severe damage to the vehicles suggested the Sonata was traveling at a high rate of speed (more than the 40-mph posted speed limit) when it impacted the passenger side of the Mustang. The speed of the Sonata became an important consideration, and a forensic engineer was asked to determine it.

At the time of the forensic investigation, the collision vehicles were no longer available for crush damage assessment. But the police file did include a very good photograph set of the damaged vehicles from all sides, and the NHTSA crash test studies of the exemplars (same make, model, and year as both vehicles) afforded comparative crush damage analysis. The list below summarizes the data used for this EDA.

- Intersection geometry
- Point of impact (POI) and vehicle final resting positions (FRPs)
- General collision type (i.e., front-to-side or T-bone)
- Vehicle curb weights and gross vehicle weights (GVW)
- Good crash scene photographs (postcollision vehicles inclusive)
- Police crash scene survey measurements (survey of horizontal and vertical relief)
- Weather data (site-specific road conditions determined from photos and confirmed with local National Weather Service data)
- NHTSA exemplar crash test data
- *Momentum–energy dissipation* (MED) analysis

The following analysis is conservative and determines a conservative estimate of the Sonata collision impact speed. Figure 6.3 depicts the collision scene, the vehicles' approaches to the intersection, and their postcollision trajectories. Number sequence 1 to 4 relates to sources of energy dissipation identified in Table 1, which is included as an inset within the figure. Energy dissipation sources 1 and 2 occurred simultaneously at the POI, after which a period of inelastic collision occurred with the cars coupled (see Fig. 6.3 enlargement at upper right), followed by vehicle separation and Mustang change in direction (energy dissipation source 3). Energy was also dissipated (source 4) with the Sonata's final impact with the signal pole.

Analysis

What follows is a step-by-step quantification of the sources of energy dissipation (ED) for the collision.

Source 1 ED. Sonata front impact crush, average KE_{crush} based on two 2006 Sonata exemplars:

Crush kinetic energy is

$$KE_{crush} = \frac{1}{2}mv^2$$

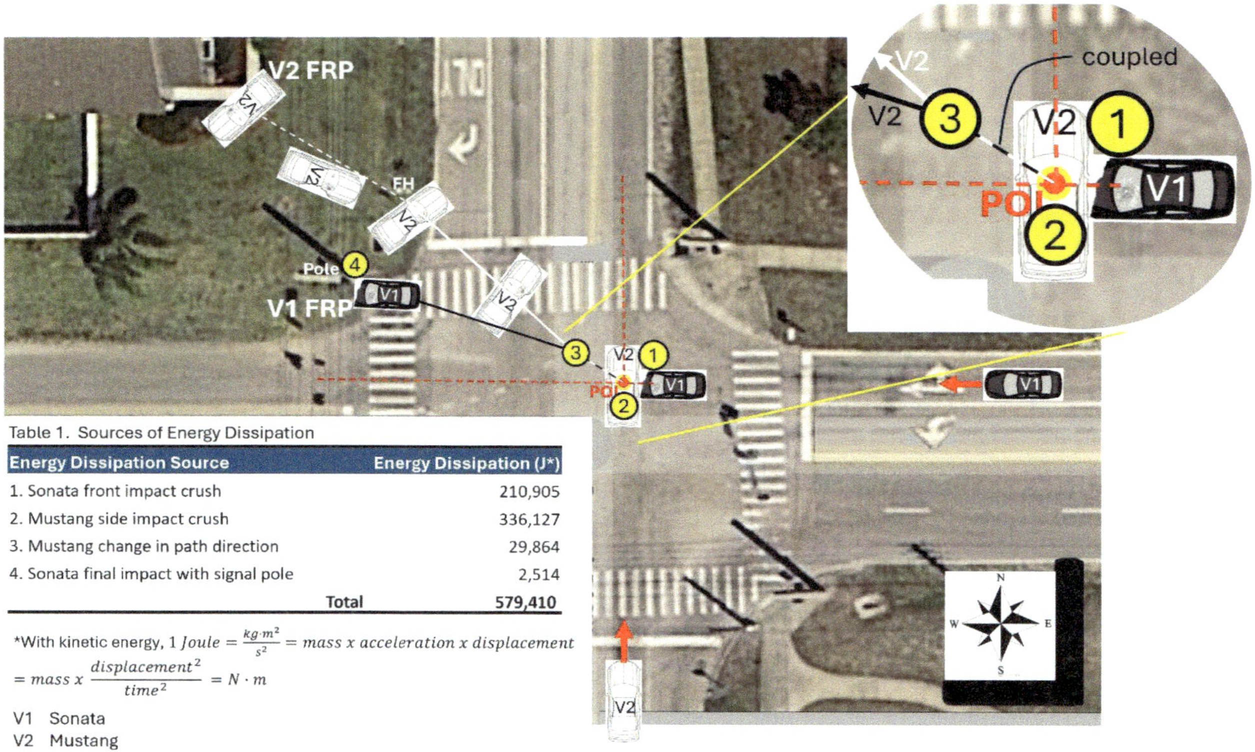

Table 1. Sources of Energy Dissipation

Energy Dissipation Source	Energy Dissipation (J*)
1. Sonata front impact crush	210,905
2. Mustang side impact crush	336,127
3. Mustang change in path direction	29,864
4. Sonata final impact with signal pole	2,514
Total	579,410

*With kinetic energy, 1 $Joule = \frac{kg \cdot m^2}{s^2} = mass\ x\ acceleration\ x\ displacement$

$$= mass\ x\ \frac{displacement^2}{time^2} = N \cdot m$$

V1 Sonata
V2 Mustang
FH Fire Hydrant
FRP Final Resting Position

FIGURE 6.3
Collision scene energy dissipation sequence and analysis summary (Table 1 – inset).

where the respective masses and test speeds for exemplar Sonata 1 and Sonata 2 are $m_1 = 1710\ kg$ (car + driver) and $m_2 = 1718\ kg$ (car + driver + accessories), and $v_1 = 15.7\ m/s\ (35\ mph)$ and $v_2 = 15.67\ m/s$. The corresponding crush kinetic energies are $KE_{c1} = 210{,}748\ J$ and $KE_{c2} = 210{,}926\ J$, the average of which is $KE_{c(avg)} = 210{,}837\ J$.

Source 2 ED. Mustang side impact crush analysis. Equivalent KE based on KE of moving deformable body (MDB) and red exemplar 2012 Mustang (NHTSA crash test).

A side-impact crush comparison of the 2012 subject (white) and NHTSA test (red) Mustangs is performed using a photogrammetric technique that requires a good set of damage photographs taken from multiple angles of the subject vehicle (Fig. 6.5, upper frames). The NHTSA crash test on the 2012 exemplar Mustang is illustrated in Figure 6.4. NHTSA uses an MDB crash vehicle that is a wheeled steel chassis fitted with a deformable barrier

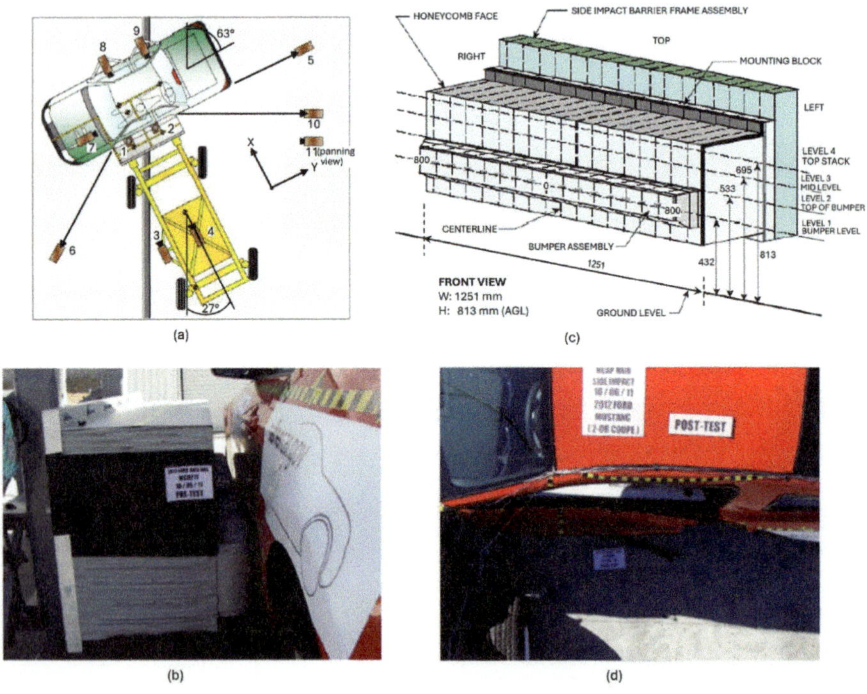

FIGURE 6.4
NHTSA crash test on a 2012 Ford Mustang. (Courtesy of NHTSA.): (a) Illustration of impact test (MDB yellow; 2012 Mustang test vehicle green); (b) Pre-test right side view of MDS positioned against side of 2012 Mustang test vehicle at ideal impact point; (c) Schematic of MDB showing dimensions face/bumper levels; and (d) Post-test top view of side impact damage to 2012 Mustang test vehicle (35 mph impact).

shaped with dimensions that mimic a generic vehicle front end (Fig. 6.4(c)). The weight of the MDB is adjustable and was similar to that of the 2006 Sonata strike vehicle, and its strike speed was 35 mph (56.3 kph) for the test. An image of the postimpact damage to the test Mustang (Fig. 6.4(d)) attests to the relatively light crush damage associated with the 35-mph impact.

Crush is next compared to that of the subject white Mustang (Fig. 6.5, upper frames). It should be obvious that the crush damage is greater for the subject Mustang. Notably, its chassis is significantly more deformed. A top view of the comparative damage profiles is applied to the exemplar with

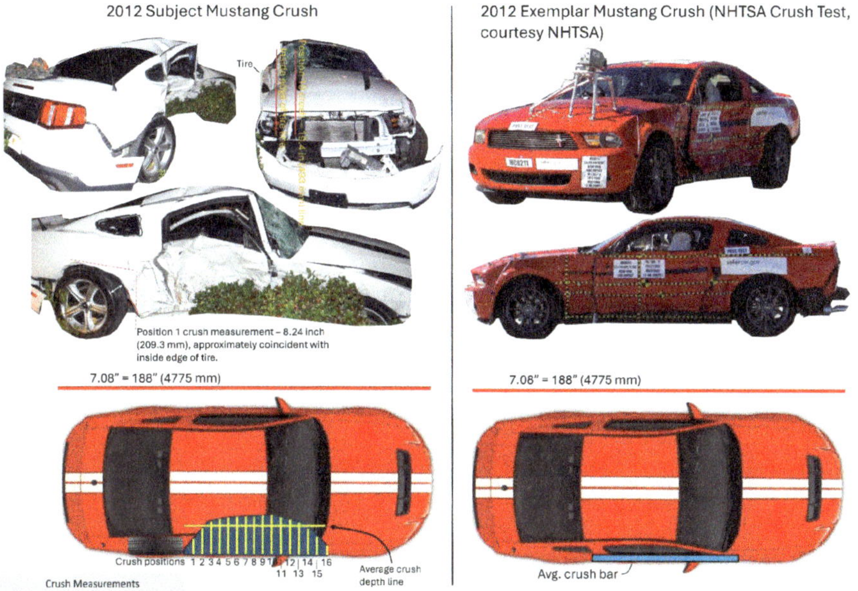

Crush Measurements

Position	Scale (in)	Actual (in)	mm
1	0.31	8.24	209.1948
2	0.48	12.75	323.9146
3	0.6	15.94	404.8932
4	0.65	17.27	438.6343
5	0.68	18.07	458.8790
6	0.69	18.33	465.6272
7	0.71	18.86	479.1236
8	0.71	18.86	479.1236
9	0.73	19.39	492.6201
10	0.73	19.39	492.6201
11	0.68	18.07	458.8790
12	0.65	17.27	438.6343
13	0.6	15.94	404.8932
14	0.48	12.75	323.9146
15	0.33	8.77	222.6913
16	0.19	5.05	128.2162
Average	0.58	15.31	388.8662

Scale factor: $\frac{188}{7.08} = 26.554$

FIGURE 6.5
2012 subject white Mustang crush damage comparison with the 35-mph 2012 exemplar red Mustang crash test.

the bottom frames. The average crush bar (blue) on the exemplar (lower right image) is based on NHTSA report crush data. For the subject Mustang, note that the Position 1 crush depth of 209 mm in the upper left image is approximately that of the tire width, and maximum crush occurred at position 10 deep within the passenger door at a depth of 492.62 mm (Crush Measurements table). Comparison shows that the subject Mustang's average crush depth is four to five times that of the exemplar, using a *crush factor* of 2.5 for conservative analysis. We next quantify EDs and explain the crush factor.

Equivalent momentum – Sonata and MDB

$$KE_{total} = KE_{equiv}(CF)$$

where $KE_{equiv} = \frac{1}{2}(m_{sonata})(v_{equiv})^2 - KE_{MDB,crush}$

$m_{sonata} = 1548.6\ kg$ (use curb weight[2] for most conservative analysis)

$v_{equiv} = 15.17\ m/s$ (equivalent Sonata velocity assuming momentum MDB equals momentum Sonata (see calculation below))

$m_{MDB} = 1828\ kg$

$$\frac{1}{2}m_s v_s^2 = \frac{1}{2}m_{MDB}v_{MDB}^2$$

$$v_s = 35\ mph = 15.646\ m/s$$

$$v_{MDB} = \sqrt{\left(\frac{m_s}{m_{MDB}}\right)v_s^2} = \sqrt{\left(\frac{1718}{1828}\right)(15.646)^2} = 15.17\ m/s$$

$$KE_{Sonata} = \frac{1}{2}(1548.6)(15.17)^2 = 178,188\ J$$

$KE_{MDB,crush}$ = crush strength of MDB (N/m^2) × crush area (m^2) × avg. crush depth (m)*

$$= 310,000\frac{N}{m^2}(0.937m^2)(0.15\ m)$$

$$= 43,571\ J$$

* Parameters from NHTSA report #MC0211

KE_{equiv} can now be computed:

$$KE_{equiv} = 178,188 - 43,571 = 134,617\ J$$

The relative crush depth of the subject Mustang to that of the exemplar Mustang is now determined:

$$CF = \left(\frac{C_{subj\ Mustang}}{C_{exemplar}} \right) CB$$

where $C_{subj\ Mustang}$ = estimated average crush depth in the subject Mustang = $C_{avg} \cdot S_{A1} \cdot S_{A2}$

C_{avg}= 0.58 in, avg. crush depth taken from multiple measurements from scaled damage photographs (see Fig. 6.5 Crush Measurements table) of subject Mustang.**

S_{A1} = 26.554, scale factor to convert photo measurements to actual

S_{A2} = 25.4 mm/in (scale factor to convert from in to mm)

$$C_{subj\ Mustang} = (0.58\ in) \cdot 26.554 = 15.401\ in\ (\text{or } 391.2\ mm)$$

** C_{avg} is determined from measured crush when collision vehicles are accessible. When not available, damage photogrammetry can be used with a good photograph set. Ideally, vehicle damage should be photographed from several angles for the most accurate portrayal. Measurements are then estimated based on known vehicle dimensions.

$C_{exemplar} = 73.7794\ mm$, avg. crush depth taken from crush measurements in NHTSA report #MC0211)

$CB = 0.5$, conservative bias (to ensure Sonata impact crush factor CF is conservative)

$$CF = \left(\frac{C_{subj\ Mustang}}{C_{exemplar}} \right) CB = \left(\frac{391.2}{73.7794} \right) 0.5 = 2.65$$

Use 2.5 (more conservative)

Finally,

$$KE_{total} = KE_{equiv}(CF) = 134{,}617\ J(2.5) = 336{,}544\ J$$

Source 3 ED. Mustang change in path direction. This is the energy required to reroute the subject Mustang during engagement with the Sonata.

$$KE_{Mustang(reroute)} = F \times d$$

where $F = m_{Mustang} \cdot a \cdot f$

$m_{Mustang} = 1604\ kg$ (conservative)***

$a = 9.81 m/s^2$

$f = 0.5$ (coeff. of friction, tires on dry pavement, conservative)
$d = 3.8\ m$ is the lateral distance the subject Mustang was pushed during engagement with the Sonata (diagram below at right). The diagonal distance from POI to POD (point of disengagement) was measured in the field at 5.89 m by the police during their field survey.

$$KE_{Mustang(reroute)} = \left[(1604kg)\left(9.81\frac{m}{s^2}\right)(0.5)\right]\times(3.8m)$$

$$= 29{,}898\ J$$

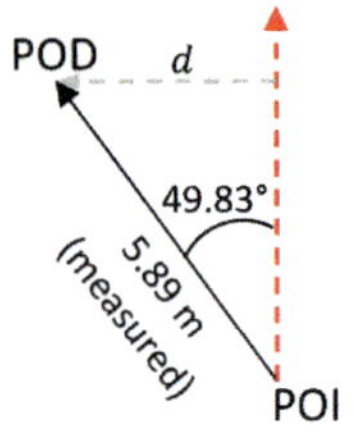

$d = 5.89 \cos 49.83° = 3.8\ m$

***For this calculation, we use Mustang curb weight 1604 kg and low f for conservative analysis.

Source 4 ED. Energy dissipated by Sonata during final impact with signal pole.

The fluid splatter pattern on the signal pole can be used to estimate the final impact speed of the Sonata when it was stopped by the pole. This speed can then be used to compute the last source of ED in stopping the Sonata. The text box below shows an image of the FRP of the Sonata and a fluid splatter trajectory diagram used for the calculation.

$$KE_{final} = \frac{1}{2} m_{Sonata} \left(v_{f(Sonata)}\right)^2$$

where

$m_{Sonata} = 1548.6\ kg$ (conservative)

$v_{f(Sonata)} = 4.03\ mph$ (see text box below)

$d_h = d_{horiz} = 0.27m$, splatter horizontal displacement

$d_v = d_{vert} = 0.11m$, splatter vertical displacement

(displacements estimated from police photographs and field confirmed)

$$KE_{final} = \frac{1}{2}(1548.6\ kg)\left(1.802\frac{m}{s}\right)^2 = 2514\ J$$

Fluid splatter trajectory:

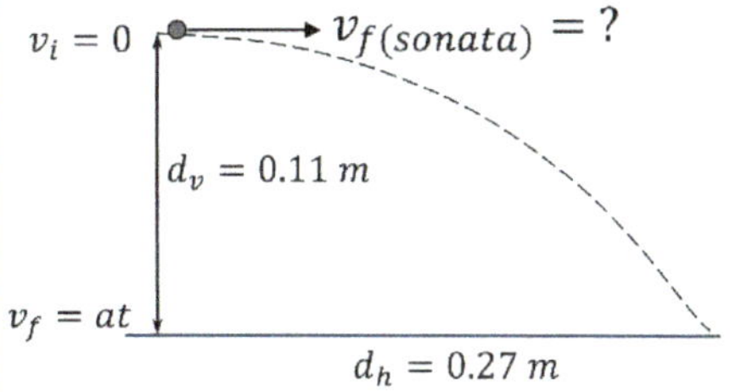

$$d_v = \overset{0}{\cancel{v_i}}t + \frac{1}{2}at^2$$

$$t = \sqrt{\frac{2d_v}{a}} = \sqrt{\frac{2(0.11m)}{9.81\ m/s^2}} = 0.1498\ s$$

$$d_h = v_f \cdot t = v_{f(Sonata)} \cdot t$$

$$v_{f(Sonata)} = \frac{d_h}{t} = \frac{0.27\ m}{0.1498\ s} = 1.802\ \frac{m}{s}\left(\frac{2.23694\ mph}{1\ m/s}\right) = 4.03\ mph$$

Total ED: Sum the total energy dissipated from the four sources.

The total energy dissipated through the collision sequence is

$$KE_{total} = KE_1 + KE_2 + KE_3 + KE_4 = 210,837 + 336,544 + 29,898 + 2,514 = 579,793\ J$$

The corresponding Sonata impact velocity ($v_{i(Sonata)}$) can now be estimated from this total dissipated energy and a unit conversion $\left(1J = 1\ N \cdot m = 1kg \cdot m^2/s^2\right)$:

$$KE_{total} = \frac{1}{2}m_{Sonata}\left(v_{i(Sonata)}\right)^2$$

Rearranging and solving for $v_{i(Sonata)}$ yields

$$v_{i(Sonata)} = \sqrt{\frac{2KE_{total}}{m_{Sonata}}} = \sqrt{\frac{2\left(579,793kg \cdot m^2/s^2\right)}{1548.6\ kg}} = 27.36\frac{m}{s}\ (\text{or } 61.2\ mph)$$

This impact speed is conservative and based on the Sonata curb weight and $CF = 2.5$. Had we used $m_{Sonata} = 1710\ kg$, the estimated Sonata speed at impact would be 58.3 *mph*. Further, *CF* is most likely greater, as the side crush depth of the subject Mustang is more than 2.5 times that of the exemplar Mustang (see Fig. 6.5). Our analysis is conservative throughout, and a good way to report the findings is to state the Sonata's collision speed was greater than 58 *mph* (the lowest speed determined with conservative parameters). Regardless, the Sonata's impact speed well exceeded the 40-mph posted speed limit.

When using analytical or numerical methods for EDA, it is best to be cautious and conservative. Engineers performing accident reconstruction can often make the point with underestimated results, and causation proved through conservative means is more convincing to a jury.

Some key points about collision energy dissipation:

1. Crush-velocity relationship is linear for a speed range of 10–60 mph.[3],[4]
2. Crush energy dominates the energy balance: Total E ≅ Crush E + Slide E + Heat E*

   *(Crush E is nearly all Total E with head-on collisions, and up to 85% with T-bone collisions depending on postcollision trajectory (e.g., slide length, airborne departure, secondary impacts).[1],[4]

6.3 CDR and EDR: A Modern Standard for Accident Investigation and Reconstruction

Millions of passenger cars and light trucks on the road today around the world are equipped with an electronic *on-board diagnostic* (OBD) system and ***event data recorder*** (EDR), sometimes referred to as the automotive *black box.* More than 200 million registered vehicles are equipped in the United States and Canada alone, and approximately 98% of new vehicles sold in the United States and Canada today have an EDR.[5]

EDR data is used for vehicle crash research, accident investigation and reconstruction, product liability litigation, safety system performance analysis, and insurance claims. Users include police, forensic engineers, *original equipment manufacturer* (OEM) engineers, insurance professionals, and universities. EDR is the standard for modern accident investigation and reconstruction.

Professionals perform ***crash data retrieval*** (CDR) using specialty tools like the Bosch CDR Tool to retrieve EDR data from crashed vehicles. Data is usually retrieved through a vehicle's OBD connector. However, for severe crashes with damaged electrical systems or inaccessibility to the OBD connector, data can be retrieved via *direct-to-module connection* with the *airbag control module* (ACM), *electronic control unit* (ECU), or *powertrain control module* (PCM).

The forensic expert routinely involved with automotive collisions is an ***accident reconstructionist***. Northwestern University Center for Public

Safety (NUCPS) is a preeminent US institution that offers comprehensive training and an internationally recognized certification program for police and professionals seeking to become an accident reconstructionist. Accreditation Commission for Traffic Accident Reconstruction (ACTAR) is another renowned organization that offers accident reconstruction (AR) certification. Add a P.E. (Professional Engineering) license and advanced training and certification as a CDR specialist from a reputable third-party collision/crash organization (there are several in the United States, Canada, and Germany) and you have an excellent set of credentials for a career in modern AR.

How PCMs and EDRs Work

A PCM (or ECM), the *"brain"* of a vehicle's engine control systems, integrates the functions of the engine and transmission control, with other modules, recording performance information. This component regulates more than a 100 factors from acceleration-throttle interaction, fuel injection, ignition timing, and air-to-fuel ratios, to automated traction (braking), drivability, and on-board climate controls. The common experience of the lit *"Check Engine"* indicator on the car instrument panel (Fig. 6.6) is the PCM in action, which records *Diagnostic Trouble Codes* (DTCs) for the auto-repair mechanic to troubleshoot.

FIGURE 6.6
Check engine indicators are common to automobiles.

Sensors monitor the various systems and send data to the PCM for storage. Some PCMs can store up to 6 minutes (360 seconds) of data. It is common for PCMs to store up to 20 seconds of predeployment data and 5 seconds of postdeployment data before shutoff after a restraint deployment (e.g., airbag) event. In the event of a crash or other critical event, the EDR (or *black box*) is triggered to record a data imaging file in the ACM, ECU, or PCM. CDR retrieval software can plot critical parameters of this data (e.g., vehicle speed, acceleration, engine RPM, and braking) against relative time in seconds as a graphic representation of the event. The accident reconstructionist then interprets the related data file and graph. Often, it is that last 25 seconds of crucial information that aids understanding of a crash event. The following example provides insight into the world of AR and CDR.

Black Box Time Capsule

It was a nice spring afternoon on the Florida coast when a man left happy hour and drove away in his red truck. He drove west across the island before entering the causeway for the mainland.

Two couples were paired on Harleys, easy riding behind a blue SUV in the right lane as they approached the causeway. Seeing the fast-approaching truck in their mirrors, they moved to the left lane to pass around the SUV and avoid it. They were side by side in the lane next to the SUV when the truck rear-ended it with such force that the bumpers locked as the impact crushed and pushed the SUV downward. Now coupled and skidding out of control, one Harley went down with a controlled slide into the grassy median; the other was not so lucky.

Swerving left across the lanes, the coupled SUV-truck took out the second motorcycle before careening into the median and separating. At separation, the truck flipped end-over-end and landed upside down in the near eastbound lane. The SUV spun once maybe twice, before coming to rest in the far eastbound lane, not far beyond the truck. How the young female motorcycle passenger came to rest curled up under the front end of the SUV without it hitting her was nothing short of a miracle. Her man lay seriously injured next to his hog back on the other side of the road in a westbound lane.

Three years passed before a state attorney's office (SAO) hired a forensic engineer. Review of the SAO file, including eyewitness accounts and excellent police photo-documentation and measurements of the accident scene, provided the evidence for the recreation (Fig. 6.7). However, a twist in the case challenged the prosecution: the driver of the red truck claimed the truck was a *runaway*. Having been struck by lightning and subsequently repaired just days before, truck acceleration and throttling reportedly became sudden and uncontrollable during the moments leading up to the collision. An interview with the auto repair mechanic confirmed that a lightning strike

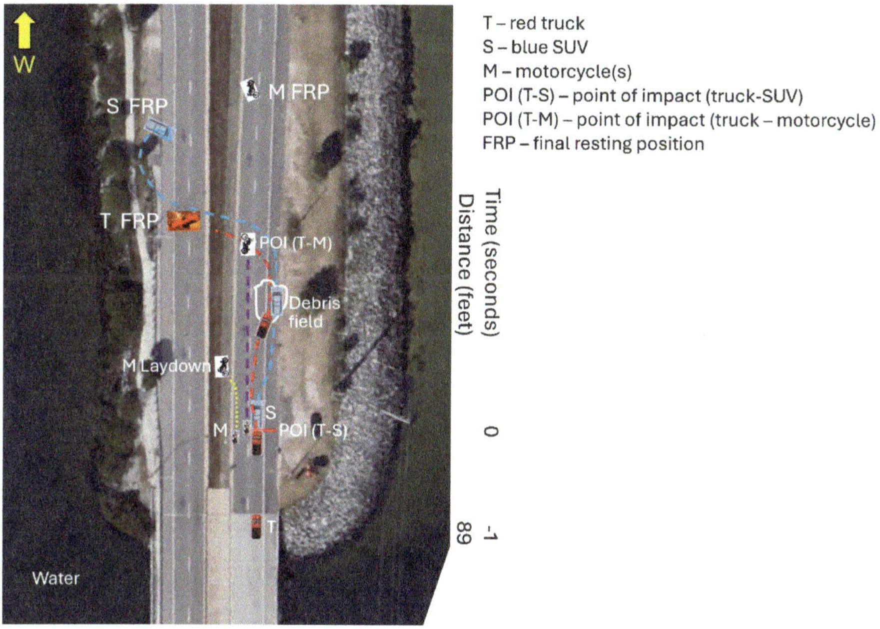

FIGURE 6.7
Recreation of causeway collision scene. (Base image courtesy of Google Maps.)

destroyed dashboard wiring harnesses and a couple of diagnostic modules, but that the truck's PCM and EDR remained intact. Crash data recovery was possible.

Crash Data Retrieval

The red truck was accessed in an impound yard for CDR and OBD[3] diagnostics. CDR indicated an intact PCM circular buffer with approximately 200 seconds of relevant trip data, the last 20 seconds of which were most important to evaluating truck performance on approach to the crash. Results from this analysis are depicted in Figures 6.8 and 6.9. Twelve columns of recorded performance data from T-minus 6 seconds to 3.2 seconds after collision are shown in Figure 6.8. Additional precrash data was considered for the truck approach sequence depicted in Figure 6.9.

First, note that from T-minus 15 seconds until SUV impact, the truck was speeding from the last island intersection until it hit the SUV (Fig. 6.9, Frame 1). The speed limit at the island intersection was 45 mph, transitioning to 60 mph across the relief bridge. The truck accelerated to 87 mph before decelerating at the relief bridge as it approached the SUV. CDR data indicated accelerator pedal release and engine throttle reduction but no braking at T-minus 6 seconds (Fig. 6.8, uppermost yellow highlighted data), followed by

Abrupt accelerator pedal deceleration and engine throttle reduction; no significant slowing

Brakes applied

Brief accelerator pedal increase while braking

Brakes off, shift into neutral

Impact

Restraint deployed

Vehicles coupled and slowing

Torque reduction

Brief torque Increase then neutral coast

Buffer Address (Hex)	Relative Time (calc.) (Seconds)	Restraint Deployment Signal (Received/ Not Received)	Speed, Vehicle Indicated (MPH [kmh])	Accelerator Pedal, % Full (%)	Engine Throttle, % Full (%)	Brake Switch (On/Off)	Brake SC, De-activate (On/Off)	ABS (Active/ Not Active)	Trans-mission Neutral (Neutral/ Not Neutral)	Trans-mission Reverse (Reverse/ Not Reverse)	Speed Control (On/Off)	Engine RPM (RPM)	Driveline Torque Commanded (N-m)	Driveline Torque Active (N-m)	Traction Control (Active/ Inactive)	Stability Control (Active/ Inactive)	Key On Timer, 63.75 max (sec) (Seconds)
EA003D30	-6.0	Not Received	80 [129]	44.5	33	OFF	OFF	Not Active	Not Neutral	Not Reverse	OFF	3394	326	303	Not Active	Not Active	63.75
EA003D40	-5.8	Not Received	79 [127]	6.5	17.5	OFF	OFF	Not Active	Not Neutral	Not Reverse	OFF	3210	128	109	Not Active	Not Active	63.75
EA003D50	-5.6	Not Received	80 [129]	5.5	17.5	OFF	OFF	Not Active	Not Neutral	Not Reverse	OFF	3217	84	66	Not Active	Not Active	63.75
EA003D60	-5.4	Not Received	79 [127]	21.5	18.5	OFF	OFF	Not Active	Not Neutral	Not Reverse	OFF	3196	73	55	Not Active	Not Active	63.75
EA003D70	-5.2	Not Received	79 [127]	14.5	17.8	OFF	OFF	Not Active	Not Neutral	Not Reverse	OFF	3185	98	80	Not Active	Not Active	63.75
EA003D80	-5.0	Not Received	79 [127]	4.5	15	OFF	OFF	Not Active	Not Neutral	Not Reverse	OFF	2904	82	68	Not Active	Not Active	63.75
EA003D90	-4.8	Not Received	79 [127]	9.5	13	OFF	OFF	Not Active	Not Neutral	Not Reverse	OFF	2471	95	61	Not Active	Not Active	63.75
EA003DA0	-4.6	Not Received	78 [126]	2.5	11.5	OFF	OFF	Not Active	Not Neutral	Not Reverse	OFF	2178	83	45	Not Active	Not Active	63.75
EA003DB0	-4.4	Not Received	78 [126]	5	10.5	OFF	OFF	Not Active	Not Neutral	Not Reverse	OFF	2180	90	22	Not Active	Not Active	63.75
EA003DC0	-4.2	Not Received	78 [126]	7.5	10	OFF	OFF	Not Active	Not Neutral	Not Reverse	OFF	2142	29	7	Not Active	Not Active	63.75
EA003DD0	-4.0	Not Received	77 [124]	5	10	OFF	OFF	Not Active	Not Neutral	Not Reverse	OFF	2151	31	9	Not Active	Not Active	63.75
EA003DE0	-3.8	Not Received	77 [124]	2.5	9	OFF	OFF	Not Active	Not Neutral	Not Reverse	OFF	2127	24	4	Not Active	Not Active	63.75
EA003DF0	-3.6	Not Received	77 [124]	14	12	OFF	OFF	Not Active	Not Neutral	Not Reverse	OFF	2109	10	-4	Not Active	Not Active	63.75
EA003E00	-3.4	Not Received	77 [124]	19.5	17	OFF	OFF	Not Active	Not Neutral	Not Reverse	OFF	2205	72	37	Not Active	Not Active	63.75
EA003E10	-3.2	Not Received	76 [122]	15.5	15	OFF	OFF	Not Active	Not Neutral	Not Reverse	OFF	2124	132	77	Not Active	Not Active	63.75
EA003E20	-3.0	Not Received	76 [122]	16.5	14.5	OFF	OFF	Not Active	Not Neutral	Not Reverse	OFF	2107	109	82	Not Active	Not Active	63.75
EA003E30	-2.8	Not Received	76 [122]	21.5	17	OFF	OFF	Not Active	Not Neutral	Not Reverse	OFF	2101	120	68	Not Active	Not Active	63.75
EA003E40	-2.6	Not Received	76 [122]	32	20.5	OFF	OFF	Not Active	Not Neutral	Not Reverse	OFF	2070	181	110	Not Active	Not Active	63.75
EA003E50	-2.4	Not Received	76 [122]	26.5	19.5	OFF	OFF	Not Active	Not Neutral	Not Reverse	OFF	2093	200	124	Not Active	Not Active	63.75
EA003E60	-2.2	Not Received	76 [122]	30	19.5	OFF	OFF	Not Active	Not Neutral	Not Reverse	OFF	2078	190	122	Not Active	Not Active	63.75
EA003E70	-2.0	Not Received	75 [121]	38.5	22	OFF	OFF	Not Active	Not Neutral	Not Reverse	OFF	2085	202	124	Not Active	Not Active	63.75
EA003E80	-1.8	Not Received	75 [121]	25.5	10.5	OFF	OFF	Not Active	Not Neutral	Not Reverse	OFF	2084	248	154	Not Active	Not Active	63.75
EA003E90	-1.6	Not Received	75 [121]	0	10.5	OFF	OFF	Not Active	Not Neutral	Not Reverse	OFF	2076	127	74	Not Active	Not Active	63.75
EA003EA0	-1.4	Not Received	74 [119]	0	9.5	ON	OFF	Active	Not Neutral	Not Reverse	OFF	2081	60	16	Not Active	Not Active	63.75
EA003EB0	-1.2	Not Received	73 [117]	0	8.5	ON	ON	Active	Not Neutral	Not Reverse	OFF	2012	20	2	Not Active	Not Active	63.75
EA003EC0	-1.0	Not Received	69 [111]	0	8	ON	ON	Active	Not Neutral	Not Reverse	OFF	1870	13	-2	Not Active	Not Active	63.75
EA003ED0	-0.8	Not Received	66 [106]	0	7	ON	ON	Active	Not Neutral	Not Reverse	OFF	1777	4	-7	Not Active	Not Active	63.75
EA003EE0	-0.6	Not Received	64 [103]	15	7	ON	ON	Active	Not Neutral	Not Reverse	OFF	1631	5	-7	Not Active	Not Active	63.75
EA003EF0	-0.4	Not Received	42 [68]	0	7.5	OFF	ON	Not Active	Neutral	Not Reverse	OFF	1000	112	94	Not Active	Not Active	63.75
EA003F00	-0.2	Not Received	39 [63]	0	8	OFF	ON	Not Active	Neutral	Not Reverse	OFF	1060	58	41	Not Active	Not Active	63.75
EA003F10	0.0	Not Received	41 [66]	0	7.5	ON	ON	Active	Neutral	Not Reverse	OFF	1196	21	5	Not Active	Not Active	63.75
EA003F20	0.2	Received	43 [69]	0	6.5	ON	ON	Active	Neutral	Not Reverse	OFF	1194	13	-4	Not Active	Not Active	63.75
EA003F30	0.4	Received	43 [69]	0	7	ON	ON	Active	Neutral	Not Reverse	OFF	1032	17	0	Not Active	Not Active	63.75
EA003F40	0.6	Received	42 [68]	0	7	ON	ON	Active	Neutral	Not Reverse	OFF	948	15	-2	Not Active	Not Active	63.75
EA003F50	0.8	Received	41 [66]	0	6.5	ON	ON	Active	Neutral	Not Reverse	OFF	880	20	1	Not Active	Not Active	63.75
EA003F60	1.0	Received	41 [66]	0	6	ON	ON	Active	Neutral	Not Reverse	OFF	818	18	0	Not Active	Not Active	63.75
EA003F70	1.2	Received	38 [61]	0	6	ON	ON	Active	Neutral	Not Reverse	OFF	644	20	10	Not Active	Not Active	63.75
EA003F80	1.4	Received	37 [60]	0	6	ON	ON	Active	Neutral	Not Reverse	OFF	500	23	5	Not Active	Not Active	63.75
EA003F90	1.6	Received	34 [55]	0	6	ON	ON	Active	Neutral	Not Reverse	OFF	425	20	1	Not Active	Not Active	63.75
EA003FA0	1.8	Received	31 [50]	0	6.5	ON	ON	Active	Neutral	Not Reverse	OFF	322	33	14	Not Active	Not Active	63.75
EA003FB0	2.0	Received	19 [31]	0	0	ON	ON	Active	Neutral	Not Reverse	OFF	340	41	23	Not Active	Not Active	63.75
EA003FC0	2.2	Received	17 [27]	0	4	ON	ON	Active	Neutral	Not Reverse	OFF	342	58	37	Not Active	Not Active	63.75
EA003FD0	2.4	Received	15 [24]	0	5	ON	ON	Active	Neutral	Not Reverse	OFF	329	60	89	Not Active	Not Active	63.75
EA003FE0	2.6	Received	13 [21]	0	4.5	ON	ON	Active	Neutral	Not Reverse	OFF	380	70	96	Not Active	Not Active	63.75
EA003FF0	2.8	Received	9 [14]	0	4.5	ON	ON	Active	Neutral	Not Reverse	OFF	364	82	100	Not Active	Not Active	63.75
EA004000	3.0	Received	8 [12]	0	6	ON	ON	Active	Neutral	Not Reverse	OFF	387	72	96	Not Active	Not Active	63.75
EA004010	3.2	Received	7 [11]	0	5	ON	ON	Active	Neutral	Not Reverse	OFF	306	85	105	Not Active	Not Active	63.75

FIGURE 6.8
EDR data from T-minus 6 seconds through 3.2 seconds postcollision (end of collision event).

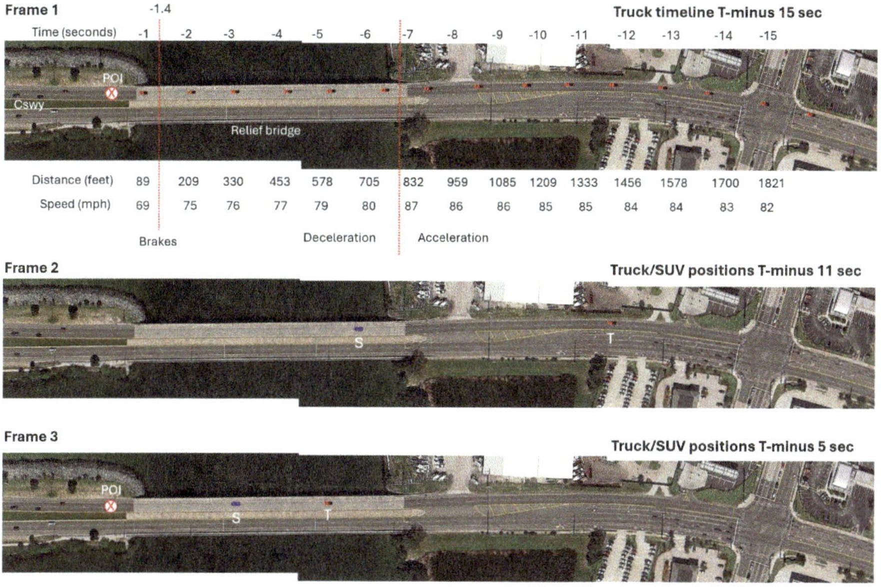

FIGURE 6.9
Precrash timeline from T-minus 15 seconds. Frame 1 depicts the truck's approach to the causeway as recreated from the EDR data. Note the high rate of speed and acceleration of the truck as it approached the relief bridge. Frames 2 and 3 show the relative positions of the truck (T) and SUV (S) at selected instances before collision at the POI. Note truck deceleration (Frame 1) on the bridge as it approached the SUV (Frame 3).

a couple of seconds of coasting before light braking at T-minus 1.4 seconds (red highlighted data) before impact. Most interesting is the instance of brief accelerator pedal increase while braking at T-minus 0.6 seconds (again, yellow highlight), an occurrence suggesting a foot across both pedals or both feet on the pedals. Further, at T-minus 0.4 seconds the truck shifted into neutral without braking (green highlight) before impact with the SUV four-tenths of a second later.

> Had the truck throttle stuck fully open causing the truck to race at high speeds for 15 seconds or more, wouldn't an attentive driver have attempted to stop the vehicle?

Scrutiny of the PCM data indicated a pattern of high throttle/rapid acceleration from intersection stops and rapid acceleration as the truck cleared traffic as it traveled across the island. Table 6.1 summarizes the instances with explanations. The challenge was to definitively explain acceleration anomalies and rule out *sudden involuntary acceleration* (i.e., a runaway truck).

The "89 Blip"

The *"89 Blip"* refers to instances in the EDR data where Engine Throttle % Full (ET%F) exceeded Accelerator Pedal % Full (AP%F), and ET%F had a value of 89% full over several seconds. Instances occurred at times T-minus 69, 28, and 12 seconds in Table 6.1. First, the typical operating range for ET%F for this truck is 5% to 90%. ET%F can reach 99.5% to 100% when the accelerator pedal is fully depressed (i.e., *floored*), as occurred at T-minus 153s for six seconds earlier in the drive.

There were several instances in the PCM data where ET%F exceeded AP%F when the truck driver heavily depressed the accelerator pedal. The corresponding RPM (revolutions per minute) and driveline torque increases indicated that the PCM responded to the sudden throttle changes before throttling back down into the nominal upper 20s N-m. This was the normal response to a sudden drastic AP%F change. For example, at T-minus 69 seconds (Table 6.1), following an automatic shift from second to third gear, the driver was heavy on the accelerator pedal (62.5% full) when ET%F was 89% full and engine RPM reached a high of 4044. As the engine RPM dropped to 2457 with the shift, the driver continued to depress the accelerator pedal. This is abnormal driving behavior as recognized by the PCM and *Keep Alive Memory* (KAM), a section in the PCM that retains information on vehicle performance that the computer has learned from driving behavior. The combination of low RPM (gear shift) and high (62.5%) AP%F caused the PCM to adjust ET%F – first upward from 31.5% to 89% with the gear shift and heavily depressed accelerator pedal – then held constant at 89% for 3 seconds, before

TABLE 6.1

Instances of Truck Throttle-up with Acceleration Discerned from the PCM Data

Time (T-minus) and Duration (s)	Accelerator Pedal (% Full)	Engine Throttle (% full)	Max RPM	Explanation
T-minus 153 s for 6 s	100	99.5	4467	Full throttle upon green light at intersection, no traffic
T-minus 79 s for 9 s	25.5–31.5	24.5–31.5	2587	Driving in traffic
T-minus 69 s for 3 s	62.5	89*	4044	Rapid acceleration after clearing traffic
T-minus 65 s for 39 s	45.5–48.5	26.5–34.5	2457	Driving in traffic
T-minus 28 s for 13 s	47.5–80	89*	4247	High throttle, rapid acceleration, no traffic
T-minus 15 s to 12 s	64.5–66	76.5–70	3609	High throttle through intersection, no traffic, speeds low 80s
T-minus 12 s to 9 s	65	89*	3989	Rapid acceleration, no traffic, speeds upper 70s

* See *"The 89 Blip"* discussion for a detailed explanation.

the driver began to let off the accelerator pedal and ET%F returned to a more normal correlation at T-minus 65.4. KAM learned to adjust ET%F to a new high (89%) when AP%F exceeded about 62.5%.

The next instance where ET%F > AP%F occurred at T-minus 28 seconds, when the driver again heavily depressed the accelerator pedal from 45% to 80%, when the truck was then speeding (> 65 mph) toward the last intersection before leaving the island. Again, ET%F adjusted to 89% with sudden acceleration, having maintained this level for nearly 15 seconds as the accelerator pedal was depressed. A last similar instance of rapid acceleration occurred 12 seconds before impact with the SUV. Lasting for 3 seconds, when KAM adjusted ET%F to 89% when AP%F was 65%.

> Had the accelerator pedal stuck fully depressed (i.e. floored), the prudent driver would attempt to play with the pedal by repeatedly tapping it to try and unstick it, then shift into neutral and turn the truck off, then brake – and apply the emergency brake if necessary – to slow the vehicle and pull over to the side of the road. The EDR data did not indicate attempt to do any of these maneuvers.

A final observation in Table 6.1 data is the *"Oh No"* moment. At T-minus 0.6 s to 0.4 s from impact, the driver did the *Oh No* brake and shifted to neutral before hitting the SUV.

OBD was performed to check the accelerator-throttle function. The truck was equipped with a drive-by-wire system, and accelerator position to the throttle body control assembly was communicated through the PCM. Diagnostic checks on throttle position (TP) and accelerator pedal position (APP) and APP1 and APP2 sensor correlations, voltages, and APP agreement with TAC (Commanded Throttle Actuator) position were performed from gas pedal inputs. Close-to-full-throttle events were monitored, graphed, and checked against the manufacturer's mechanical and electronic specifications. All measured parameters were within specification limits and accelerator-throttle operation was found to be normal.

Conclude or Not

At this point the reader might think that the evidence overwhelmingly supports a conclusion of reckless driving by the truck driver. However, there are many performance parameters not recorded by the EDR and qualitative analysis requires a holistic approach to data evaluation for definitive proof. As we learned, comparison of select parameters such as AP%F and ET%F is not necessarily indicative of anomalous operating parameters as KAM adjusted for new driving behavior. What is proof-positive is real-time monitoring of PCM performance during driving scenarios with an exemplar truck, starting with the one inferred by the explanations posed in Table 6.1. Such analysis would definitively establish the correlations and performance.

Notes

1. The converse of an inverse problem is a forward problem, one that predicts outcomes from known parameters.
2. Curb weight is a vehicle's weight with a full tank of fuel and all standard equipment.
3. The PCM stores specific information that gets locked into a vehicle's computer memory in the event of an RDS Event. OBD is a separate memory system that catalogs errors from vehicle sensors and other systems. It also keeps a time history of other system bad data as error codes.

References

1. Miroslaw Gidlewski, Leon Prochowski, Leszek Jemioł and Dariusz Żardecki (2019). The process of front-to-side collision in motor vehicles in terms of energy balance. *Nonlinear. Dyn.* 97:1877–1893.
2. Crash Test Database, 2026. https://www.nhtsa.gov/research-data/databases-and-software, Test No's. 7465, 5453.
3. Campbell, K., "Energy Basis for Collision Severity," SAE Technical Paper 740565, 1974, https://doi.org/10.4271/740565.
4. Andrzej Żuchowski (2015). The use of energy methods at the calculation of vehicle impact velocity. *Arch. Automot. Eng.* 68(2):85–111.

5 Canis, Bill & Peterman, David Randall. "Black Boxes" in Passenger Vehicles: Policy Issues, report, July 22, 2014; Washington D.C. (https://digital.library.unt.edu/ark:/67531/metadc462832/: accessed March 22, 2025), University of North Texas Libraries, UNT Digital Library, https://digital.library.unt.edu; crediting UNT Libraries Government Documents Department.

Case Study 6: A Notorious Drive

- Truck v. pedestrian
- Accident reconstruction
- Backing technology

Message to students:

> Automotive technology can enhance, distract and detract. Old-school mirrors or backing camera display; you decide.

A *commuter trail* is a path or road used primarily by people to travel from one place to another, often between home and work. They are sidewalks, bike paths, pedestrian walkways, or dedicated trails designed for walking, biking, or other non-motorized modes of travel. Commuter trails connect neighborhoods, businesses, and other destinations in a way that avoids traffic congestion and encourages active exercise.

DOI: 10.1201/9781003623311-14

FIGURE CS6.1
(a) The notorious drive between Lots 1A and 1B, depicting driveways, public access and utility easements, and the truck backing path to the point of impact (POI) with the pedestrian. Inset (b) is the aerial survey that defined the Plan B reconstruction area.

Fifty years ago, hillside homesteaders used a commuter trail through the woods to reach local businesses and a community center in the valley below. Progress brought subdivisions and kids' soccer, and a portion of the trail became a common property line, utility easement, public access easement, and driveway for four property owners. The intention was to preserve the commuter tradition. This portion of the old trail became *the drive,* a link between two sides of subdivided land that formed a winding loop through the new subdivision. Dog walks, runs, bicycle rides, and occasional vehicles, all appreciated the winding way. However, county planners called this a *"notorious drive"* (Fig. CS6.1) because of the potential conflicts that could arise.

The Incident

One sunny summer day in 2022, a property owner backed a 2019 F350 truck over a pedestrian who was talking on a cell phone while walking in *the drive.* Ran over twice, concussed by the truck's rear differential, and two years of rehabilitation later, the walker survived. Then came the legal battle; people chose sides, and "No Trespassing" signs were posted on the drive amidst the preliminary investigation. The experts would have to go to Plan B: accident reconstruction.

Plan B

Plan B required locating a site with terrain similar to the incident scene, acquiring an exemplar F350, and using a model of similar stature to the victim. The site, a gravel pit free of visual obstructions, was laid out with cones (Fig. CS6.2) using the aerial information from Figure CS6.1(b). An exemplar truck (same year and model) with the same backing equipment – side and interior mirrors and backing camera with interior visual display – was acquired. The 5′2″ model was clothed in a dark top to mimic the victim's appearance at the time of the incident. Reenactment used a driver of similar stature to that of the defendant driver, interior photography to image mirrors and backing camera displays, and ground-level and drone videography to document backing trials. Positions 1 through 4 (Fig. CS6.2, upper left) represent truck backing positions for the representative images in Figures CS6.3 and CS6.4.

F350 Backing

Mirror images are contrasted with those of the video display screen (Figs. CS6.3 and CS6.4) at the four backing positions along the truck's 150-foot backing track. Noteworthy is the backing camera video display field of view (FOV) with enhanced park aid (EPA) and alert zone (AZ). The 80-degree FOV

FOV is backing camera field of view; AZ is backup zones; RA is reconstruction area; positions 1 – 4 for truck backing photographs.

Backing position 4

FIGURE CS6.2
Accident reconstruction site depicting exemplar truck backing track (L, aerial view) and truck-to-pedestrian model photographs on the right.

FIGURE CS6.3
Camera display and mirror images from truck backing positions 1 and 2.

first detects the model at position 2 when the truck is approximately 90 feet away from the model in the backing turn. The black centerline of the backing track is visible within the AZ in the display images, and the EPA is the truck icon at the upper right corner of the display screen (Fig. CS6.5).

The model became visible in the rear-view and right-side mirrors at position 3, when the truck was straightening out from the backing turn about 50 feet away from the model. Partially obscured by the passenger headrest

FIGURE CS6.4
Camera display and mirror images from truck backing positions 3 and 4.

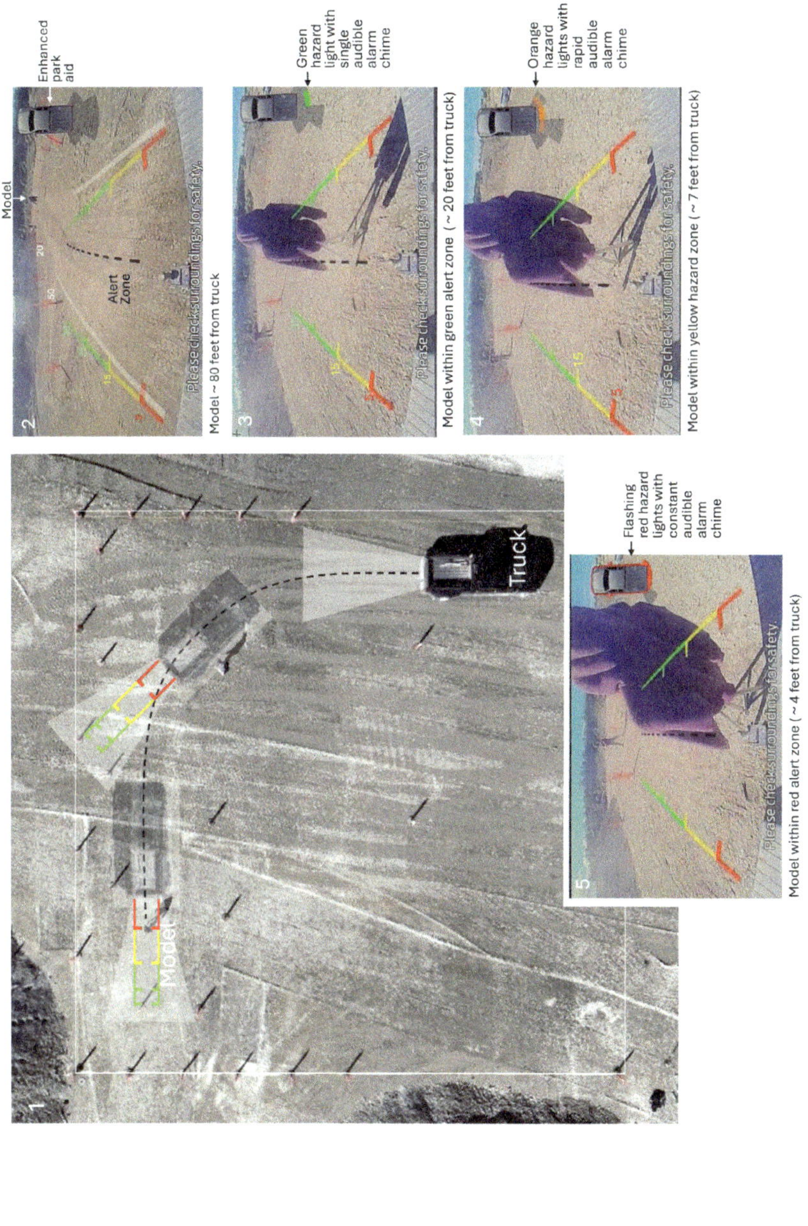

FIGURE CS6.5
Images of the model within backing camera display as the truck approaches from 80 feet away.

in the rear-view mirror, the side mirror and camera display provided the clearest views of the model. At 30 feet away (position 4), only the head of the model remained visible in the rear-view mirror while the entire model was clearly visible in the center of the display screen.

Shifting focus to the camera display as the truck approached from 80 feet away (Fig. CS6.5), attention is now given to the EPA and AZ features. From this distance, these important backing features are intended to alert the driver of an obstacle within the drive path. Looking at images 2 through 5, the EPA engaged with a green hazard light and single audible chime when the model entered the AZ at 25 feet from the back of the truck. Images 4 and 5 show the hazard light sequence and increase in alert chime frequency as the truck closed on the model. The EPA and chime alerts are designed to warn the driver visually and audibly about an approaching conflict within the AZ.

Automotive Backing Technology and Manual Checks

Most drivers spend less than 1% of their driving time in reverse, yet national statistics indicate that about one-quarter of all collisions occur while backing[1]. The Texas Department of Transportation reported that for 2019, 10,335 of its drivers cited backing as a factor in a vehicle crash, and of those, there were 36 serious injuries and eight deaths. Further, *back over* incidents kill an estimated 200 people annually nationwide and injure more than 12,000[2].

A consensus among state transportation safety boards is that automotive backing technology (ABT) has improved safety and convenience. Backing incident rates and related injuries are expected to decrease as a growing number of vehicles incorporate rear-vision camera systems, automatic rear braking, and other safety technology features. However, significant problems with ABT persist (Table CS6.1) and the problem of *Overreliance and User*

TABLE CS6.1

Summary of Automotive User Interface (UI) Problems with Backing Technology

Problem Type	Description
Overreliance	Drivers may trust cameras too much, ignoring mirrors and blind spots.
UI Complexity	Steep learning curve, lack of tactile feedback (vibrations or audible clicks), and visual overload.
System Integration Conflicts	Delays or failures when switching between internal and external infotainment systems.
Distraction	Screen multifunction can divert attention from safe backing.
Technical Failures	Camera/image issues (e.g., blurred images, blank screens, delayed camera activation) resulting from hardware, wiring, or software problems.

Interface Issues has the attention of psychology. *Technology and Psychological Distraction* is a problem with real consequences of multitasking and *dual-task processing*[1] while driving[3]. New tech bells and whistles lead to new habits, and sometimes unintended ramifications.

A century of driving has included manual checks with backing; mirror checks and head checks (head rotations left and right) of the blind spots. Commercial truck drivers must perform walk-arounds of their rigs and inspections of their surroundings, looking for pedestrians and other potential hazards before driving. Should automation preclude or replace manual checks? How could an F350 truck driver and a pedestrian, both familiar with the drive and its commuters, fail to perceive each other on a clear sunny day?

Overreliance on Automotive Backing Technology

Overreliance on ABT (or backup cameras) can lead to a false sense of security as their sole source of information when backing. Backing camera field of view (FOV) can range from 80 to 170 degrees depending on the vehicle, and the camera cannot capture everything behind or around the vehicle. The greater the FOV the greater the potential for exclusive camera use, neglect of mirrors and physical checks of blind spots, and increased risk of collisions with obstacles and pedestrians. Also, backing cameras linked to a multifunctional display screen can divert driver attention away from safety while reversing. Drivers may interact with the screen by checking between cameras while backing or looking at nonessential backing information such as climate controls, navigation, or entertainment features. Further, there are also several UI problems with backing technology (Table CS6.1).

Technology and Psychological Distraction

Technology can be a significant source of *psychological distraction* due to its constant availability, notification systems, and multitasking facilitation, which can lead to decreased focus and attention spans. Also, AI (artificial intelligence) is thinking for us when we write and perform virtual transactions, and unannounced automatic software updates interrupt and frustrate users. The developers are trialing AI at user expense. The automotive industry is competing to offer the best user navigation and infotainment experience. These distractions and interruptions impact work, education, social interactions, driving, and other aspects of daily life.

Further, a constant stream of notifications from social media and messaging apps interrupts focus, makes it difficult to concentrate on tasks, and otherwise annoys. Technological multitasking may *seem* efficient, but research suggests it hinders performance, cognition, and memory retention. Ease of access to technology for the sake of immediate gratification for some is a significant distraction from education, productivity, and driving.

Platted Driveway and Driver Multitasking

A *platted driveway* is a legal driveway. Officially recorded on a subdivision plat, this means it is part of the official land division and planning for a particular area, including easements or rights of way that may be associated with the driveway.

Land that is subdivided must afford access from all lots to the nearest serviced road via an access easement and/or driveways. The site of the truck-pedestrian collision was in the drive that joined three subdivided lots: 1A, 1B, and 1C (Fig. CS6.6). The county surveyed and platted a common driveway for lots 1A and 1B to Windy Way, a county-maintained cul-de-sac. This driveway was also deemed a public access easement (PAE) for commuters as the western portion of the drive that connected with Bush Lane at the east side of the subdivision. Four properties had common property lines and a common corner in the drive near its east end, and this portion of the drive was within a public utility easement (PUE). This platting and recording made the drive between Windy Way and Bush Lane accessible to commuters and property owners alike. It also meant for vehicles from lots 1A and 1B to depart/return via Windy Way, which was a maintained road. Bush Lane to the east was an unmaintained dirt trail of a power line before being improved to a single-lane gravel drive for access to new subdivided properties: lots 1207, 1242, 1244, and 1206 as shown in Fig. CS6.6. Bush Lane connected with another maintained road at its southern end.

A neighbor working outside approximately 300 feet away heard the F350 start in the driveway of Lot 1B when the pedestrian would have been closer, walking in the drive from the east. At the house, the truck was approximately 150 feet from the drive. Both driveway and drive had flat gravel surfaces and the area encompassing them was devoid of vegetation and obstructions. There was nothing to impede driver and pedestrian visibility that fateful day.

Deposition testimony by the truck driver indicated his wife and three young children were also in the truck, the children fastened in the back seat. There was conversation and music tuning going on as the truck backed out

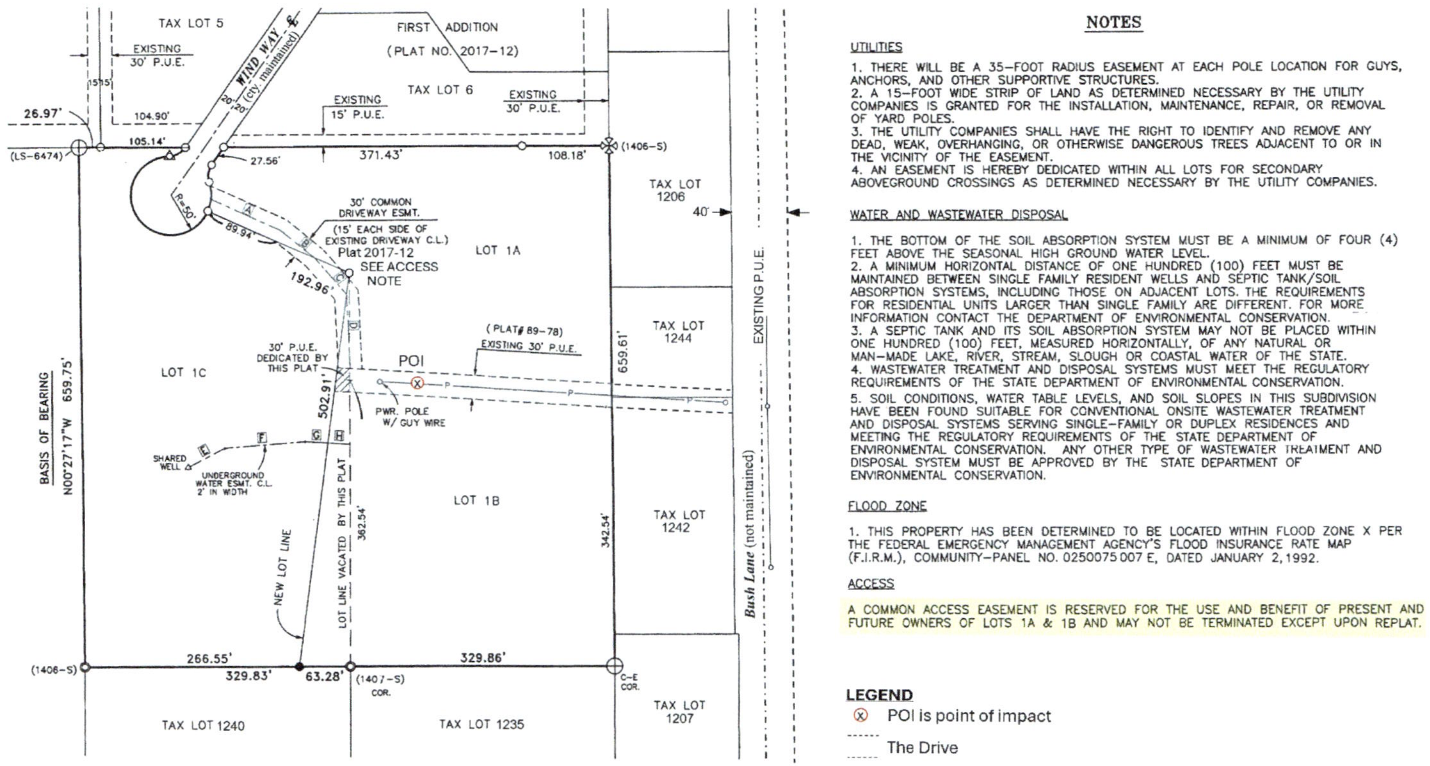

FIGURE CS6.6
The drive and platted common driveway (shaded light blue) from Wendy Way to Lots 1A and 1B and Bush Lane.

of the driveway. The pedestrian was talking on her cell phone as she walked in the drive. Who was responsible for this tragic accident?

> The tragedy of life is what dies inside while we live.
>
> **Albert Schweitzer**
> *French German theologian, writer, philosopher, and physician*

Note

1. Dual-task processing as it relates to driving involves a motor task and cognitive task, or two cognitive tasks that a driver must simultaneously process. Psychologists study dual-task paradigms and effects of divided attention and interference on cognitive and motor performance.

References

1. L.R. Robinson (2021). "Backing Up Happens." Online Defensive Driving Modules, National Safety Council. https://youtube/kj1b_v_17qM. Accessed 13 April 2025.
2. Statewide Crashes by Unit Contributing Factor, Texas Department of Transportation, 2019. https://ftp.txdot.gov/pub/txdot-info/trf/crash_statistics/2019/29.pdf. Retrieved 11 May 2025.
3. David L Strayer (2015). "Is the technology in your car driving you to distraction?" *Policy Insights Behav.* Brain Sci. 2(1):1–9.

7

Structure Failures

CONTENTS

CONCEPTS	KEY TERMS	ADDITIONAL READING
Structure Failures	Engineering disasters	Exercise 13
Bridge Redundancy	Load paths	Exercise 14
Threshold Inspections	Punching shear	
Milestone Inspections	Effective depth	Case Study 7
Best Practices Forensic Investigation	Failure chain	
Climate Impacts on Northern Infrastructure	Failure theories	
Failure Analysis in Design		

Engineering disasters are failures in engineering design, construction, or operation, which result in loss of life, injury, and significant damage. The focus of this chapter is structure failures. First, some historical background about how centuries of failures eventually led to formal building codes and professional engineering and architectural requirements. Next, failure causes are considered. Then some notable engineering disasters and subsequent forensic investigations that led to regulatory changes are presented. The chapter would not be complete without also considering climate impacts on northern infrastructure and investigation challenges. Finally, the case is made to include failure analysis in structural design curricula. Although the tragic examples presented have led to improvements in engineering codes, safety standards, and risk management protocols, the frequency of engineering disasters has not diminished in a hundred years of modernized design and construction.

DOI: 10.1201/9781003623311-15

7.1 Structure Failures

History records amphitheater failures in antiquity, and bridges, towers, and church failures throughout the Renaissance. Collapses at wooden, aged, and poorly maintained Roman amphitheaters killed thousands at Fidenae and Rome in AD 27 and AD 141, respectively. Many churches and retaining structures throughout Europe and a few dams experienced failures during the 15th to 18th centuries. Quite a few bridges failed in Europe in the 1800s, and the Broughton and Yarmouth suspension bridge collapses and the Dee Bridge disaster between 1831 and 1847 in the United Kingdom emphasized the need for modernized standards for engineered structures.

The United Kingdom established the first systematic national building standard with the Metropolitan Buildings Act 1844, the purpose of which was to improve housing and business standards and regulate activities deemed harmful to public health. Among its provisions:

- Builders were required to give surveyors two days notice before building.
- Regulations were established regarding the height of rooms, the thickness of walls, and the materials used in repairs.
- Divided and classified existing buildings.
- Design and placement of fireplaces, chimneys, and drains; and streets had to be built to minimum requirements.

In London, the Metropolitan Buildings Office was established to regulate construction and building use, and surveyors were empowered to enforce building regulations.[1]

The United States also experienced a series of failed bridges, dams, and buildings during the latter half of the 1800s. Notably, the partial collapse of the Pemberton Mills textile factory in Lawrence, Massachusetts, in 1860, killed 145 and seriously injured another 166 workers. Also, the Virginia State Capitol disaster of 1870, when an overcrowded courtroom on the third floor collapsed during a Supreme Court hearing, killing 62 and injuring another 251 people (Fig. 7.1). Pemberton Mills was only seven years old when the partial collapse occurred; substandard construction and second-floor equipment overload were the causes. This disaster led to administrative and regulatory improvements in industrial construction and workplace safety.

Baltimore established the first building code in the United States in 1891. Substantive changes were made after the Great Baltimore Fire of 1904, and the updated code was published as the *Handbook of the Baltimore City Building Laws* before being formalized as a permanent building code in 1908.[2] Then came the Great Molasses Flood of 1919 (Fig. 7.1 at right), which prompted the Boston Building Department to require engineering and architectural calculations to be signed and filed with subsequent construction projects.[4]

"Ruins of the Pemberton Mills, at Lawrence, Massachusetts, the Morning after the Fall," *Harper's Weekly* 4, no. 160 (January 21, 1860), 33. (Courtesy of the New England Historical Society.)

1870 Virginia State Capitol disaster: "The Richmond Calamity – Interior of Hall of Delegates - Getting out the dead and wounded. From a sketch by W. L. Sheppard," *Harper's Weekly*. AP2 .H32. Special Collections, University of Virginia, Charlottesville, Va.

Before (above) and aftermath (below) of the Great Molasses Flood in Boston, 1919, one of the strangest disasters in American history.[4]

FIGURE 7.1
Nineteenth- and early twentieth-century structure failures from the archives.

Following this lead, US cities and states began requiring registered professional engineers to sign off on plans for major infrastructure projects.[3]

Fast forward to today, and a hundred years of failures have helped improve our design codes. What cost 25 cents for 13 pages in 1923 is now the $250, 2432-page, 16th edition of the *American Institute of Steel Construction (AISC) Steel Construction Manual*. In theory, improved design codes should build better and reduce the frequency of failures, right? The first quarter of the 21st century has had its share of major engineering disasters. The I-35W highway and FIU pedestrian bridge collapses, the Champion Towers South condominium disaster, and the Oroville Dam spillway collapse to name a few (Fig. 7.2).

7.2 Causes of Structural Failure

Forensic investigations have revealed numerous causal factors for structural failures:

- design flaws
- design processes
- design complacency
- insufficient design review
- inadequate planning
- construction malpractice
- faulty installation
- miscommunication
- software
- overloading
- overstressing
- deferred maintenance
- environmental influences
- inclement weather
- substandard materials
- material stress and fatigue
- insufficient material
- ethical breaches (fraud, corruption)
- breach of contract
- negligence
- sabotage
- poor standards of care
- freeze/thaw
- frost heave

FIGURE 7.2
Modern structure failures and a failed roof cover. The National Transportation Safety Board (NTSB) is a nongovernmental organization that investigates aviation, highway, marine, railroad, and pipeline transportation incidents, including accidents involving the transportation of hazardous materials. NIST is the National Institute of Standards and Technology, a government agency that investigates building collapses.

This list is not exhaustive, and combinations of these factors often work together to fail structures. Further, design flaws (e.g., erroneous calculations, incorrect use of design factors, overreliance on software analysis) and deferred maintenance are common themes with modern failures, and it is difficult to quantify contributions from designer ego, litigiousness, and jurisdiction relative to managerial and administrative decision-making.

7.3 Bridges and Redundancy

Bridge redundancy is about having multiple ways for loads to distribute (or *load paths*) within a bridge. If one part of the bridge fails, the remaining parts can still carry the load, preventing a catastrophic collapse.

FIGURE 7.3
Load path redundancy in a pretensioned concrete girder bridge span with a steel-reinforced deck. (a) Pier 2 can compensate for the loss of pier 1, and (b) piers 1 and 3 together compensate for the loss of pier 2.

Redundancy is safety, ensuring that a bridge remains functional even in a damaged condition. For example, a cable-stayed bridge provides multiple load paths with its cables that connect the deck to the towers. A reinforced concrete bridge that includes structural indeterminacy, load path redundancy, and internal redundancy (pretensioned rods and steel reinforcement) has three layers of protection against failure. In Figure 7.3, we see how such a bridge remains functional when a support fails and load transfers to adjacent piers. Redundancy means safer, more reliable, and damage-tolerant bridges.

7.3.1 1967 Silver Bridge Collapse

The Point Pleasant Bridge (popularly known as the ***Silver Bridge***, Fig. 7.4 at left) was built in 1928 as the first steel ***eye-bar suspension bridge*** of its type in the United States. It crossed the Ohio River as US Route 35 from Point Pleasant, West Virginia to Kanauga, Ohio. Silver Bridge collapsed into the river on December 15, 1967, during rush-hour traffic.

Key facts about the bridge:

- **Design**: Low redundancy, high-strength eye-bar suspension
- **Designer**: J. E. Greiner Company, of Baltimore, MD, a civil engineering firm specializing in bridge design. Founded 1908.
- **Builder**: American Bridge Company of Coraopolis, PA, now 125 years old. Builder of bridges, marine structures, the Chrysler Building and the Empire State Building.
- **Opened**: May 30, 1928
- **Design Life**: 75 years
- **Functional Life**: 39 years

Silver Bridge drawing above photograph of collapsed bridge. Courtesy of the West Virginia Herald-Dispatch.

High redundancy eye-bar suspension bridge. Inset is detail of nine parallel sister eye-bars pinned. Images courtesy of Jet Lowe.

FIGURE 7.4
Silver Bridge look before (above left) and after (below left) collapse into the Ohio River. Example of an eye-bar suspension bridge with high redundancy is at right. The more eye-bars pinned in parallel, the higher is redundancy.

- **Cause:** A single eye-bar in one of the suspension chains failed. Specifically, a cleavage fracture in the lower half of the head of eye-bar 330 in joint C13N, together with a ductile fracture on the opposite side of the eye from the cleavage fracture, resulted in the complete separation of eye-bar 330 from the C13N joint. Sister [redundant] eye-bar 33 had slipped from the eye-bar pin at the same joint. Once the north eye-bar chain separated at joint C13N, total collapse was imminent since the towers were supported on rockers.
- **Impact:** Total collapse due to low structural and load path redundancy.
- **Casualties:** 46 people died, and 9 others injured.
- **Investigation:** Forensic, three years, National Bureau of Standards (NBS) assists the NTSB. See Special Publication 925 (NBS SP 925) and NTSB Report HAR 7101.
- **Aftermath**: **Congressional Action:** Congress passed the National Bridge Inspection Standards in 1968. Two similar eye-bar suspension bridges were subsequently inspected; one was demolished, the other had sufficient redundancy

(four eye-bars to a chain instead of two) and remained in service until 1993.

Anecdote: Aging infrastructure remains a problem in the United States and routine inspections do not always detect problems. In 1983, the Mianus River Bridge in Greenwich, Connecticut collapsed, causing three deaths. The 2007 collapse of the I-35W Mississippi River bridge in Minneapolis killed 13. Inspections documented problems, but deferred maintenance delayed repairs. In September 2009, emergency repair of a failed eye-bar in the San Francisco-Oakland Bay Bridge occurred during a scheduled closure.

7.4 Threshold and Milestone Inspections

Coastline condominiums are still the rage from the Texas Panhandle to Massachusetts' Cape Cod. Currently, preconstruction demand for units in Texas, Florida-Alabama, and the South Carolina barrier islands are high, and coastal mainland demand in the southeast remains steady.

Numerous design errors led to the 1981 collapse of the five-story Harbor Cay Condominium while under construction in Cocoa Beach, Florida. Workers were pouring concrete on the fifth floor when the structure collapsed. The rumble, ground shake, and choking gray cloud are forever etched in my mind. On the pile, looking and listening for signs; hair was the unmistakable sign of one buried in the rubble. Eleven workers died, and 23 others were injured on that fateful day. This disaster was totally preventable.

Key facts about the disaster:

- **Design**: Five-story flat plate concrete residential condominium project.
- **Developer**: Towne Realty, Milwaukee, Wisconsin
- **Builder**: Univel Inc., a family-owned Cocoa Beach construction company, under the supervision of Dynamic Construction Company.
- **Collapse**: March 27, 1981 (during construction).
- **Functional Life**: 0 years.

- **Investigation:** The OSHA (Occupational Safety and Health Administration) inspection team and US Department of Labor arrived on site within one hour of the collapse and began an investigation. An NBS team arrived a day later to initiate a formal forensic investigation. This investigation included a thorough design review and materials testing.
- **Causes**: Numerous design and construction errors. Primary design flaws cited by NBS[6]:
    1. Undersized floor slabs were 200 mm (8 in) thick instead of the American Concrete Institute's Building Code minimum requirement of 280 mm (11 in).
    2. Steel reinforcement chair spacers were 110 mm (4.25 in) high, too short, and coupled with too thin slabs provided insufficient *effective depth* for shearing.
    3. No design calculation checks for punching shear. This omission supports the case for no shear reinforcement consideration. Thin slabs without shear reinforcement increase the risk of punching shear failure at load-bearing columns.

        Department of Commerce report stated, "*The analysis showed that shear stresses in the slab at many column locations on the fifth floor exceeded the nominal shear strength. Thus, punching shear failure at one of the columns precipitated a progressive failure of the slab throughout the entire fifth floor.*"
    4. Inadequate reshoring.
    5. Poor quality concrete. Fifth-floor concrete strength was determined to be 85% of the required 4000 psi (27,600 kPa) compressive strength.
- **Impact:** Total collapse due to insufficient floor slab design.
- **Casualties:** 11 workers dead, 23 injured.
- **Legislative Action:** Florida Threshold Inspection Law of 1982.

Despite knowing about the problems associated with barrier island infrastructure projects (e.g., erosion, damming and dredging, hard levees), and NOAA predictions that climate change will worsen barrier island erosion, money talks, insurers run the risk, and failures will keep inspectors and forensic engineers busy.

Some disasters that made the news and led to legislation intended to lessen the potential for certain types of structural failures are highlighted here. Case study 7 at the end of this chapter presents an engineer's inspection of a third, and how band-aid repairs may be the prelude to disaster despite the new regulations.

1981 Harbour Cay Condominium Collapse

Punching Shear

Punching shear in flat-plate reinforced concrete slabs is a major concern that usually governs design. This phenomenon arises due to the relatively small thickness of the slab and its direct support on columns, resulting in high shear stresses transferred between the slab and columns.

Primary causes #2 and #3 are linked by *effective depth*,[1] the critical parameter for punching shear calculations. Recall the floor slabs were undersized by three inches (80 mm); eight inches thick instead of the required 11 inches (280 mm). The computation below determines effective depth (d_{eff}) for punching shear for both required and undersized slabs with the use of Figure 7.5 and #5 rebar for longitudinal and transverse reinforcement. For the calculation, h is slab thickness, c is concrete cover, and $d_{s,y}$ and $d_{s,x}$ are diameters of longitudinal and transverse rebar (i.e., 16 mm for both).

For the 11-in (280 mm) thick slab:

$$d_x = h - c - d_{s,y} - \frac{d_{s,x}}{2} = 280 - 20 - 16 - \frac{16}{2} = 236\ mm$$

$$d_y = h - c - \frac{d_{s,y}}{2} = 252\ mm$$

$$d_{eff} = \frac{d_x + x_y}{2} = \frac{236 + 252}{2} = 244\ mm\ (9.6\ in)$$

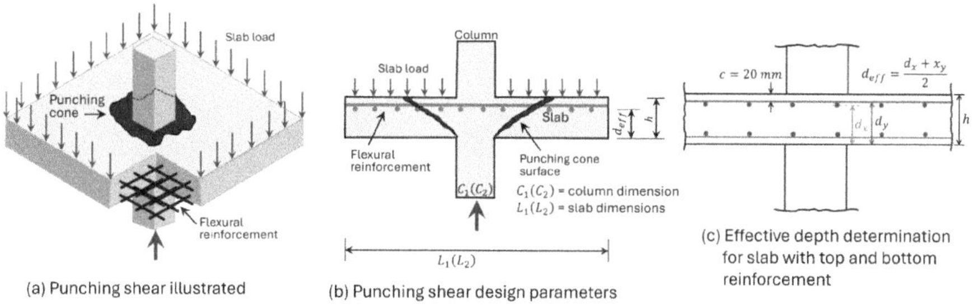

(a) Punching shear illustrated

(b) Punching shear design parameters

(c) Effective depth determination for slab with top and bottom reinforcement

FIGURE 7.5
Illustration of concrete floor slab punching shear at a load-bearing column.

For the 8-in (200 mm) thick slab:

$$d_x = 156\ mm,\ d_y = 172\ mm,\ \text{and}\ d_{eff} = 164\ mm\ (6.47\ in)$$

The undersized slab resulted in a 32.8% reduction in effective depth. Had the engineer of record then evaluated punching shear for the undersized slab with and without shear reinforcement, the slab failure should have been realized.

Threshold Inspections

This structural failure highlighted the need for additional oversight during the construction of certain types of buildings to ensure compliance with design and construction standards. Florida established the *Threshold Inspection Law* in 1982. This law requires critical inspections to be performed by certified special inspectors on buildings over three stories or 50 feet (15.2 m) tall, or those with assembly occupancy greater than 5000 sf (465 m^2) and occupancy greater than 500 people during construction to prevent structural failures and ensure public safety. The threshold inspector inspects structural elements like slab/column dimensions and embedded steel type and placement, spacing, and concrete cover during construction to ensure compliance with permitted drawings.

Subsequent structure failures such as the 2019 New Orleans Hard Rock Hotel[2] and 2021 Champlain Towers South Condominium collapses would lead to additional regulations.

2021 Champlain Towers South Condominium Collapse

Champlain Towers South (CTS), a 12-story beachfront condominium in the Surfside suburb of Miami, Florida, partially collapsed on June 24, 2021 (Fig. 7.6), killing 98 and injuring 11. An engineering report documented long-term concrete degradation, water penetration, and corrosion in the 40-year-old south tower in 2018. NIST sent scientists and engineers to the site on June 25, 2021 to initiate a full investigation into the collapse; a primary goal was to establish *best practices* to prevent similar disasters in the future. The CTS homeowner association hired an engineering firm to independently investigate the collapse.

Key facts about the disaster:

- **Designers**: William M. Friedman & Associates Architects, Inc. (architectural drawings) and Breiterman Jurado & Associates (project engineering).
- **Builder**: Nattel Construction in partnership with Champlain Towers South Associates, a group of 15 different development companies.

FIGURE 7.6
Partially collapsed Champlain Towers South. (Courtesy of Miami-Dade Fire Rescue Dept.)

- **Designer**: Breiterman Jurado & Associates
- **Built**: 1981–82
- **Collapse**: June 24, 2021
- **Design Life**: 75 years
- **Functional Life**: 40 years
- **Investigation:** NIST's comprehensive forensic investigation of many potential causes. Release of findings expected in 2025.

Causes: PRELIMINARY[7][8]

The CTS collapse resulted from a combination of factors, including structural deficiencies, inadequate maintenance, and potential effects of nearby construction. A breakdown of the contributing causes follows.

1. Structural Deficiencies and Design Failures:
    - **Under-designed pool deck slab.** The deck slab was not strong enough to withstand the loads it was subjected to.

- **Cracking and spalling.** Significant concrete deterioration was observed in the pool deck and parking garage.
- **Corrosion of reinforcing steel.** NIST materials testing revealed extensive and various types of corrosion within the building, which weakened the structure.
- **Compromised columns.** Column supports exhibited distress; interior concrete and reinforcing steel corrosion are suspect.
- **Punching shear.** Pool deck slab experienced punching shear failures, where concentrated loads from columns caused local slab failure.

2. Neglect and Lack of Maintenance:
   - **Water damage.** Pool leaks, rainfall, and potential saltwater sources contributed to the degradation of concrete and reinforcing steel.
   - **Lack of inspections.** Required inspections designed to assess building structural integrity would have detected problems sooner.
   - **Unnecessary additional weight.** Planters with palm trees and other additions to the pool deck added unnecessary weight and potentially increased stress on the deck slab.
3. Potential role of Nearby Construction:
   - **Vibrations:** Vibrations from construction on nearby luxury condominiums may have contributed to the structural damage.
   - **Ground settlement:** Possibility of ground settlement due to the weight of the nearby construction.

Preliminary Failure Chain: Engineering firm WJE Associates (WJE) conducted an independent investigation that identified similar issues and concluded that punching shear failures in the pool deck (columns punched through the deck slab) caused the failed deck to then apply loads to the building, which caused it to fail. Videography sequencing from multiple CTS and private camera footages support this theory. NIST is nearing completion of its comprehensive investigation; findings are expected to be released in 2025.

Milestone Inspections

The CTS collapse spurred the *Florida Milestone Inspection Law* in 2022, requiring mandatory structural inspections (***milestone inspections***) for aging condominiums and cooperative buildings. Specifically, applicable buildings three or more stories tall must undergo a milestone inspection to assess structural integrity at 30 years of age. Subsequent inspections are required every 10 years thereafter. Buildings within three miles of a coastline must be inspected at 25 years of age, and every 10 years thereafter. The primary goal of these inspections is to identify and address structural deterioration to ensure the safety of the buildings. In 2023, the law was amended with the *"Glitch Bill"* for further clarification and allowed local jurisdictions to modify the initial 25-year inspection requirement for coastal buildings based on environmental factors.

The milestone inspection process involves two phases: a Phase One inspection to assess the overall structural condition, followed by a Phase Two inspection if substantial structural deterioration is identified. The inspections are performed by a licensed professional engineer or architect, who then issues their report of findings to the condominium or cooperative association. Associations, which have a fiduciary duty to prioritize the long-term health and safety of the buildings, then manage any required repairs or retrofits in a timely manner. The law emphasizes the importance of timely inspections and reserve funding (the states have reserve funding laws) to avoid deferred maintenance.

NIST (Best Practices) Forensic Investigation

The scope of the NIST forensic investigation is unique in that it includes government, university, and private industry experts working across six technical projects for the most comprehensive investigation of its kind. The technical projects are summarized:

1. ***Building and Code History***: Assess the entire history of the building from design through partial collapse, including relevant codes and standards, design plans and specifications, construction records, inspections, maintenance, renovations and alterations relevant to additional loads, and environmental conditions.
2. ***Evidence Collection and Preservation***: Innovative tagging and data collection methods.
3. ***Remote Sensing and Data Visualization***: Analyze all available 2D and 3D data in establishing a geographical information system (GIS) model designed to support other investigation projects.
4. ***Materials Science***: Evaluate strength, appropriateness, uniformity, and deterioration of construction materials used.
5. ***Geotechnical Engineering***: Evaluate building foundation design, as-built construction, and current subsurface conditions (soil and groundwater).

6. ***Structural Engineering***: Emphasis given to load testing and generation of computer models to simulate failure initiation and progression.

It will be interesting to study the outcomes of this investigation once they are reported.

7.5 Climate Impacts on Far North Infrastructure

A colleague once described a 19th-century book about the history of the Netherlands which featured a sequence of coastal cartography maps of Europe's Atlantic coastline that, when thumbed through, elicited the perception of coastal changes with sea level rise. One can Google *'Beringia,'* click Wikipedia's entry, and watch the *Paleoenvironmental Atlas of Beringia* timelapse of the *Bering Land Bridge* from 21,000 years ago to the present and observe sea level rise. Climate has, is, and always will change. According to Working Group I (WGI) of the Intergovernmental Panel on Climate Change (IPCC), subarctic and arctic regions of the Earth will experience the greatest warming and impacts to infrastructure during this century[9].

Forensic engineers have investigated *frost heaves, frost jacking,* and *permafrost degradation* phenomena associated with infrastructure built in and on frozen ground. Heated buildings on ice-rich soil, roads through permafrost areas, and utilities in frost-susceptible soils (Fig. 7.7). Designers and builders

FIGURE 7.7
Engineering and contending with northern landscapes.

must contend with *drunken forests*, *thermokarst sinkholes*, and construct with *thermosyphons*, insulated foundations, and raise buildings on gravel pads, shallow piles, and skids, techniques used to preserve permafrost. We do not want a repeat of Siberia's *'doorway to the underworld.'*

Infrastructure management moving forward is sure to include small footprints, specialized geotechnical engineering for intermittent and discontinuous permafrost areas, better foundation designs for *thaw consolidation*, and moveable structures and enforced decommissioning of infrastructure removed from Arctic tundra and taiga. Wherever mankind endeavors in the far north, the forensic engineer will be challenged with the investigation of failures.

7.6 Making the Case for Failure Analysis in Design

This chapter tells a story of how building failures and forensic investigations have led to legislation designed to reduce failures. Building codes have been upgraded, new types of inspections have been added, and existing structures are undergoing repairs and upgrades. Yet recent constructions have failed, and some young structures predate the new inspection laws. (Case Study 7 is about a young condominium complex that fits this scenario, and what band-aid repairs might lead to.)

We hope to learn much from the outcomes of NIST's best practices forensic investigation of CTS. NIST, NTSB, and private industry are repositories of lessons learned. ***Failure theories***[3] in materials science and engineering are limited to specific stress conditions. Deferred and inadequate maintenance is a problem that has contributed to failures (e.g., I-35W bridge collapse). DOT budgets are dependent on annual appropriations that are subject to political and economic cycles, and condominium associations do not like to give bad news of special assessments for structural repairs.

Why not take what we know and develop new practical strategies to prevent structural failures?

- Harness what we have learned from failures and integrate practical failure analysis into design.
- Integrate *Performance Evaluation* considerations.

- Create infrastructure banks. Plan for the worst - tax appropriations (or lottery proceeds) invested in well-managed growth strategy portfolios - and prepare for proactive repairs and rehabilitation.
- Grant programs for qualified applicants for structural repairs to help offset reserve depletions.

Notes

1. *Effective depth* in the context of concrete slabs and punching shear refers to the distance from the compression face of the slab to the center of the longitudinal [tension] reinforcement bars (Fig. 7.5(b)). This depth is crucial for calculating shear strength and punching shear stress for load-bearing support columns.

2. The New Orleans Hard Rock Hotel was under construction when it partially collapsed on October 12, 2019, resulting in three worker deaths and dozens more injured. The building was subsequently demolished, government officials continued to debate actions against potential responsible parties in 2025, and the project's future remains undecided. Intrigue with this disaster involves an OSHA investigation, lack of NIST – OSHA cooperation, and 23M taxpayer dollars. Why NIST did not investigate and the grand jury decision from a related lawsuit are worth the read.

3. *Failure theories* (FTs) in materials science and engineering help engineers design structures and components that can withstand *expected* loads without failing. Examples are Tresca's Maximum Shear Stress Theory, Von Mises Criterion (a distortion energy theory), and Saint-Venant's Maximum Principal Strain Theory. Designed to help predict when a material will fail under different loading conditions with consideration for stress, strain, and material properties, such theories cannot well account for environmental influences, unforeseen additional loads, and consequences of deferred maintenance of structures in dynamic settings.

References

1. Institute of Historic Building Conservation (2022). *A Brief History of Building Regulations*. www.npt.gov.uk, Accessed 12 December 2024.

2. Sherry H. Olson (1997). *Baltimore: The Building of an American City*. Johns Hopkins University Press, Baltimore (Md.), 248 p. ISBN 0-8018-5640-X.

3. Stephen Puleo (2004). Dark Tide: The Great Boston Molasses Flood of 1919. Beacon Press. ISBN 0-8070-5021-0.

4. Tikkanen, A. (2025, June 6). *Great Molasses Flood. Encyclopedia Britannica*. https://www.britannica.com/topic/Great-Molasses-Flood, Accessed 22 May, 2025.

5. National Transportation Safety Board (December 16, 1970). Highway Accident Report: Collapse of U.S. 35 Highway Bridge, Point Pleasant, West Virginia, December 15, 1967 (PDF). Report No. NTSB-MAR-71-1, Retrieved 15 March 2025.

6. Hai S. Lew, Nicholas J., S. George Fattal and Martin F. Batt (1982). Investigation of Construction Failure of Harbour Cay Condominium in Cocoa Beach, Florida, *NBS Building Science Services 145*, Washington, D.C., 135 p.

7. Morabito Consultants (October 8, 2018). Champlain Towers South Condominium Structural Field Survey Report (PDF). Retrieved 24 May 2025.

8. J. Huergo (2025). *Champlain Towers South Investigation Team Provides Update to Structural Engineers*, https://www.nist.gov/news-events/news/2025/04/champlain-towers-south-investigation-team-provides-update-structural, Accessed 25 May 2025.

9. Paola A. Arias, et al. (2021). *Technical Summary. In Climate Change 2021: The Physical Science Basis Contribution of Working Group I to the Sixth Assessment Report of the Intergovernmental Panel on Climate Change* [Masson-Delmotte, V., P. Zhai, A. Pirani, S.L. Connors, C. Péan, S. Berger, N. Caud, Y. Chen, L. Goldfarb, M.I. Gomis, M. Huang, K. Leitzell, E. Lonnoy, J.B.R. Matthews, T.K. Maycock, T. Waterfield, O. Yelekçi, R. Yu, and B. Zhou (eds.)]. Cambridge University Press, Cambridge, United Kingdom and New York, NY, USA, 33–144 pp. doi:10.1017/9781009157896.002.

Case Study 7: As the Roofs Go, So Go the Buildings

Message to students:

> Two most important components of a building: the roof and the foundation. They keep the elements at bay.

Florida is the retirement capital of the United States, offering warm weather, excellent healthcare and senior living centers, and year-round recreation. Approximately 25% of its population were retirees in 2025, many of whom prefer coastal living.

Elysian Condominiums is an upscale coastal common interest development with areas and facilities owned "in common" by its members. Three 12-story and one 5-story buildings comprise the condominium towers built in 2005–2007. Community *members* are condominium unit owners, and the Elysian Condominium Association (ECA) is the volunteer elected *Board* that serves along with a maintenance team and a property management company (PM) to maintain the buildings and common elements using *reserves*.

DOI: 10.1201/9781003623311-16

Monthly assessments are collected from each member to fund the Reserve Account (RA). The ECA is responsible for maintenance decisions and prudent fiscal management of the RA. The Elysian maintenance team is responsible for property security, grounds keeping, other light routine maintenance, and scheduling contractors. Property management collects assessments, manages annual budgets, procures bids, and informs and makes recommendations to the Board.

This case study is about failed building roofs, *Mr. Band Aid*, and how Board mismanagement has *kicked the bucket down the road* in dealing with looming structural problems. Every Board has a leader, and deference to one with a *band-aid mentality*[1] can be costly.

Background

A timeline of events (Table CS7.1) is key to understanding the backstory to the forensic engineering that defined the roof problems and building leaks at Elysian. The condominium buildings are numbered 1, 3, 5, and 7 as indicated in the cover image. The reader will recognize how the ECA was misled by roofers, which led to ECA fiscal mismanagement of reserves, which triggered a state investigation of its finances. Further, the Board discarded one engineer for another, who was then ignored after confirming serious roof problems and identifying leaks into the buildings. Mr. Band Aid and property management demonstrated little regard for professional contractors and their work products.

TABLE CS7.1

Timeline of Events at the Elysian Condominiums

Date	Event	Description
2005 2006	Phase I Development	Building 7, Clubhouse/Pool/Tennis & Basketball courts built
2007	Phase II	Building 5 built
2008	Phase III	Buildings 1 and 3 built
	Phase IV	Marina built
Aug 2008	Roof inspection	Roofing contractor identified *"blistering and delamination of foam roofs all bldgs."* and recommended resurfacing all bldgs. with new roofs.
Summer 2010	Roofs replaced	The four condominium bldgs. resurfaced with new *"elastomeric coating systems."* Roofer Z hired to manage roofs on a 5-yr resurfacing plan.
2012	Reserves discrepancy complaint	Filed with Florida Department of Business and Professional Regulation (DBPR)

(Continued)

TABLE CS7.1 (*Continued*)

Timeline of Events at the Elysian Condominiums

Date	Event	Description
2012-13	DBPR investigation	Investigation revealed underfunding of Roof Reserve since 2011. Required reestablishing Roof Reserve funding at 2010 level.
Nov 2012	Engineer 1 Roof Inspection Report	Bldg. 1 roof failed, needs new roof. Bldgs. 3, 5 and 7 roofs suspect (recoating may be necessary). Roof study recommended but not approved by ECA.
Jan 2013	Engineer 1 Engineering Inspection Report	Structural inspection: all buildings in good structural condition with exception of roof systems. Recommended roof study to discern problems and solutions. ECA rejected roof study.
Mar 2013	Bldg. 1 roof resurface	Roofer Z proposal to resurface Bldg. 1 roof rejected by ECA, citing 5-yr warranty. Roofer Z also issued Limited Warranty letters for all roofs, indicating only 5-yr warranty period. ECA rejected letters.
Sept 2013	Roof warranty extension agreement	Roofer Z agreed to Extended Maintenance Warranty with annual roof inspection and maintenance for 8 years (through 2020).
Sept 2014	Roof coatings analysis	Coatings Specialist hired to analyze building roof cover membranes for thickness and remaining useful life for bldgs. 3, 5 and 7 roofs. Reported membrane thicknesses and remaining useful lives as: Bldg. 3: 25–32 dry mils* thickness, 3–4 yrs useful life Bldg. 5: 30–32 dry mils, 3–4 yrs useful life Bldg. 7: 25–32 dry mils, 3–4 yrs useful life (15 dry mils was the industry standard minimum for surface coating thickness in 2014) Base foam (SPF) estimated useful life 15–20 yrs. (No testing and no basis for this assertion).
Oct 2015	Engineer 2 Reserve Study (RS)	Documents review discovered reserves discrepancy and prior roofs history. See *Engineer 2 Reserve Study Report Comments* below this table for details.
Dec 2015	Roof Conditions Assessment	Tasks performed: roofs core sampling and testing, conditions mapping, and moisture detection survey. Results reported to the ECA.
Dec 2015	Roof Corrective Action Plan (CAP)	CAP not approved by ECA.
Late Oct 2021	Request for new RS	Engineer 2 was contacted by new property management (PM2). PM2 requested new RS by Nov. 15. Engineer 2 requested updated information on status of condominium roofs and RS deadline extension to Dec. 1. No reply from PM2.

* *Dry mils* (or *Dry Film Thickness*) is the thickness of a coating after it has fully cured, after solvents or water ingredients have evaporated. It represents the final solid thickness of the coating expressed in mils (thousandths of an inch).

Engineer 2 Reserve Study Report Comments:

1. *"A significant change in the Roof reserve has negatively impacted reserves. Roof reserve funding decreased since 2011 and was discontinued in 2014. We understand that the Board was led to believe that the SPF roofs were viable, and that roof deficiencies could be managed with the existing reserve and a 5–10 year 'scarf and recoat' plan. However, new information has become known, and the roof systems are not viable.* ***All condominium roofs leak and water has impacted rooftop structures. A comprehensive investigation of the roofs and structural impacts is imperative.*** *Consequently, the ECA should reestablish roof reserve funding at $45,000 for the 2016 contribution, a level approximately 20% higher than the 2012 funding level. The roofs investigation will lead to a change in the roof management strategy, and significant rehabilitation funds may be needed in the short term. We recommend the Board consider an aggressive funding strategy for the roofs at this time to avert special assessment if possible."*
2. *"Condition assessment of the roof component (four condominium towers) is incomplete currently. The roofs are known to be leaking and in need of comprehensive investigation for sufficiency and impacts to building structures. The ECA has been made aware that the condition assessment of the roofs is subject to investigation findings and that this reserve study should be amended with a Roof Addendum upon investigation completion."*

Mr. Band Aid

What remained unclear through the timeline was the relationship between Mr. Band Aid, a roofing contractor, and the initial property manager, ECA understanding of the magnitude of the roof problems, and Board deference to Mr. Band Aid's decisions. Association board members are volunteers; oftentimes few have a building science background, and sometimes only one or a couple are active members.

Roof Investigation

Reserve studies require a high level of understanding of building systems, construction, and financial planning. An RS, usually performed by a real estate expert or engineer with the necessary expertise, should be updated every three to five years depending on individual state requirements. Elysian's previous RS was performed by a real estate professional in 2010.

Engineer 2 was hired to perform the next RS in 2015. A document review provided drawings of the buildings (i.e., plans) and previous professional reports.

Documents Review

Important information about the condominium roofs was revealed during review of the ECA documents. Engineer 1's inspection report described significant problems with the roofs, summarized and paraphrased below:

1. General conditions of the three 12-story condominium building roofs:
   - Roof covers are SPF (spray polyurethane foam) with elastomeric cover membranes.
   - Cover membranes appear weathered, discolored, and exhibit significant oxidation and chalking in some areas. This high level of degradation is inconsistent with the young life of the membranes.
   - Roof cover slopes were highly variable, producing undulated surfaces, which impede drainage to some drains and scuppers. Water ponding was evident and extensive. SPF mounding impedes water discharge at some scuppers and allows local ponding.
   - Some degree of blistering on all roofs. A few blisters were collapsed with cracking.
   - Testing is warranted to determine roof cover compositions.
2. Building 1 roof cover was different from that of the other three buildings. It had a top surface of heavily broadcasted brown granules with *"…as many granules as could possibly stick in the coating. The granules completely obscure the coating beneath, making it difficult to observe except where granules are worn away or holes exist. Other problems observed are widespread foam delamination (estimated 20–30% of roof area) and blistering. Blisters range from a few inches to several feet across; a couple of blisters had collapsed. This roof cover is a complete failure and water penetration is certain."*
3. The ECA competitively bid to re-coat all the buildings' roofs, specifying a silicone-based membrane coating. Bid specifications called for a protective silicone membrane and included several manufacturer's data for their silicone products. Three roofing contractors replied with bid estimates. A contract was awarded to a roofing contractor. At some point *after* procurement, a change was made for a polyurea coating system. This change was not formally documented. Also, the contractor provided the ECA with warranty documentation for an *"acrylic"* product.

Engineer 1's report and photo-documentation were good and thorough, but he was not approved to test roof materials. Instead, the ECA retained the services of a *"coatings specialist"* to assess the remaining useful life of the roof membranes. The specialist attributed three to four years of remaining useful life to the membranes but offered no basis for their conclusion. The new RS included a thorough inspection of the buildings and recommendation for a Roof Conditions Assessment to be added as an addendum. The ECA approved Engineer 2's scope of work.

Roof Conditions Assessment

Engineer 2 photo-documented roof conditions (Fig. CS7.1), mapped degradation features (Figs. CS7.2 to CS7.5), and collected core samples from each roof system for laboratory testing to determine material compositions.

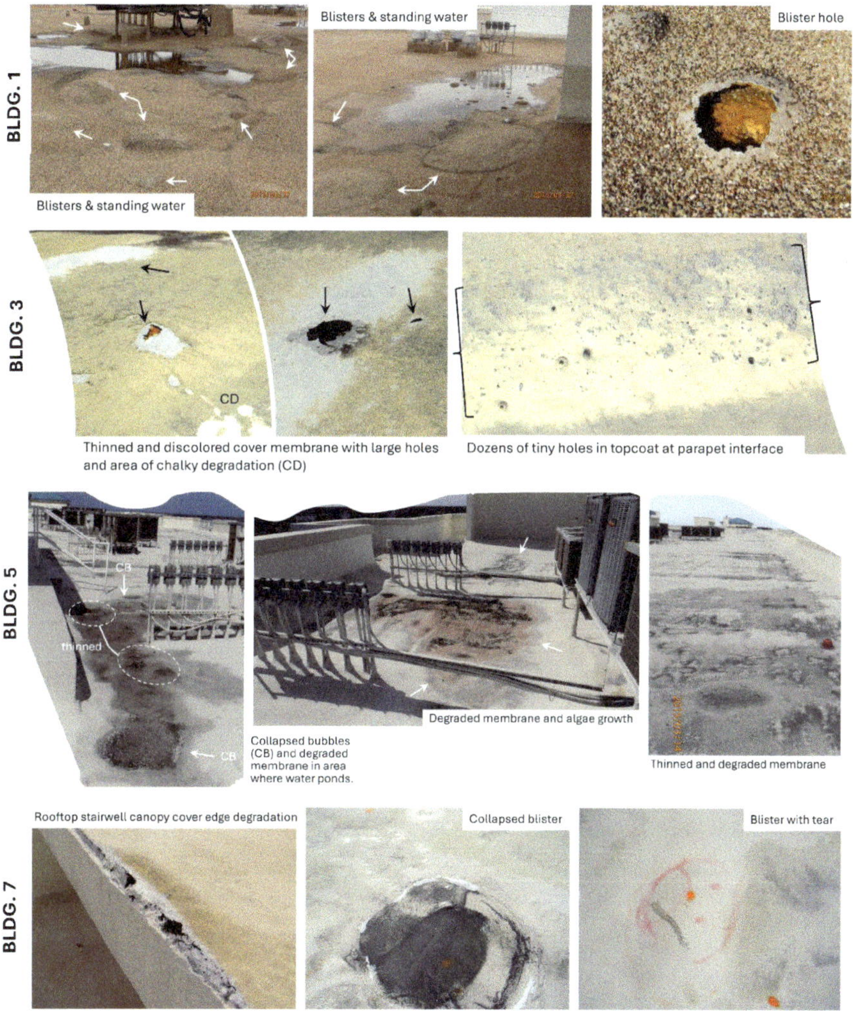

FIGURE CS7.1
Representative examples of roof degradation symptoms. Note the many problems with the different types of roof cover on Bldg. 1, and significant discoloration of the Bldg. 3 cover.

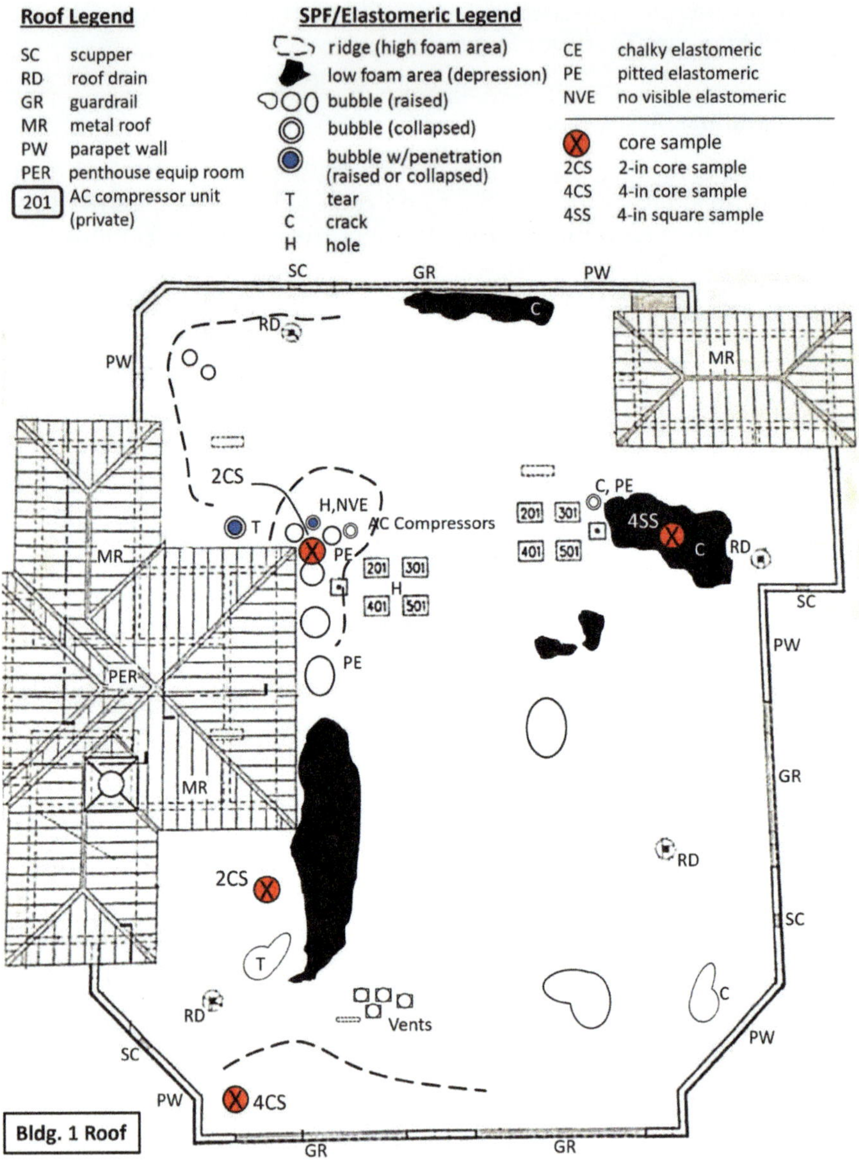

FIGURE CS7.2
Building 1 roof map of degradation features and sample locations.

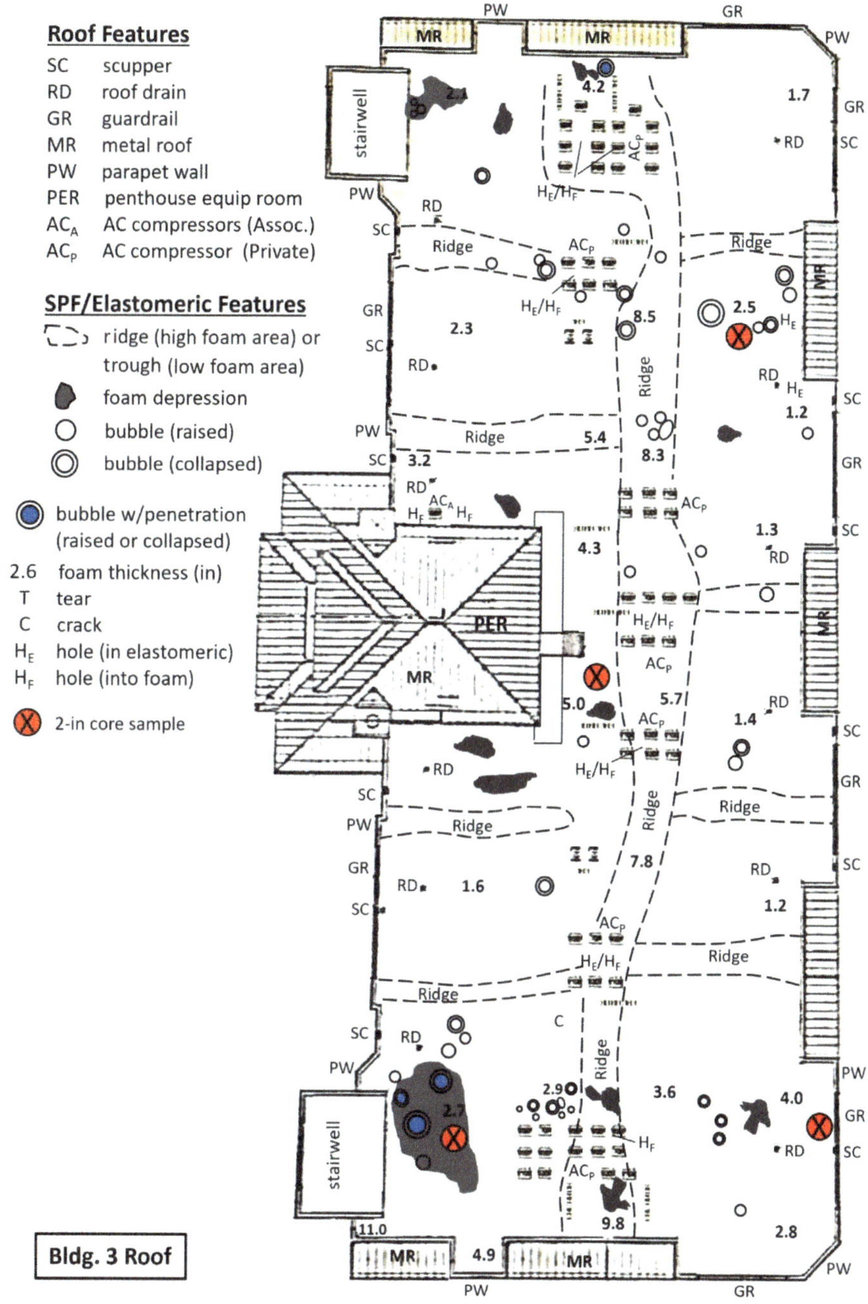

FIGURE CS7.3
Building 3 roof map of degradation features and sample locations.

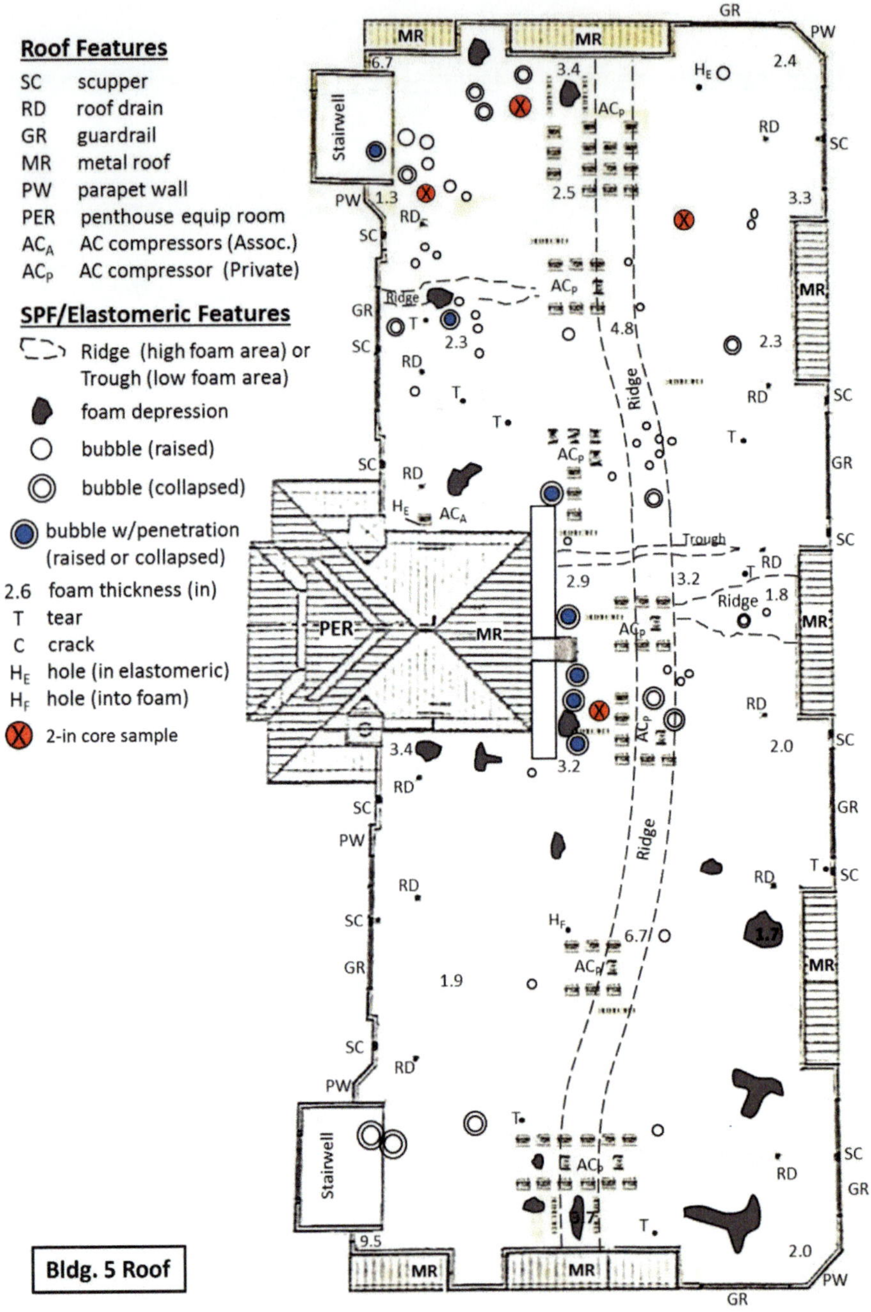

FIGURE CS7.4
Building 5 roof map of degradation features and sample locations.

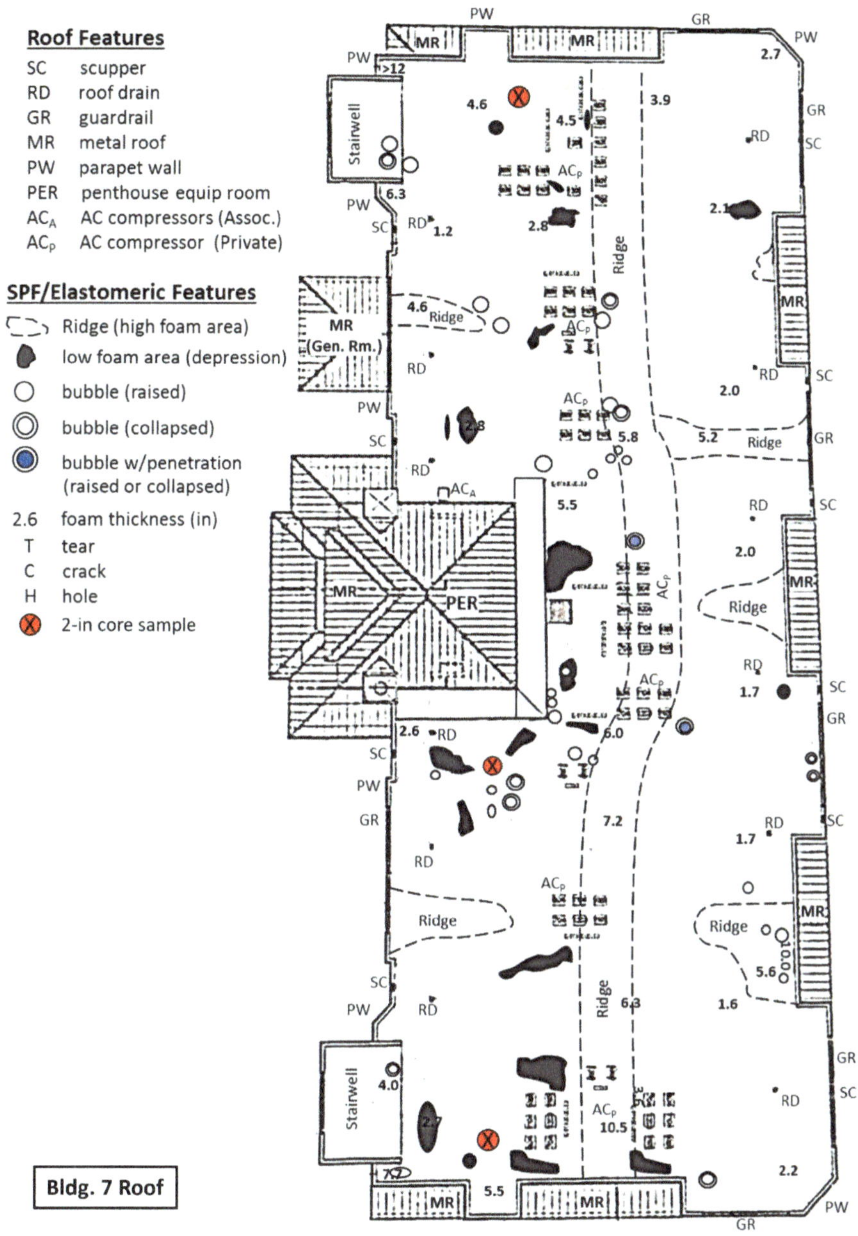

FIGURE CS7.5
Building 7 roof map of degradation features and sample locations.

A subcontractor performed moisture detection surveys to assess water beneath the roof systems. Results indicated water penetration to the concrete deck was pervasive on all four condominium building roofs.

Chemical analysis revealed the roof systems comprised different materials. All had SPF, but each was finished with a different topcoat of either silicone, polyurea, acrylic, or combination thereof. Specifically,

- Building 1 roof was finished with a granule-impregnated silicone topcoat. The grey coating between the foam and granules was a silicone-based membrane that was applied directly to the SPF. No primer (or basecoat) was present.
- Buildings 3 and 5 roof topcoats were typically two-ply (basecoat and topcoat) compositions of two different water-based acrylics or silicone layers. Typically, the basecoat was of low tensile strength elastomer and that of the topcoat much higher tensile strength as the more durable layer for foot traffic.
- Building 7 roof membrane was polyurea applied directly to the SPF. Although touted as strong, durable, and compatible with SPF, *"Polyurea is used as the first component in SPF-based roofing systems, followed by an elastomeric coating for UV protection"*[1]. Also, two foam samples from this roof contained *"resin-rich"* SPF, which implied improper mixing of the two-part product.

Not all elastomeric coatings are created equal. The SPFA (*Spray Polyurethane Foam Alliance*) offers contractors elastomeric product specifications and guidance for commercial and industrial roofing applications. Also, they will aid contractors with choosing a product specifically designed for SPF foam and remind them to adhere to the manufacturer's instructions as a prudent standard of care[2].

Seeps

Sealing roof decks under SPF covers is recommended. This protects against moisture entering the roof system and superstructure. However, a project may specify concrete roughing to increase SPF adherence – all the more reason to finish a roof system with a quality flexible, durable, waterproof, and UV-resistant membrane. Further, an elastomeric cover is typically a two-pass system comprising a basecoat and topcoat (Fig. CS7.6). The basecoat adheres to the SPF, and the topcoat is designed for weather protection and durability.

When rain or other moisture from the roof penetrates the superstructure of a concrete building, it can go undetected for a long time. Concrete is porous and seepage can be terribly slow. *Seeps* are a telltale sign of water intrusion, and it is important to understand seep manifestations (e.g., location, seepage rate, leachate, and coloration of stains).

Water that permeates concrete can lead to leaching due to a chemical reaction that occurs when penetrated moisture dissolves calcium hydroxide

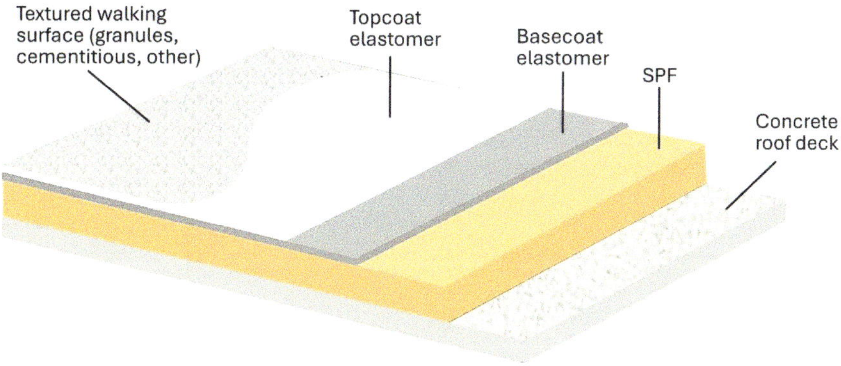

FIGURE CS7.6
Typical SPF-elastomeric roof system for condominium buildings.

($Ca(OH)_2$) in the concrete. This precipitate then reacts with carbon dioxide (CO_2) in the air to form calcium carbonate ($CaCO_3$), which is then carried to the concrete surface and deposited, resulting in stains[2]. This process is called *concrete leaching* or *efflorescence*. Although efflorescence itself is not a structural problem, its presence can indicate leaks in the concrete. This is why seep location and stain color matter.

Examples of different types of seeps are highlighted in Figure CS7.7. The Building 1, third-floor-level exterior wall seep is a significant concern. When

FIGURE CS7.7
Example water seeps from walls at various locations in the buildings. Rust-colored stains from seeps in Bldg. 1 (L) indicate corroding steel reinforcement in the concrete; something is leaching black to taint seep water from the Bldg. 7 roof deck-parapet interface (M); and vertical efflorescent seeps in Bldgs. 3 and 5 walls denote concrete leaching.

water seeping from a wall leaves rust-colored stains, concrete reinforcement corrosion is the likely source. Two adjacent parapet seeps (middle image pair) that produced black stains down the walls may be indicative of deck sealant degradation. The far right vertical seeps in the figure represent where efflorescent leachate runs down the wall under the paint, leaving a distinct white trail. Where efflorescence occurs, concrete is weakening.

Deferred Maintenance as an Operating Strategy

The Elysian condominium towers were built with bad roofs, a problem that should have been detected with inspections during construction. However, this was not the case, and the new ECA took over managing control of the property. Then bad roofs were replaced by bad roofs in 2010, by a roofing contractor who installed "elastomeric coating systems" and managed to land a 5-yr roof-maintenance contract. (Manufacturers of elastomeric roof products tout 25 to 50 years of amazing longevity.) Thinking they now had lasting roofs, the ECA reduced and then ceased funding their roof reserve to channel monies elsewhere. This is called *reserves manipulation*, a violation of state law governing condominium reserves. Further, transfer of reserves between line items requires majority approval from membership. This ensures the community is aware of and supports prioritizing repairs when necessary.

A condominium association has elected officers: President, Vice President, Secretary, and Treasurer. Mr. Band Aid was Vice President (VP) when he controlled the decision-making through the roofs debacle. The Board president was elderly and sickly, the treasurer traveled a lot, and the secretary often deferred to the VP. And property management took its marching orders from the VP. Thus, the band-aid mentality ruled. What happens when the ruler encourages the hire of a roofing contractor that installs bad roofs? Deferred maintenance with band aids never fixes anything for long and is a costly operating strategy that can lead to failure.

> At some point, you just pull off the Band-Aid, and it hurts, but then it's over and you're relieved.
>
> **John Green**
> *American author and YouTuber*

Notes

1. A *band-aid mentality* refers to addressing problems with temporary, superficial solutions rather than tackling the root causes. It's about quick, cheap fixes that alleviate symptoms in the short term, with little concern for addressing the

underlying issues that manifest over the long term. This strategy leaves the problem to resurface or worsen later.

2. An industry-standard guide for elastomeric roof coatings does not exist as a single, monolithic document. Instead, industry practices and recommendations are found in various resources such as product datasheets, application handbooks, and professional roofing guides. These resources offer guidance on product selection, surface preparation, application techniques, and safety measures.

References

1. Spray Polyurethane Foam Alliance, *SPFA-102 A Guide for Selection of Elastomeric Protective Coatings Over Exterior Spray Foam* (2015).
2. Randall K. Noon (2001). *Forensic Engineering Investigation*. CRC Press, 78–84 pp. ISBN: 0-8493-0911-5.

8

Hydrologic Control, Floods, and Coastal Barriers

CONTENTS

CONCEPTS	KEY TERMS	ADDITIONAL READING
Water	Hydrologic control structures	Exercise 15
Hydrologic Control	Pressure head	Exercise 16
Hydrostatic Pressure	Boils	
Bed Degradation	Foundation undermining	Case Study 8
Underseepage	Submarine groundwater discharge	
Hydrologic Study		

Water is a dynamic force. Since the agricultural revolution of about 5000 years ago, independent societies in ancient Egypt, Mesopotamia, the Indus Valley, and northern China began mastering the hydraulics for controlling water from rivers and streams for irrigation. However, whatever we create to control or harness the power of water has potential consequences, and sometimes our structures fail. In this chapter, students recall what they learned in Fluid Mechanics and look at how forensics helps us understand some interesting hydrologic events.

8.1 Hydrologic Control

Hydrologic control means managing water for useful purposes. We do it with *control structures* (Fig. 8.1) engineered to manage the flow and storage of water within a specific area, including directing, controlling, or preventing water movement. These structures are used for various purposes like flood control, irrigation, water supply, and navigation. Common examples such as

DOI: 10.1201/9781003623311-17

FIGURE 8.1
Examples of hydrologic control structures.

dams, reservoirs, canals, culverts, weirs, levees, gates, and coastal barriers are briefly described:

- **Dams**: Structures built across a river or stream to impound water and create a reservoir for various purposes.
- **Reservoirs**: Artificial lakes formed by dams, used for storing water, regulating flow, and energy production.
- **Channels and Canals**: Man-made waterways used for irrigation, navigation, and water supply.
- **Culverts**: Structures that allow water to flow under a roadway, railroad, or other obstruction.
- **Weirs**: Structures built across a waterway to raise the water level and control flow, often used for measuring water flow or diverting water.
- **Levees**: Engineered embankments built along river courses to protect adjacent lands from flooding.
- **Gates**: Movable barriers used to control water flow, often in conjunction with dams, canals, and reservoirs.
- **Pumps**: Mechanical devices used to move water, often used in conjunction with levees and other structures to manipulate water levels or move water to other locations.
- **Manufactured Coastal Barriers**: Structures like seawalls, breakwaters, and groins built to protect coastlines from erosion, flooding, and other coastal hazards.

8.2 Floods and Coastal Barriers

Hydrostatic pressure forces and conservation of flow energy are topics covered in a Fluid Mechanics (or *Fluids*) course, where engineering students learn that water pressure increases linearly with depth in the ground, a reservoir behind a dam, or the ocean. A good Fluids book includes some reservoir problems, and a favorite is the hydroelectric dam problem with its large amount of water in the reservoir behind the dam. Students are challenged with converting the water's potential energy to electrical power output as it passes through the turbine-generator equipment (or *gen set*) in the powerhouse 100 feet (30.5 m) below on the other side. They learn that a large amount of water can produce electrical power.

A levee along a rising river is a similar problem, only without the power generation. As the river rises and creates a large amount of water (or ***pressure head***[1]) against the levee, water pressure increases in the soil beneath. Should the river reach flood stage, seepage under the levee – called *underseepage* – can become problematic for infrastructure on the landside of the levee. The levee was built to protect against floods. But like a dam, the levee too has its limits.

What follows are some water events that required the services of forensic engineers. From a far north river flood to an epic midwestern flood, then to hurricane alley, and north again to an arctic coastal barrier wall with good intentions but unintended consequences, students will learn how water challenges infrastructure.

Tanana Flooding Undermines Foundation

Heavy rains in July 2008 caused the Tanana River to flood in Fairbanks, Alaska. Foundations in a residential subdivision were undermined when two feet (0.6 m) of water inundated homes. An insurance company hired a forensic engineer to investigate the magnitude of the problem at several properties. What follows summarizes the investigation and findings at one flooded home.

The two-story wood frame home was built on a constructed soil pad and supported by a concrete slab-on-grade foundation with footer walls. Located about a tenth of a mile (0.16 km) from the river, gray river silt stained trees two feet (0.6 m) above the ground (Fig. 8.2) to mark the high water line in the surrounding woods. At the house, the high water line was approximately ten inches (25.4 cm) on the outer walls, and sandbags did little to prevent flood water intrusion. ***Boils***, small circular muddy depressions along the edge of the foundation (Fig. 8.2 C), were the telltale sign of pressurized water having eroded fines out from under the foundation during the flood. Once flood water receded and hydrostatic pressure under the slab decreased to ambient, the boils stopped bubbling out their discharge and collapsed, leaving small

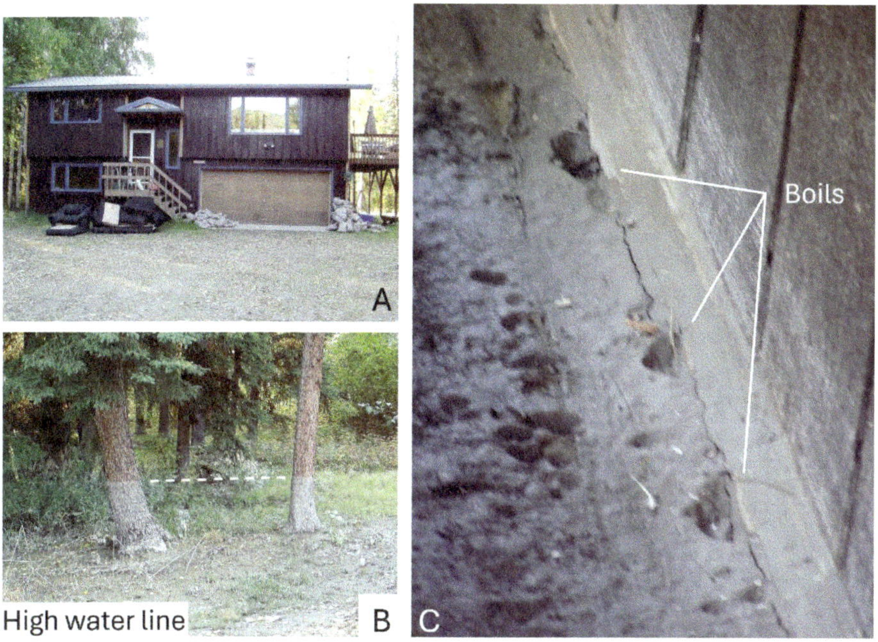

FIGURE 8.2
Flood evidence at the subject property. Note the high water line on the trees (image B) and structure wall (image C), and boils along the edge of the foundation.

craters. Determining the extent of ***foundation undermining*** was key to establishing effective repairs.

Test pits dug on two sides of the house confirmed good foundation soil for support and drainage (Fig. 8.3), a reason differential settlement was not an issue. Also, with good compaction, soil erosion under the house should be minimal.

Flooring inside the first floor of the house and the bottom couple of feet of sheetrock had been removed to prevent mold, always a good idea immediately after a flood. This also exposes the foundation and wall cavities for inspection.

A broomstick or hammer and a good ear is one way to assess voids beneath a concrete foundation. The trained ear can discern the difference between

FIGURE 8.3
Test pits dug next to the house indicate 0–2 ft (0.6 m) of sand and rounded riverine gravel to 2 ft (0.6 m) depth, confirming good foundation soil.

compacted soil and voids (or cavities) beneath the concrete by sound. A dull thud (compacted soil under concrete) versus a higher pitch hollow ring (cavity) is the audible difference. The forensic engineer applied this *sounding* technique to map voids beneath the foundation as shown in Figure 8.4. Foundation cracks were also inspected to distinguish old from new and upwelling of fines with the pressurized flood water.

Five cracks were observed in the foundation. Two longitudinal cracks, denoted Cracks #1 and #2 in Figure 8.4, spanned the length of the foundation from north to south. Crack #1 was an old crack, which was painted over in the living room, and which grew in length with flooding through the utility room and into the garage. Crack #2 was a new hairline crack, as were three short transverse cracks (perpendicular to cracks #1 and #2) in the garage. Portions of cracks denoted *"active"* were filled with sediment that upwelled into the garage and guest bedroom during the flood. The bottom left image in Figure 8.4 (arrow) shows how one can easily discern a crack that was *active* during flooding by sweeping away the upwelled sediment on one side of it.

Voids mapping indicated that the pair of longitudinal cracks acted as pressure relief once the foundation soil became saturated. This is a good thing; hydrostatic pressure relief minimized erosion and precluded differential

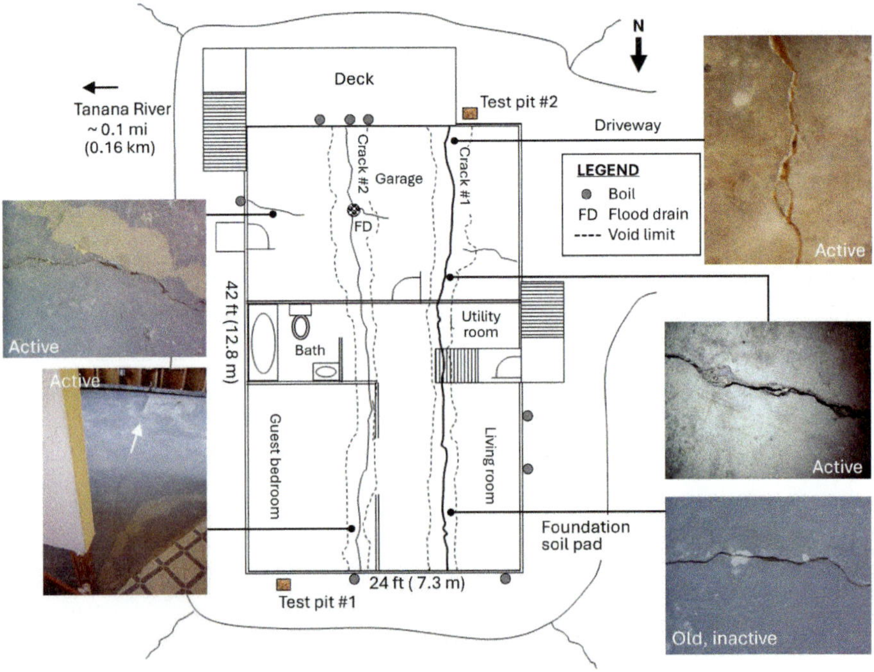

FIGURE 8.4
Mapped voids under the foundation and images of cracks that were active and inactive during the flood.

settlement. Further, injection grouting and crack sealing were inexpensive remedies that restored foundation integrity.

The Great Flood of 1993

The Great Flood of 1993 was a devastating Midwestern United States (US) natural disaster that caused an estimated $15B US in damages. Unseasonably dry cool air from the northwest converged with warm moist air from the south over nine states in May. From Wyoming to Michigan and Minnesota to Oklahoma, this ~400,000 sq mi *convergence zone* parked itself over the mighty Mississippi and Missouri Rivers for a couple of months. Rain fell for 200 days, heavy steady rains for 40 of 61 days in June and July, and hundreds of levees failed. The Missouri River crested at a record 48.9 ft (14.9 m) in Kansas City, within inches of the top of the levee that protects the North Kansas City (NKC) industrial district (Fig. 8.5). Depending on which statistical model one has faith in, this epic flood was a once in a 100- to 1000-year event[1].

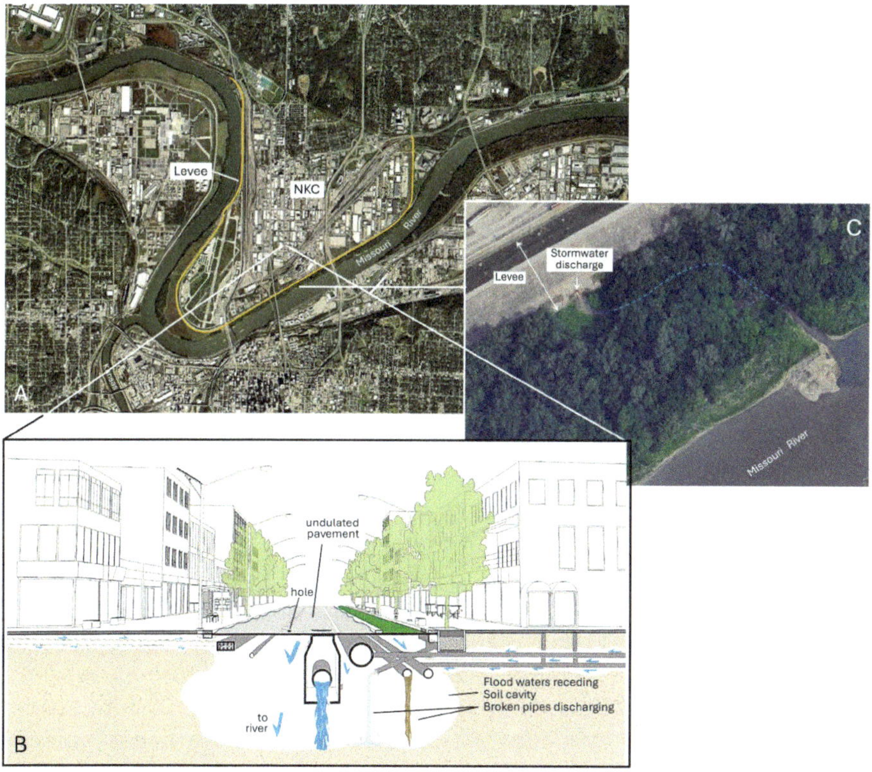

FIGURE 8.5
Consequences of underseepage and discharge points for receding water from the Great Flood of 1993. (Satellite photographs courtesy of Google Maps.)

Flood-fighting activities at the NKC levee included pumping to divert water, which prevented flooding of the Kansas City downtown airport (visible on the point in Fig. 8.5, image A). Designed to protect against a 500-year flood, sinkholes developed on its landward side, a telltale sign of ***underseepage***, when river water pushed through a layer of sand beneath the levee[2][4]. Flood mitigation efforts prevented the river from overtopping the levee, but underneath and hidden lurked hydrostatic pressure forces beyond what flood-weakened soils could withstand. The forensic engineer is certain to be awestruck and challenged by the variety of impacts to infrastructure such a calamity causes.

Imagine pavement floating on liquified soil that undulates while walking on it, and upward-burst foundations that invoke thoughts of *Tremors*, that 1990 film about a bored trio from small-town Perfection, Nevada seeking desert adventure when they happen across mysterious deaths in nowhere USA and must survive giant, prehistoric, worm-like underground monsters hungry for humans.

Tasked with performing ground-penetrating radar surveys to discern the underground, the engineer must first *recon* the area. Walking toward an intersection, a tiny hole is noticed in the pavement near a utility hole cover. Using a flashlight, the engineer peers into a cavernous underground devoid of soil and with broken and draining pipes (Fig. 8.5, image B). Instinctively dropping to all fours, the engineer slowly backs away from the only thing supporting the pavement, a single four-ft-diameter utility vault in the center of the intersection. The least path of resistance for retreating flood waters back to the river is via underground utility alignments and discharge points in the levee (Fig. 8.5, image C). Investigation of the NKC levee concluded an *underseepage and riverbank instability* failure mode.

Major floods on the Missouri River system are usually caused by a series of short-duration, high-intensity storms following a prolonged period of rains and snowmelt from Wyoming, Montana, and the Dakotas that fill tributaries. Landward, rainfall significantly reduces the infiltration capacity of the soil, which further contributes to greater than normal flows in the stream channels. The system is particularly vulnerable in early summer. Before the '93 flood, there were the floods of 1951–52. How might these floods have altered river channels and impacted levee protection?

The US Army Corps of Engineers (USACE) in cooperation with the Mid-America Regional Council conducted a major *bed degradation*[2] *study* of the lower Missouri River system between 2014 and 2017 to address these and other questions. The study included a geotechnical analysis component that evaluated levee corridor soils with regards to flood-risk management. Channel degradation and slope instability, underseepage, and landside slope instability (Fig. 8.6 (a)) were problems that warranted restoration. This study led to the $529M Kansas City Levees Program that included pump station upgrades, levee and floodwall raises to five ft (1.5 m), and seepage and stability improvements[3][4]. Relief wells, underseepage berms, riprap, and revet mattress are examples of the improvements (Fig. 8.6 (b)).

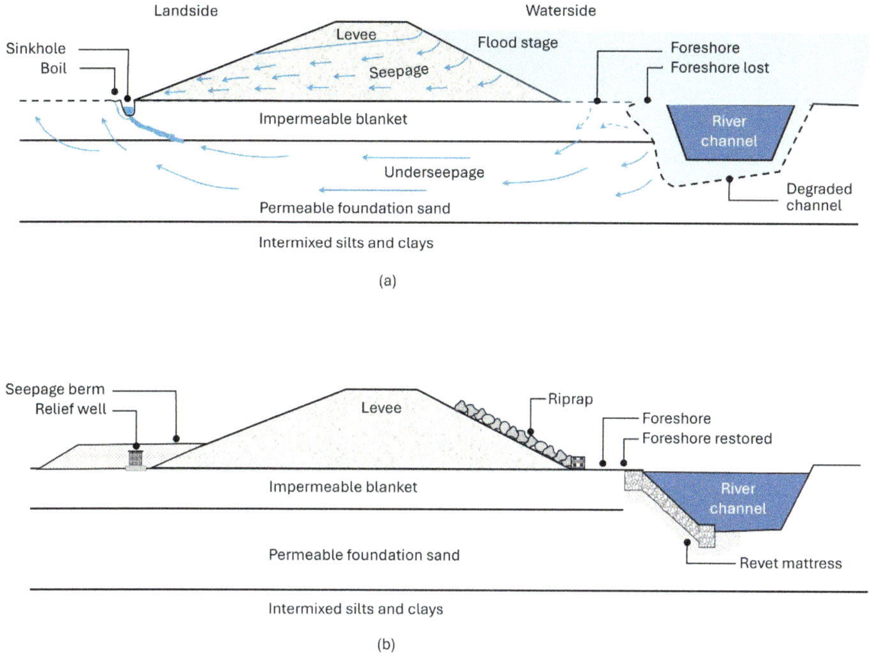

FIGURE 8.6
Levee (a) at flood stage with underseepage and landside instability, and (b) strengthened with waterside improvements and landside seepage mitigation.

Hurricane Damage Distinctions

Hurricane season in the southeastern US is from June 1 to November 1 each year. From Category 1 glancing blows to powerful Category 5 super storms, they bring torrential rains, high winds, tornadoes, and destruction. The insured will encounter hurricane deductibles and coverage is dependent upon flood or wind damage, as if that matters to those with loss.

The 2022 hurricane season brought a pair of hurricanes to rural riverfront property on a lazy river in Florida. Hurricane Ian made landfall as a Category 4 storm with 150 mph before weakening to Category 1 as it passed through in late September. It dumped fifteen inches of rain in the area before heading to South Carolina. Then came Hurricane Nicole during the early morning hours in early December. Little rain, and downgraded to a tropical storm, she passed over the property with a tornado warning in effect. Daylight brought destruction: a partially ripped-off roof from the house and a destroyed dock (Fig. 8.7). Tree branches lay strewn across the yard between the two structures.

A claim was filed, photographs sent, and an insurance adjuster told the homeowner the roof was covered but not the dock: *"The dock sustained flood damage, and as your policy states that since you live in a flood plain, it is not covered."* The homeowner requested the adjuster visit the property and *see* the

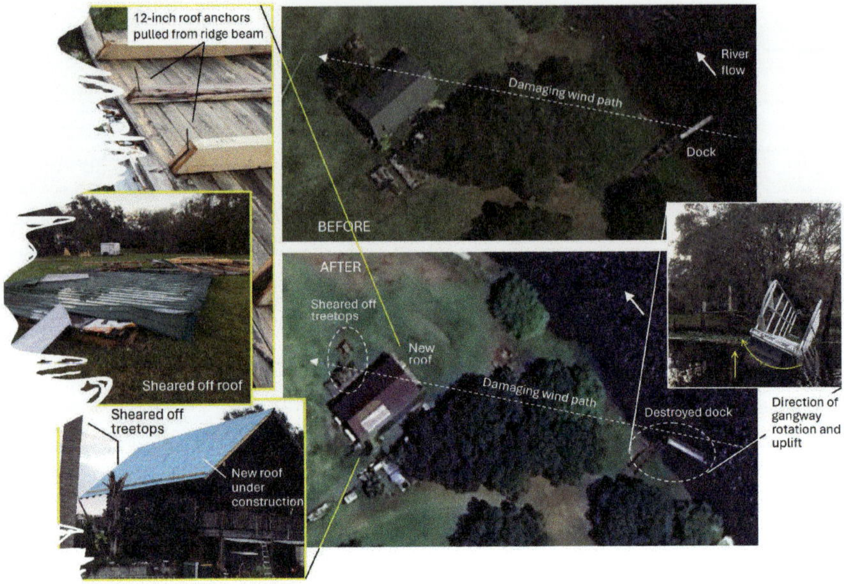

FIGURE 8.7
Distinguishing wind from flood damage. (Satellite photographs courtesy of Google Maps.)

damage firsthand; the adjuster declined based on *"you sent good photographs."* The property owner then hired a forensic engineer to investigate.

Investigation required use of a canoe, camera, Google Maps (or drone), some common sense, and review of local hydrologic data for the river. The canoe afforded the river perspective, which revealed more about the damage than from land observations. Google Maps provided the before-and-after satellite images used in Figure 8.7. One can even see the decapitated trees next to the house and the upriver orientation (a clue) of the dock gangway in the AFTER image. The disposition of the gangway, together with ***United States Geological Survey (USGS) Hydrological Data (USGS-HD)*** from the nearest river monitoring station, did the convincing.

Gangway[3] twist and deformation at its connection with the main dock, and pattern of dislodged piles suggested high rotating winds racked the dock during Hurricane Nicole. The gangway was lifted and twisted counterclockwise (CCW) to the direction of river flow, a rotation that then twisted the end of the wooden dock, pushing it downward and to the right (against the direction of water current) to its final deformed state. The torque caused dock piles to dislodge and bend in two directions. Engineering deduction surmised (1) the gangway would have *rolled under* and shifted downstream (not upstream) with river flow had it significantly increased, and (2) only a strong wind could lift the heavy gangway and counter-rotate it to an upstream position and disorient piles in multiple directions. The USGS-HD would support this assertion.

USGS-HD from a nearby upstream hydrologic station was reviewed for nominal river flow characteristics and those pertaining to the two hurricane events. Average flow speed of the lazy river was 0.3 mph (0.44 ft/s, 0.13 m/s). The river crested at flood stage on October 3rd after Hurricane Ian, with a water level at 11.37 ft (3.47 m) above NAVD 1988[4] and peak velocity of 0.34 mph (0.5 ft/s, 0.15 m/s). The water level and speed had dropped to well below flood stage to 7.49 ft (2.3 m) and 0.2 mph (0.29 ft/s, 0.09 m/s), respectively, by November 9, the day before Hurricane Nicole passed over. Nicole's influence on the lazy river was minimal, having increased the water level by only 0.35 ft (0.11 m) and river speed by 0.07 mph (to 0.4 ft/s, or 0.12 m/s) on November 10[5]. This data indicated minimal influence on river flow conditions caused by Hurricane Nicole. Further, the river was well under flood stage on November 10, and its flow speed was simply too slow to be destructive to marine docks. This information, together with the landward line of destruction to the house – downed tree limbs and sheared-off roof and treetops – was conclusive. With a phone call and report from the forensic engineer, the insurer reversed its decision and covered the dock loss.

Climate Change, Arctic Coastal Erosion Plan, and Unintended Consequences

Kotzebue lies at the tip of an Arctic *barrier spit*, a narrow point of land that projects from the Baldwin Peninsula into Kotzebue Sound, in northwestern Alaska (Fig. 8.8). Composed of sands and gravels transported by

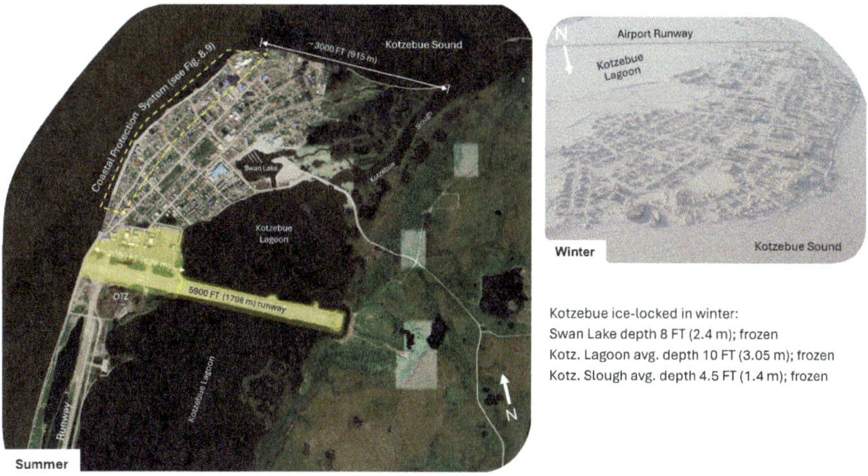

FIGURE 8.8
Kotzebue in summer and winter. Yellow highlight is the airport underlain with densely compacted soil. Winter image provides perspective of the town locked in ice. (Base satellite image at left courtesy of Google Maps.)

longshore currents, the spit is constantly changing due to erosion, accretion, and storms.

Initially known as Qikiqtaġruk in Inupiaq, the site was a central trading hub for native groups for hundreds of years. Then Russian explorer Otto von Kotzebue *"discovered"* Kotzebue Sound in 1818, and trade established a reindeer station and community there in 1897. Then came the post office two years later, which needed a name, so the community adopted Kotzebue from the sound. Then, in 1958, the U.S. Air Force established a base in Kotzebue (thus the airport) and a long-range radar (LRR) station. The base operated until 1983 before becoming a commercial airport; the LRR station remains an important NORAD[5] facility.

Kotzebue grew to become a regional service, shipping, transportation, and health center for Northwest Alaska. Today, with a population slightly over 3000, it is a major transportation and supply hub for the Northwest Arctic Borough. Known for its subsistence hunting and fishing, tourism, and government services, it has most recently taken center stage in the climate change debate.

Floods damaged Kotzebue about every eight to ten years. In the early 1980s, seaside erosion protection measures included galvanized wire gabion baskets filled with rock and gravel, a groin composed of gravel-filled 55-gallon steel barrels, and various types of plastic or fabric bags filled with sand and gravel along the beachfront. However, these installations lasted only a few years before being destroyed by storms and surges.

In 1984, the Alaska Department of Transportation and Public Facilities (AKDOT&PF) conducted an erosion control study to address beach erosion along a mile of Shore Avenue between the airport and Crowley Marine Dock at the north end of the spit. Storm surges, storm-driven waves, and ice action were eroding the beach at a rate of one to three feet (0.3 to 1 m) per year[6]. Infrastructure threatened by erosion included beachfront residences and businesses along Shore Avenue, outbuildings, sheds, water tanks and fuel tanks, utility poles and underground utilities, a boat launch, boat storage, and repair structures, schools, clinics, churches, other public buildings, airport facilities and a runway, and sites of significant cultural and archeological value.

The worst flood event on record was during a 1990 storm when abnormally high waters flooded the airport, lift stations, sewer mains, and a few residences. Subsequent storms in fall 2002 and 2004 produced damaging storm surges and significant beach erosion; the 2002 storm caused major damage to Shore Avenue. In 2009, an AKDOT&PF project repaired and widened Shore Avenue, and installed a riprap revetment and a 4400-foot (1340 m)-long sheet pile wall for erosion protection (Fig. 8.9). An existing sheet pile wall and ACB mat[6] at Crowley Marine Dock was working well to control erosion up at the north end of the spit[6].

What could go wrong with this modern coastal erosion protection system? Read Case Study 8 to find out.

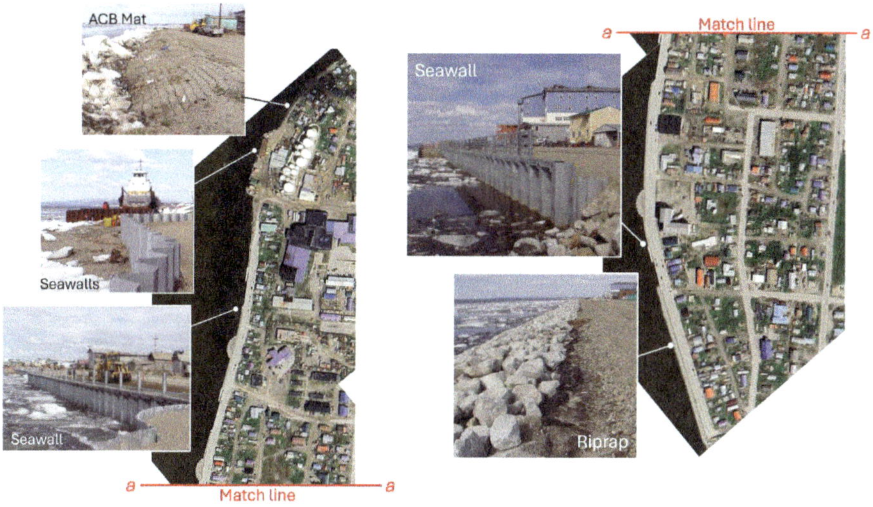

FIGURE 8.9
Erosion control features of the seaside coastal protection system. Features include articulated concrete block (ACB) mats, seawalls (old docking section and new 4400 FT (1340 m) beach wall), and riprap. (Satellite images courtesy of Google Maps.)

8.3 Hydrologic Study

Designers of hydrologic control structures know there are always consequences with whatever is built to control water. For example, with the weirs shown in Figure 8.1, seepage and uplift pressure under the structure are consequences that could undermine the structure itself if not carefully evaluated as a design consideration.

The Oroville Dam spillway blowout in February 2017 was a near-catastrophic event as operators were trying to keep the reservoir from overflowing the dam during unusually high rainfall that winter. They first used the spillway, then the emergency overflow spillway, once the primary chute failed. A forensic investigation determined that cracks in the chute allowed water to seep under and cause tremendous hydrostatic uplift forces on the concrete spillway. A hole opened, and the deluge eroded an estimated 1.7 million cy (1.3 million m^3) of bedrock. Review of the geological records revealed that much of the spillway was built on moderately to severely weathered bedrock [7]. An emergency overflow spillway saved the day, but not before also causing severe bedrock erosion. The downstream city of Oroville, California, was spared a disastrous flood. Someone or persons made the decision to build the dam on weak bedrock. A *hydrologic study of bedrock scour* would have been a prudent design consideration.

The 1967 Fairbanks flood, one of the worst disasters in Alaskan history, led to the Chena River Lakes Flood Control Project in North Pole, Alaska. The project includes a dam and its impoundment. Since completion in 1979, the project has caused impacts on groundwater:

- Elevated groundwater levels downstream.
- Altered water quality in drinking water wells.
- Potential contamination from sulfolane and PFAS (per- and polyfluoroalkyl substances, or *forever chemicals*) from a former oil refinery.

Additionally, the contamination requires careful management of excavation dewatering in the area. A hydrologic impacts study would have been a prudent design consideration.

Seawalls like the one installed at Kotzebue to fight against coastal erosion disrupt the natural ***submarine groundwater discharge*** (SGD) of fresh groundwater from the spit to the ocean. Water bodies surrounding the spit freeze up in winter, and a wall sunk deep into the seabed along the ocean side is now a barrier to SGD. Should the city water supply spring a leak, where might that water go? (Another teaser for Case study 8.) Further, SGD can impact the salinity, temperature, and overall health of coastal habitats. A *hydrologic and ecosystem impacts study* would have been a good idea when the coastal erosion plan was designed.

The point is that ***hydrologic study*** should always be integral to the design of hydrologic control structures. To ignore this important consideration is to risk legal actions and have the forensic engineer pay a visit later.

Notes

1. In the context of fluids, *pressure head* refers to the height of a body of liquid that would generate a specific pressure. It is a way to express pressure as a vertical distance or *"head."* For example, 45 feet (13.7 m) of water against a levee will cause a pressure at the base of the levee equivalent to a column of water 45 feet tall, or 19.5 psi (134 kPa) gauge pressure.
2. *Bed degradation* refers to erosion or downcutting of river or stream channels. Floods can magnify channel erosion and undermine protective barriers.
3. A gangway is a narrow passage with handrails used to board or disembark boats.
4. The North American Vertical Datum of 1988 (or NAVD 1988) is the vertical datum for orthometric heights established for vertical control surveying in the United States in 1991.

5. NORAD (North American Air Defense Command) is a defense network of radar installation sites established across Alaska and the Canadian Arctic during the *Cold War* era. Now known as North American Aerospace Defense Command, NORAD remains a vital defense system for North America.

6 An *articulated concrete block* (ACB) system comprises concrete blocks spaced out from each other and interlocked with a high-strength geogrid across the beach slope. They can endure storm surge and work best where wave action is light.

References

1. Bruce D. Malamud, Donald L. Turcotte and Christopher C. Barton (1996). The 1993 Mississippi river flood: A one hundred or a one thousand year event? *Envir. Eng. Geosci.* 2(4):479. doi:10.2113/gseegeosci. II.4.479.
2. United States General Accounting Office (1995). Midwest Flood: Information on the Performance, Effects, and Control of Levees, GAO-RCED-95-125, pp. 85.
3. USACE (2017). Missouri Riverbed Degradation Feasibility Study: Technical Report. Kansas City District. Retrieved from www.nwk.usace.army.mil/projects/MoRiverDegradation/Documents/MORiverBedDegradationStudy/TechnicalReport.pdf
4. USACE (2009). Missouri Riverbed Degradation Reconnaissance Study, Section 905(b) Analysis. Kansas City District. Retrieved from www.nwk.usace.army.mil/projects/MoRiverDegradation/Documents/MORiverBedDegradationReconStudy.pdf
5. United State Geological Service (2022). Hydrologic Data, Reston, VA, http://waterdata.usgs.gov/ monitoring-location/, Accessed 12 November 2022.
6. USACE (2008). Alaska Baseline Erosion Assessment: Erosion Information Paper – Kotzebue, Alaska. Alaska District. Retrieved from https://www.poa.usace.army.mil/Portals/34/docs/ civilworks/BEA/Kotzebue_Final%20Report.pdf
7. Independent Forensic Team (2018). Report of Oroville Dam Spillway Incident. Retrieved from https://damsafety.org/sites/default/files/files/Independent%20Forensic%20Team%20Report%20Final%2001-05-18.pdf

Case Study 8: Nowhere To Go But Up

- Winter Flood
- Groundwater Mounding
- Property Damage
- Insurance Claim

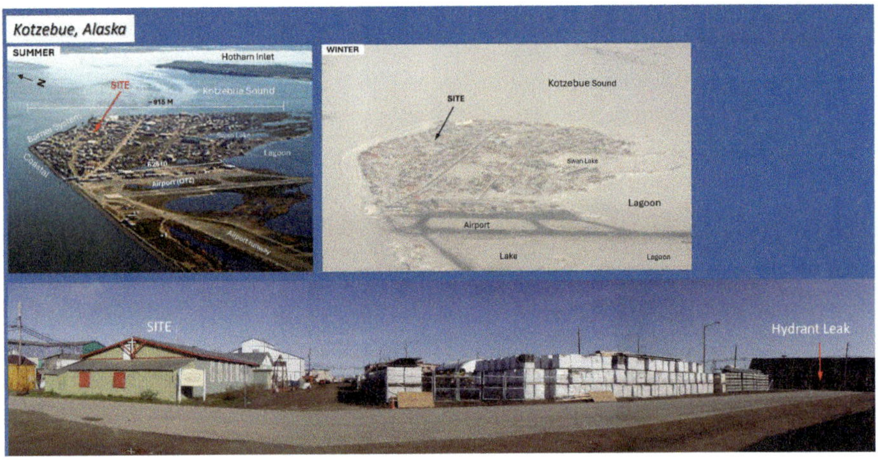

Message to students:

> When investigating events involving hydrologic control structures, ask to see the hydrologic study.

Incident

In early March 2011, the basement of a church in Kotzebue, Alaska, flooded for nine weeks. Early speculations led to several possible causes: a late February storm surge, influence from a new coastal erosion protection system (CEPS), church sewer-line freezeup, or a leak in the city's water distribution system. Meetings with municipal and state transportation department (the CEPS management agency) officials garnered support from the city

 DOI: 10.1201/9781003623311-18

utility provider, which investigated the underground utilities servicing the church. The utility provider excavated and inspected the church water and sewer lines, tested flood water from the basement, and concluded that neither city water nor a frozen sewer line was related to the basement flooding. Consequently, the church filed a claim with their insurance carrier and hired a forensic expert to determine the cause of the winter flood. Employing consilience[1] with a thorough investigation, a unique combination of factors was determined to have caused the basement flooding.

Hydrogeologic Setting and Coastal Protection

Kotzebue lies at the northern tip of a narrow spit between Kotzebue Sound and a tidal lagoon, on the Baldwin Peninsula. Kotzebue Sound is very shallow around the spit; the maximum depth of the tidal lagoon is about 3.1 m, and June Creek – a freshwater stream that originates from Devil's Lake – seasonally empties into the lagoon at its southern end beyond the airport (Fig. CS8.1(b)).

The Baldwin Peninsula is dotted with tundra lakes and ponds amid grassy and mossy tundra lowlands, a terrain known as polygonal patterned ground[2]. Lowlands east of the peninsula are drained by the Kobuk, Selawik, and Noatuk Rivers, which contribute large amounts of freshwater to Hotham Inlet and Kotzebue Sound[1].

A 1929 aerial photograph of the undeveloped spit shows shallow-relief topography of striated beach ridges at a time when only a few buildings existed along the shoreline dune ridge (Fig. CS8.1(a)). Natural surficial soils are generally described as dark gray and mottled silty loams and silty alluvium of marine and glacial origin[2,3]. Although groundwater is available above and below permafrost in Kotzebue, the local aquifers have not been developed for drinking water due to inadequate supply and saltwater influence.

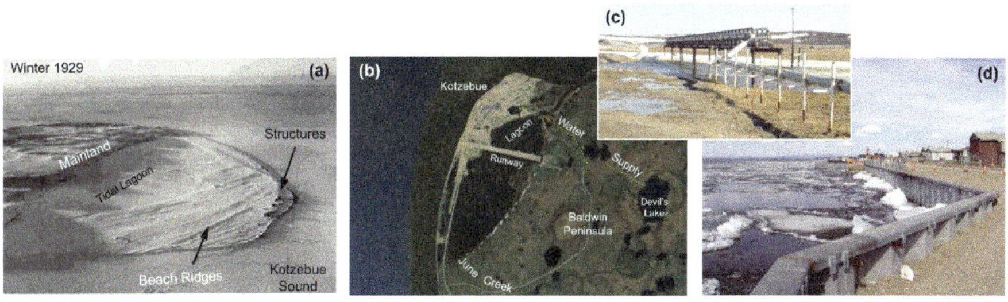

FIGURE CS8.1
(a) 1929 aerial photograph of early Kotzebue; (b) satellite image of Kotzebue[6] annotated with city water supply and June Creek features; (c) Kotzebue fresh water supply pipeline from Devil's Lake; and (d) new seawall. (Base photographs (a) and (b) courtesy Alaska Historical Society and Google Maps, respectively.)

The mean minimum, annual, and maximum air temperatures for Kotzebue are –25°C, –6°C, and 15°C, respectively, and mean annual snowfall and precipitation are about 1.1 m and 0.23 m. Ground thawing occurs between June and mid-September each year, and the active layer is about 1 m[4]. Kotzebue Sound is ice-free from early July until early October annually; from November to May the spit is locked in ice.

A documented history of storm surges, storm-driven waves, and ice action has caused beach erosion and localized flooding in the Kotzebue area since 1963. According to a 2008 US Army Corps of Engineers (USACE) *Erosion Information Report*[5], approximately 1.4 km of the western bank of Kotzebue was eroding at a rate of about 0.3 to 1 m per year. A 1990 storm caused the worst erosion and flooding on record when waters rose to 2.2 m above normal high water and flooded the airport, lift stations and sewer mains, and several residences. Fall storm surges in 2002 and 2004 caused significant damages that took two years to repair after each event.

Consequently, the Alaska Dept. of Transportation and Public Facilities earmarked funds for the CEPS. A geotechnical report for the project characterized soils along Shore Avenue (a major street along the shoreline and near the site) as a meter of sandy-gravel and gravel-sand fill above natural beach sand/loam deposits. (Sandy-gravel *pit run* from the area of what is now the Ralph Wien Memorial Airport was used to construct foundation pads and roadbeds in Kotzebue as it developed.) Seasonal frost depths to 2.1 m were encountered within beach borings, indicating the beach-side active layer averages about 2 m. Free groundwater was encountered above permafrost at between 1.6 and 2.1 m in the eight borings. The system of revetments and a barrier wall was installed during the summers of 2009-11. The steel seawall (Fig. CS8.1(d)) spans 1.04 km and is embedded 4.6 to 10.7 m deep. The project design did not include a hydrologic impact study.

Investigation

Numerous people were interviewed during the investigation to obtain first-hand accounts of the church flooding event and previous storm surge and flood events, Kotzebue historical imagery and soil and groundwater information, Kotzebue water utility plans and monitoring data and fire hydrant leak details, and design plans and reports for the shoreline CEPS.

The church was originally constructed in 1929 as a single-story wood-framed building founded on timbers. A two-story addition with concrete basement and wood-frame upper story was added in 1965. The addition slab foundation is approximately 1.2 m below the ground surface. Design plans for the addition did not show a well within the basement. A well located under the stairwell down to the basement was likely an afterthought, installed after the addition was constructed. A hole had been cut in the foundation to accommodate the 1.7 m deep rectangular well. The well was framed over in 1976, when the church was connected to city water and sewer utilities and was forgotten until the March 2011 basement flooding.

A timeline of the flood event is summarized:

- February 15 to 26: Kotzebue experienced a series of winter storms that brought blizzard conditions, heavy snowfall, and a storm surge on February 25. Property owner accounts indicated that the CEPS protected most of Shore Avenue from flooding, and that basements at four commercial properties west and nearby the site had not flooded.
- February 27: The sewer line from the church froze up. This line runs underground from the church laundry room diagonally out to Betty Street, and then up Betty Street for about 80 m to a sewer main behind the Post Office and Eskimo Building (Fig. CS8.2).
- March 6: The church basement flooded. A contractor performed flood mitigation, first having pumped flood waters from the laundry sump to the outside grounds, and then performed plumbing checks. Flood waters were not deemed septic.
- March 8: Water was observed leaking into the basement at the foundation joint with the wood-frame enclosed forgotten well. The wood enclosure was cut open and the sump pump was relocated into the timber crib well for pumping. Based on the pumping rate, the water leak rate into the basement was estimated at 10.9 m^3/day, equivalent to 2 gal/min (gpm). The church subsequently notified the city and asked that they check their water distribution system for potential problems.
- March 11: A local contractor thawed approximately 58 m of the sewer line before it became functional again. An internal camera inspection of the line revealed several breaks.
- March 23 to 25: The city utility provider collected and tested basement floodwater samples and performed leak detection of a nearby main and the water service line to the church. Leak detection was inconclusive. The utility provider subsequently excavated and tested the water line to the church for leaks; none were found. The church basement flooding continued unabated during this work. A memorandum sent to the church stated that the tested water samples contained "...no chlorines, fluoride levels were average, and salinity was elevated." Upon requests for copies of the test results, the utility provider said they did not retain the data.
- March 30: With permission, the utility provider monitored rising water in the church basement for 7 hours after the sump pump was shut off.

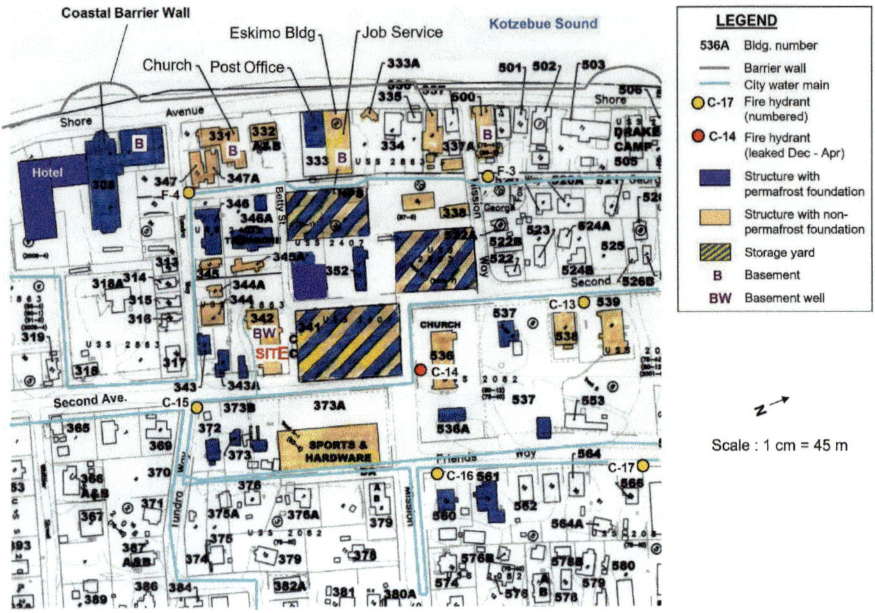

FIGURE CS8.2
Informal infrastructure survey for permafrost assessment.

- April 20 and 23: The city utility provider received a report of standing water in the snow around fire hydrant C-14, located at the intersection of Second Avenue and Mission Street (see Fig. CS8.2), approximately 82 m north of the church site. The hydrant was excavated and a leaking butt weld was repaired on April 23.
- April 1 to May 20: Basement flood monitored. At its peak, the flow rate of water from the well was 144 m^3/day (26.4 gpm) on April 20. The following church basement observations were recorded after the fire hydrant leak was repaired:

7 days after fire hydrant repair (April 30)	Water level in well and floor drains still high. Pumps not running, but on the verge (meaning the water level is just below the pumps' float switch).
10 days after hydrant repair (May 3)	Water level in well and floor drains still high. Pumps not running, but on the verge. Dropped pump down 6 inches (15.2 cm) in well and it ran continuously.
14 days after hydrant repair (May 7)	Water level in well down. Dropped pump down 4 more inches in well and it ran continuously.
18 days after hydrant repair (May 11)	Pump still operating. Raised pump 4 inches (10.1 cm) to shut it off.
20 days after hydrant repair (May 13)	Water level in well stabilized. No further pumping required. Pump set in well to kick on with 1-inch (2.5 cm) rise in water level.

- June 8: Church basement flood event end. The measured water level in the well indicated groundwater had receded to a couple of inches below the bottom of the foundation.
- June 8: A meeting was held with City of Kotzebue officials. A city water utility official stated that he became aware of the fire hydrant leak on April 20, and that utility monitoring records indicated water leaked at between 136 to 147 m^3/day (25 to 27 gpm) since December 31, 2010. Conservative calculations indicated that at least 9085 m^3 (2.4 million gal) of water leaked by March 6, when church flooding began. By April 23, when the leak was repaired, nearly 15,520 m^3 (4.1 million gal) of water had leaked. Maximum leak estimates were reported at 12,113 m^3 (3.2 million gal) by March 6 and 20,820 m^3 (5.5 million gal) by April 23.

Permafrost Survey

Kotzebue is underlain with permafrost, but the permafrost table is considered spatially variable depending on soil and ground cover. Further, frozen soil can have profound influence on groundwater hydrology. For example, the natural sand/loam beneath a meter of gravel-sand fill of a storage yard largely covered with stockpiled materials (e.g., cover figure, bottom image) would likely remain frozen year-round. Frozen sandy loam could act as a soil barrier (just like a basement) to groundwater flow. In contrast, unfrozen soil zones and underground utility trenches constructed with non-frost-susceptible soils would be amenable to groundwater flow in winter. In other words, a large influx of water above permafrost would seek the least paths of resistance for lateral spread.

Infrastructure in the vicinity of the church site was surveyed for land use and foundation type as an informal means to infer permafrost. For example, structures that are raised above the ground on piles or post-and-pads, or those with ground-level foundations underlain with thermosyphons[3] (Fig. CS8.3) attest to permafrost properties. Undeveloped properties – storage yards, for example – were assessed for surface coverage to infer seasonally frozen ground or permafrost. (Permafrost is easily created by placing a car or pallets of material on the ground and not moving them for two or more years. The car/pallets shield (or insulate) the ground from solar radiation, thereby allowing the ground to freeze and not undergo seasonal thaw.) Other parcels developed with at-grade buildings with basements and without thermosyphons were deemed non-permafrost properties. The pattern of land use around the site – structures with permafrost and non-permafrost foundations and buildings with basements – is shown in Figure CS8.2.

FIGURE CS8.3
Examples of permafrost foundations in Kotzebue: (a) steel pile frame, (b) post-and-pad, (c) triodetic foundation, and (d) and (e) at-grade building foundations augmented with thermosyphons.

City water mains and fire hydrants, and the CEPS seawall are superimposed on the figure. Examples of structures with permafrost foundations are shown in Figure CS8.3.

Groundwater Mounding Hypothesis

The general soil profile for the vicinity of the church was a surface cover of short vegetation and organics (tundra mat) over cobbles and rock fill to 0.6 m (2 ft), over gravel sands to about 3.7 m (12.1 ft), below which is stratified silt and sandy-silt-loam permafrost. This profile varied with up to 1.1 m (3.6 ft) of fill for undeveloped storage yards and parking lots above natural sandy loam. Undeveloped lots typically experience seasonally frozen ground through long winters and can exhibit intermittent or continuous permafrost at depths less than 2 m (6.6 ft). USS tract 2407, the storage yard between the site and hydrant leak down the road (i.e., cover figure, bottom image), was used to stockpile construction materials for the CEPS project since summer 2009.

With the onset of winter, ground freezes from the surface downward and water bodies freeze deep. The Kotzebue spit eventually became locked in ice (see cover figure, winter image). Typically, sea ice is 0.6 to 0.9 m (2 to 3 ft) thick, the ice in the tidal lagoon as much as 2.4 m (7.9 ft) thick (the shallow tidal lagoon has a maximum depth of 3.1 m (10.3 ft)), and ponds and lakes around town completely freeze up. Largely confined by ice and seasonally frozen ground, the groundwater in Kotzebue becomes virtually static in winter.

The water utility provider excavated the church's water service lines for inspection on March 24. The bottom of the excavation was measured at 142 cm (4.7 ft) below grade and the water service lines were buried at 117 cm (3.8 ft) deep at that location. With the excavation open for some time, groundwater equilibrated and began to freeze at a level approximately 10 cm (4 in) above the lines, or at 107 cm (3.5 ft) below grade. During this time, water was seeping into the basement from the joint around the well casement, at about 112 cm (3.7 ft) below grade. In other words, the groundwater level was about 5 cm (2 in) above the foundation of the church on March 24. Further, water was not yet flowing from the top of the well, which is 21.6 cm (8.1 in) above the leaking joint at the basement floor. This implied the flood had not yet peaked and groundwater was rising.

The patchwork of permafrost ground around the site would act to concentrate and channel groundwater. Also, the CEPS barrier wall, which disrupted the *submarine groundwater discharge* of fresh groundwater from the spit to the sea, would impede groundwater movement further west. Likewise, the frozen lagoon to the east and other local frozen water bodies would preclude groundwater movement to the east. Also, dense engineered soil founding the airport would impede groundwater movement to the south. Under these conditions, *groundwater mounding* can occur with a large influx of water. Groundwater mounding implies elevation of the water table in response to obstructions/impediments to lateral flow. Mounding can be local or extensive, depending on the magnitude of water influx and subsurface conditions.

Reasonable values for moist gravel-sand density, specific gravity, and porosity are 2000 kg/m^3 (125 pcf), 2.6, and 0.25 (or 25%). For a conservative estimate of groundwater mounding, assuming 100% of soil porosity as air voids available for water to fill, then the groundwater table could rise 0.3 m (1 ft) over 50 acres with a nearby infusion of 15,140 m^3 (4 million gal) of water. With significant permafrost barriers, the same volume of water could raise the groundwater table even higher within this area. Alternatively, without permafrost barriers to lateral spread and lower soil porosity, the same volume of water could mound groundwater by 0.3 m (1 ft) or more over a greater area. For comparison, consider that the groundwater table was 5 cm (2 in) above the church foundation on March 24. It would take less than 2 million gal of water leaked from the nearby fire hydrant to mound groundwater the additional 16.6 cm (6.1 in) to flow out of the top of the well.

Leak rates provided by the city utility provider were consistent with those measured from the well in the church basement during the April peak flood period and hydrant leak monitoring and repair. Further, the groundwater table was within an inch (2.5 cm) of the church foundation on June 9, 47 days after the hydrant leak was repaired. This attests to how long it took the groundwater table to subside, at a time breakup was contributing meltwater to the subsurface.

Determinations and Recommendations

The engineer's report was distributed to the parties for review and consideration. It was the consistency of the basement flood monitoring data with the utility provider's leak records that led to agreement about the cause of the church basement flooding. All agreed that the leaking fire hydrant was the source of flood water and groundwater mounding was the most likely flood scenario. The church insurer and city worked out a settlement agreement for damages without the need for lawyers.

One wonders if the flood would have occurred at all had the CEPS barrier wall not been installed ~100 m (~328 ft) west of the site. Excess groundwater quite possibly would have naturally discharged to the sea instead of mounding. Regardless, the Kotzebue community recognized the benefits of the CEPS and the involved parties did not involve the project owner in the settlement agreement. A recommendation was made to the CEPS project owner that a hydrologic study be integral to the design of all future CEPS.

Notes

1. Consilience in science is a principle that relies on independent lines of converging evidence to lead to strong conclusions.
2. Polygon ground, aka tundra polygons, is one of the most widespread geomorphic features associated with permafrost. Characterized as a honeycomb network of shallow troughs delineating polygons 3 to 30 m in diameter, and with many tundra ponds and lakes, the terrain generally presents with a shallow active layer underlain by thick permafrost above deep bedrock. (The active layer is the uppermost ground that undergoes annual freezing and thawing.)
3. A thermosyphon is a passively driven heat exchange device that is typically installed in the ground adjacent to a structure on permafrost. Utilizing a cooling fluid inside the cylindrical device, and the motive forces of natural convection and conduction, heat is liberated from the ground and emitted to the atmosphere above via a finned radiator that is fixed to the device above the ground. An array of the devices around the perimeter of a building liberates heat from ground beneath the structure and preserves permafrost.

References

1. Clyde Wahrhaftig (1965). *Physiographic Divisions of Alaska.* U.S. Geological Survey Professional Paper 482, 52 p.
2. Samuel Rieger, Dale B. Schoephorster and Clarence E. Furbush (1979). Exploratory Soil Survey of Alaska: U.S. Soil Conservation Service Report, 213 p.
3. David M. Hopkins (1977). Shoreline History as an Aid to Predicting Offshore Permafrost Conditions: Environmental Assessment of the Alaskan Continental Shelf, vol. 4. Principal Investigators' Report Oct.-Dec. 1976, Boulder, Colorado, Environmental Research Laboratories, pp. 459–461.
4. National Weather Service Data, National Centers for Environmental Information, National Oceanic and Atmospheric Administration. Retrieved June 7, 2011, from https://www.ncei.noaa.gov/
5. U.S. Army Corps of Engineers, January 24, 2008, Report: *Alaska Baseline Erosion Assessment: Erosion Information Paper, Kotzebue, Alaska.* Retrieved August 21, 2022, from www.alaskaerosion.com.
6. Google. (n.d.). [Kotzebue, Alaska]. Retrieved August 21, 2022, from https://www.google.com

9

Damage Assessment

CONTENTS

CONCEPTS	KEY TERMS	ADDITIONAL READING
Hurricane Damage	Storm surge	Exercise 17
Bridge Inspection	Soil piping erosion	Exercise 18
Hail Damage	Threshold hail	
Building Condition Assessment	Quick reference guide	Case Study 9
Vibration Damage	Vibration	
	Frequency/natural frequency	
	Resonance/resonance transmissibility	
	Dampening	

In this chapter, we focus on the process of going about assessing damage to infrastructure due to a variety of causes. We first discussed hurricane damage with the lazy river example in Chapter 8. Let us now apply Chapter 2 *first principles* to assessing hurricanes and a variety of other types of impacts. Hurricane damage can be caused by wind, flood, and/or [storm] surge. The idea is to distinguish what caused damage to aid decision-making with regard to repairs, replacements, and deferred maintenance. This asset management technique is then extended to hail, building condition assessment (BCA), and vibration damage situations.

9.1 Hurricane Damage

Infrastructure inspections are routine after a hurricane. Emergency responders immediately assess and repair power and communications grids, and state, federal, and contract specialists – usually engineers – inspect and recommend repairs to impacted transportation and other vital

DOI: 10.1201/9781003623311-19

systems. Further, insurance companies hire forensic engineers to assess structural damage against policy coverage. Unfortunately, some insureds use storm events to submit claims for preexisting damages, and the COVID pandemic produced new armies of *Senior Adjusters* with little to no experience and hasty training to make insurance decisions about cause and policy coverages.

Let us apply *first principles* to the inspection of hurricane damage associated with a bridge (Fig. 9.1). The same numbered photographs in the figure are related. The figure caption indicates "***storm surge***" damage. To some

FIGURE 9.1
Storm surge damage.

viewers, storm surge is evident in the images, while others might consider that wind causes storm surge. Shouldn't wind be the underlying (hint) cause? Let's find out.

Storm Surge

Figure 9.2 is a five-step process an engineer might apply to investigate and report storm damage at the bridge approaches. With Step 1, the investigator first walks and observes the approaches in their entirety, taking notes and photographs as necessary for subsequent use. The investigator then questions the underlying cause(s) for the damage observed, looking for patterns, similarities, and differences. Steps 1 and 2 often occur simultaneously with the note-taking. A hypothesis of cause – Step 3 – naturally comes with the processing of Steps 1 and 2 information. For the scenario imaged in Figure 9.1, erosion of shoreline protection barriers, walkways, and utility components, and undermined boardwalk aprons are the telltale signs of surge, storm-driven water that overwhelmed and undercut soil beneath and beyond.

Next comes Step 4, the analysis that supports the hypothesis. For example, photographs numbered 1 in Figure 9.1 relate to the damaged area of the entrance to the boardwalk. Storm surge overwhelmed the boardwalk and eroded away approximately 52 inches (132 cm) of soil thickness (lowest #1 photograph), which led to the collapse of the sidewalk and aprons shown in the middle #1 photograph. Wind does influence storm surge, but it was the water that undercut the boardwalk infrastructure to cause the damage seen in the images, a Step 5 conclusion in the report.

Bridge Inspection

Now consider inspection of the bridge structure. Post-hurricane inspections of transportation infrastructure are a routine practice that involves engineers trained to document impacts. Inspection typically involves field documentation using standardized forms with provision to comment and photograph.

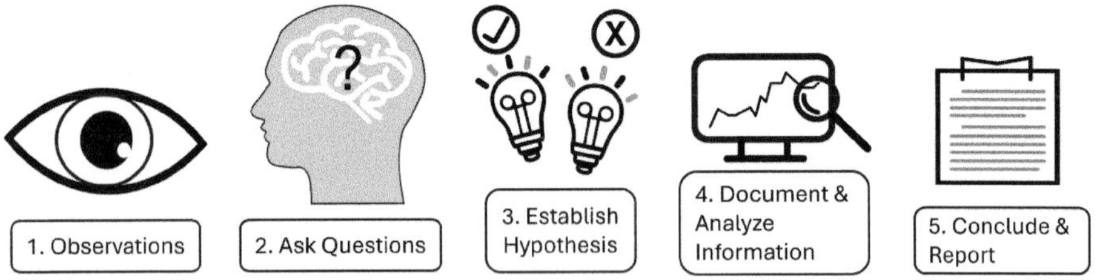

FIGURE 9.2
Investigation process for storm surge.

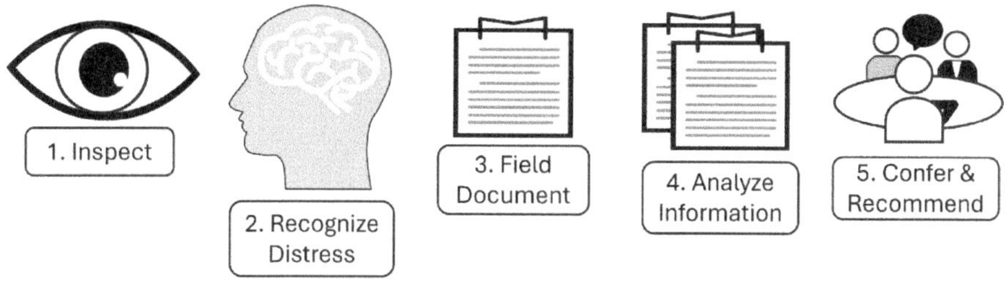

FIGURE 9.3
Inspection process of storm impacts to the bridge structure.

The information collected can be compared with that of previous inspections, and significant changes discussed with others for planning and maintenance scheduling (Fig. 9.3). Cause and effect is not part of the training in this scenario.

Damage to a bridge joint was noteworthy after the storm. Focus of Figure 9.4 is the structural components at an abutment – cornice, panels, and deck parapet. A bridge cornice is a functional and aesthetic structural component that overhangs the sides of the bridge. Designed to help protect the bridge structure from damage by rainwater and other elements and conceal irregularities, the cornice also enhances the bridge's appearance. In this instance, they are fitted to the side edges of the bridge deck at the expansion joint and under the deck parapet. The upper left corner image 1 in the figure is the subject approach and abutment focus for this discussion. More specifically, the rest of the images pertain to distresses at the south side cornice area.

Imagine you are the inspector now at Step 3 in the process (Fig. 9.3) and working with the checklist and taking photographs, as necessary. You place a finger up against the underside of the storm-shifted cornice to show how much it moved (Fig. 9.4, upper right images) when you feel the steady stream of warm moist air coming from a gap under the bottom edge of it. Placing your palm over the hole, the flow quickly moistens it. With a flashlight, you peer through the gap into a sandy void and try to make sense of what you are experiencing.

Where the approach meets the bridge span is an expansion joint, designed to accommodate nominal lateral movement (i.e., bridge expansion), but not vertical displacement. Up top, you inspect the joint and notice the strip seal detached in places along one side, but no vertical displacement. You photograph and comment that the joint needs to be resealed to prevent water and road debris from getting in. You also document other distresses in the vicinity such as rail gaps, a few cracks, and signs of reinforcement corrosion in the parapet; concrete spalling; and misalignment of a couple of panels in the MSE (mechanically stabilized earth) wall around the cornice. The air flow from the gap beneath the south cornice escaped mention as you move to the north side of the approach to document.

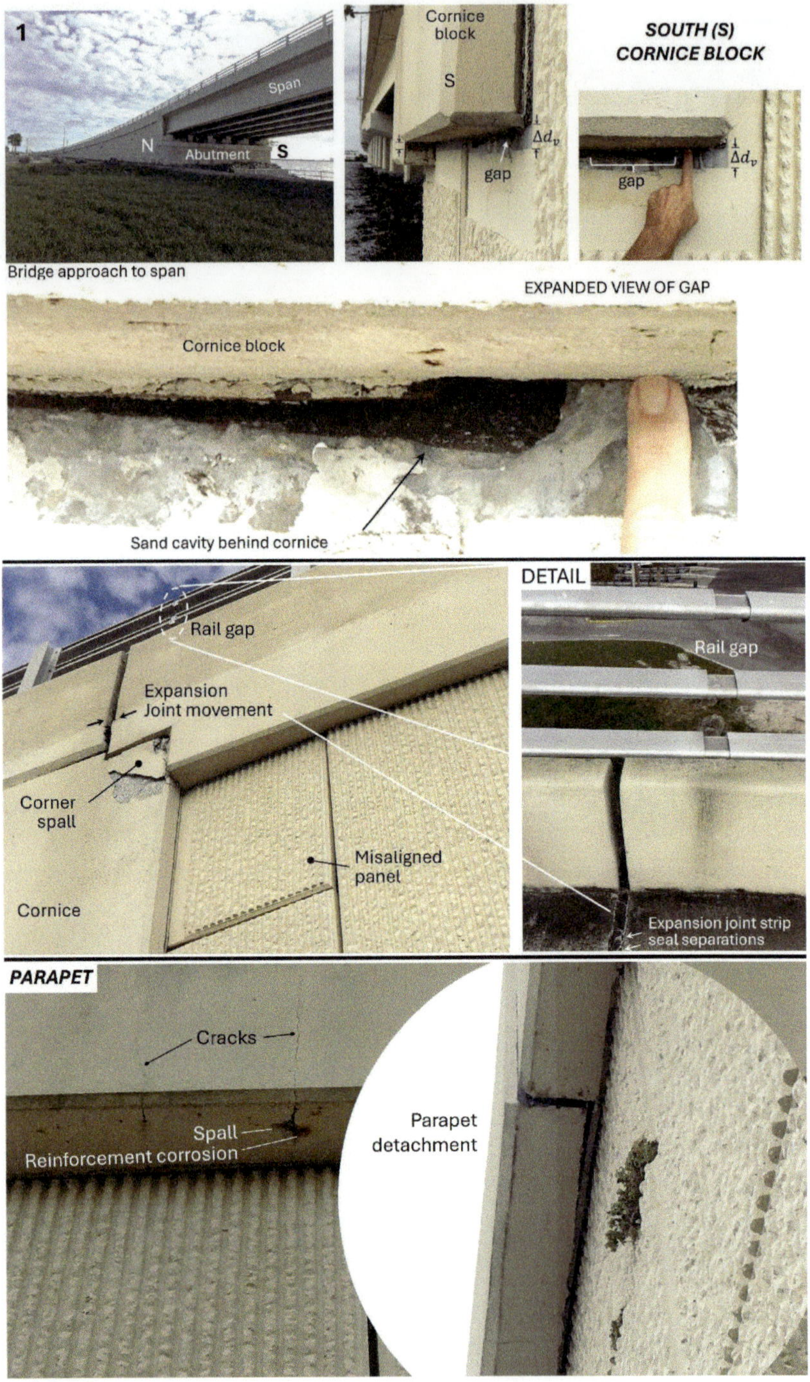

FIGURE 9.4
Distresses observed in the vicinity of the south cornice block and expansion joint.

Once back in the office, you review documentation of recent past inspections, compare notes and photographs (Step 4 in the process), and then schedule a meeting with colleagues to discuss hurricane inspection findings and recommendations for maintenance repairs (Step 5). Nothing is mentioned about the south cornice air gap, now a forgotten thought.

Moist air flowing from a wall, after a hurricane.
From a mechanically stabilized earth wall.
From an embankment of compacted soil.
That supports the road up to the bridge.

The afterthoughts that will not go away. Finally, you decide to discuss the phenomenon with your supervisor. Better late than never. The devil is always in the details, and nagging details, the least understood, need the most attention. Air moving through an MSE embankment means water got in and eroded the soil. Moist warm air means the soil is warm and giving up its moisture via convection. You learn that ***soil piping erosion*** is established in the embankment, the magnitude of which is presently unknown. In time, the road could settle. Your supervisor thanks you for bringing this to their attention and reminds you to please amend the inspection report accordingly. A course of action will soon be discussed at an upcoming meeting. The forensic engineer possesses uncanny attention to detail and understands cause and effect better than most.

Water eroded soil under the bridge.
Water eroded soil in the MSE embankment.
Water finds a way.

Roof Damage

Building roofs weather as they age; the elements degrade 24/7. Chapter 8 covers special inspections and the reserves and maintenance required by law for certain types of high-occupancy buildings. Why not schools and commercial buildings? They are also insured, have many occupants, and have roofs that weather just the same.

Owners of a charter school and owners of a commercial building filed claims for hurricane damage in 2022. The insurers hired a forensic engineer to investigate to determine if roof damages were storm-related or preexistent.

Charter School

A charter school experienced roof leaks during a hurricane. Water intrusion caused interior damage to classrooms, office space, and the parking garage, and the owners filed an insurance claim to have the roof system replaced. The building (and attached garage) had a modified bitumen (MB) low-profile roof with a front-to-back 2% slope and freefall drainage. The classroom/office portion of the building was surfaced with an MB torch down[1] small-aggregate membrane; the garage roof was MB impregnated with tightly packed large (to 3/4-in (1.9 cm)) gravels (Fig. 9.5). A metal-clad pyramid parapet system adorned the entire length of the front side of the joined roofs. Matching decorative metal-panel siding adorned the roof perimeters. The roof was supported with wood-frame truss systems. Focus of the following discussion is the roof over the garage.

The forensic engineer inspected the roofs six days after the hurricane. Observations suggested an aging roof with a couple of depressions and a waterfall emanating from one at the back edge (Fig. 9.5A). Standing water filled the space between gravels in the depressions, and runoff continued to drip from the soffit beneath the former waterfall (Fig. 9.5B and C). Soffit sag and black algae were observed at the drip (Fig. 9.5D), and the roof depression above was spongy to the touch, suggestive of underlying rot.

The engineer interviewed the school maintenance supervisor and asked him to expose the roof cavity above the garage for inspection. Wood rot was confirmed in the deck and truss members, and the drip was water coming through the roof depression. The maintenance supervisor's description of the waterfall phenomenon along the back edge of the garage roof was report-worthy: "*With heavy rain, water flows over the edge like a waterfall, and it takes about a week to drain out through the soffit.*" A report was submitted that

FIGURE 9.5
Preexistent roof rot symptoms: (A) runoff overflow; (B) roof depression above overflow; (C) drip line and algae beneath overflow; and (D) warped siding.

concluded preexisting roof damage, and that the hurricane was merely the most recent storm to expose it. The insurer worked out a deal with the school owners to prorate the cost of roof replacement.

Commercial Building

Hurricane Ian came ashore in southwest Florida as a Category 4 storm with 150 mph sustained winds on September 28, 2022. It traveled NE across the state, weakened to a tropical storm as it entered Atlantic waters, and restrengthened to a Category 1 hurricane for a second punch. This punch brought storm surge, heavy rains, and extensive flooding to US southeastern and mid-Atlantic states.

An insurance company hired a forensic engineer to investigate damage claims to commercial structures after the storm. The owner of one commercial building was adamant that the hurricane caused leaks and water-related damage to the structure and loss of supplies.

The split-level concrete block structure with a steel-frame roof system was built in 2007. A computer store occupied the two-story office and warehouse portion at the front of the building, and four small businesses occupied the single-story suites behind. The two-story section had a white low-slope corrugated metal roof that drained from front (west) to back (east) to a gutter and downspouts. A low-profile corrugated metal roof that sloped from back (north) to front (south) and drained to a trench drain behind a parapet adorned the single-story portion of the building. Decorative parapets fronted the stores, and no equipment was mounted on the roofs.

The first thing noted by the forensic engineer was numerous sloppy patches in the roof above the suites (Fig. 9.6) and problems with roof installations.

FIGURE 9.6
Commercial building roof new (2007) and with patches (2022).

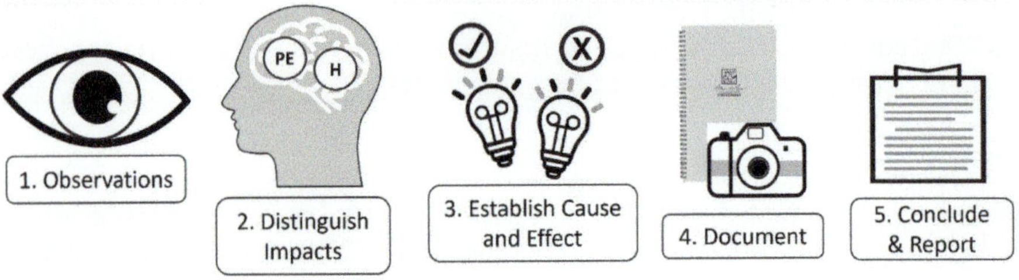

FIGURE 9.7
Investigation process for assessing storm damage. (PE is preexisting condition; H is hurricane damage.)

Important investigation aspects of this project were distinguishing preexisting defects from hurricane-caused damage and backing that up with cause and effect (Steps 2 and 3 in Fig. 9.7).

A corrugated metal roof should not be chinked[2]. Instead, the deck should be sealed with a waterproof liner, and the metal panels closed off with sealant (e.g., silicone) and tape to seal joints, seams, and fasteners to prevent water intrusion. Also, the sloped panels should sufficiently lap gutters to prevent open ends and wall seeps.

Figures 9.8 and 9.9 provide examples of the many installation problems with the roof system that existed before the hurricane. Water trails down the exterior walls emanated from leaky parapets that were improperly sealed and weathered. Water that seeped through concrete walls to the interiors penetrated the roofs through numerous unsealed gaps and where chinking (which should not have been) along roof ridges was dislodged during

FIGURE 9.8
Water trails and seeps into the building.

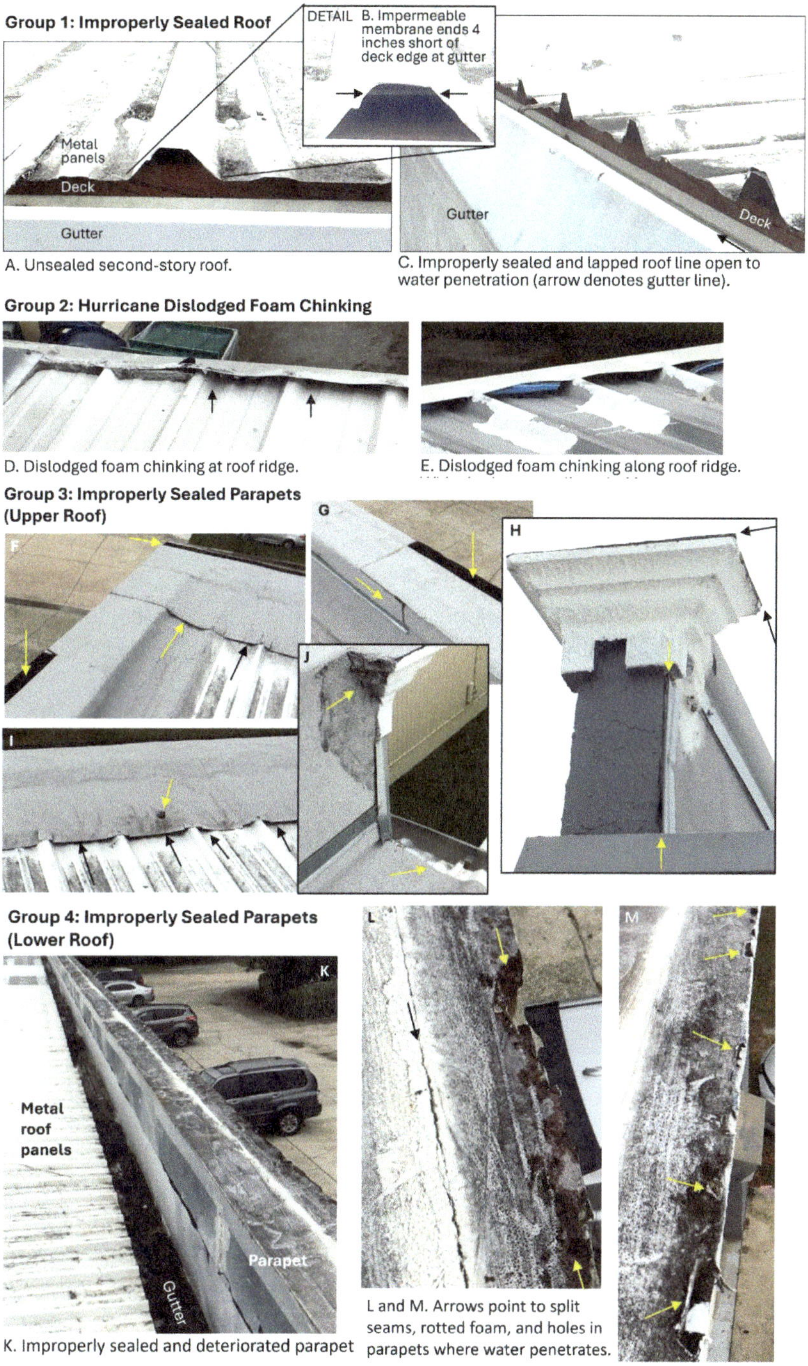

FIGURE 9.9
The many defects and weathering of poorly installed roofs.

the storm. The engineer's report opined that unprofessional roof installations were the primary cause of water intrusion damage. This opinion was backed up with representative photographs of cause-and-effect descriptions. Hurricane Ian was merely the storm that triggered roof replacement.

As the roof goes, so goes the building.

9.2 Hail Damage

Ever heard of ducks staring at an approaching thunderstorm? Geese do it too. Imagine a gaggle of geese facing the storm with their heads and bills pointed nearly vertically into the air, wings in *feather compression* mode. The phenomenon is called *sky staring,* and much like elephant *alert circles* before an earthquake, there is a purpose. The behavior is thought to protect wings and bills from hail. Waterfowl beaks are covered in sensitive nerve endings that aid eating and grooming. Bills that experience serious damage do not heal and cannot catch food. Further, a bird with a broken wing is easy prey for foxes and the like. To experience a damaged beak and broken wing is certain death. There are a few entertaining YouTube shorts of the sky-staring spectacle that are sure to bring a smile.

A June 1918 letter to a soldier in Europe during World War I tells of basketball-size hail that destroyed parts or all of buildings in a small Oklahoma town. No sensationalism or mention of "unprecedented," just a letter from mom with updates from home that every soldier at the front appreciates.

Officially, the National Weather Service honors Vivian, South Dakota with the largest diameter hailstone measured at 7.9 inches (20 cm) across, that fell during a storm on July 23, 2010[1]. Unofficial records tell of bowling-ball-sized hail in the early 20th century. A regulation bowling ball for ten-pin bowling is 8.5 to 8.595 inches (21.59 to 21.83 cm) in diameter and weighs between 10 and 16 pounds (4.54 to 7.26 kg). The size 4 youth basketball for 5 to 8-year-olds measures 8.2 inches (21 cm) across. A standard NBA basketball is between 9.43 and 9.51 inches (24.0–24.2 cm) across. Now imagine the damage a solid mass of ice the size of a basketball (or bowling ball) falling from the sky might do to a structure. Fortunately, most hail that falls during a storm is less than an inch (2.54 cm) across and does not cause serious damage[2], [3].

Recent highlights from ***threshold hail***[3] researchers are summarized:

1. According to the National Severe Storms Laboratory (NSSL) at the National Oceanic and Atmospheric Administration (NOAA), hailstones fall at varying speeds depending on their size, drag coefficient, wind conditions (updraft strength, wind speed, and direction),

and melt during fall. The NSSL provides the following general size-speed breakdown:[2]

- Small hail (less than 1 inch across): 9–25 mph (4–11.2 m/s)
- Medium hail (1–1.75 inches): 25-40 mph (11.2–17.9 m/s)
- Large hail (2–4 inches): 40–72 mph (17.9–32.2 m/s)
- Very large hail (greater than 4 inches): Up to 110 mph (49.2 m/s)

2. Shape matters. One group of researchers developed a catalog of hailstone shapes from a 6-year study of storms from Texas to North Dakota. Shedd et al.[4] found that hailstones become increasingly nonspherical the larger they are; hailstone maximum dimension tends to be 20% to 50% larger than the equivalent-volume spherical diameter; *sphericity* (how closely hailstones resemble a perfect sphere) typically ranged from 0.57 to 0.99, and corresponding drag coefficients ranged from 0.5 to 0.92. Specifically, larger hailstones and those with rougher surfaces tend to have higher drag coefficients[4].
3. Hail contact surface area is an important variable when evaluating impact forces. Hail transfers kinetic energy (KE) upon impact through a contact area defined by an indentation diameter and depression depth (Fig. 9.10). Therefore, normalizing the KE by dividing it by the contact area of the hailstone is the best way to correlate hailstone size with damage potential. What this means is that the smaller the contact area, the more likely damage will occur. Further, the literature agrees that the KE of hail increases exponentially with an increase in hailstone diameter. Because hailstone shape is highly variable and ice balls are not always spherical, it makes sense to estimate an *equivalent diameter* (D_e) for the contact area when estimating impact energy.
4. Wind speed significantly affects hail impact dynamics. Higher winds shorten impact duration, accelerate hailstones, and increase impact force, thereby enhancing the destructive potential of wind-driven hailstones by up to 18.9% according to one study[5]. Further, wind imparts a horizontal force on a hailstone in the direction of the wind. With sufficient time, the ice could achieve the same speed as the wind. Upon impact with a roof, its velocity has both a vertical and horizontal component and the *strike velocity* would be a vector addition of the two as a function

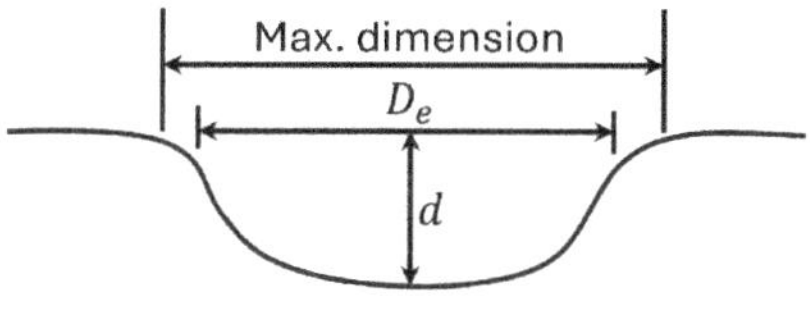

FIGURE 9.10
General cross section of hailstone indentation. Ideally, depression is defined by indentation depth and equivalent diameter.

of hail trajectory. Assuming freefall terminal velocity as the downward velocity vector and the horizontal velocity equal to that of the wind, the resulting hailstone impact energy would be maximum[6]. These wind-hail synergy studies by Dai et al.[5] and Abbott Consulting provide the investigator with useful information about hail impact energy relative to damage potential to support incident-specific assessments.

Assessing Hail Damage

What follows is information compiled from the updated information and experience as a hail ***quick reference guide*** (QRG) for damage assessment of roofs (Fig. 9.11). A fan of QRGs since my first professional job, a quality laminated and three-hole punched QRG is an excellent resource for field engineers.

Damage does not automatically mean replacement. Cosmetic damage is different from severe damage. If a 5-year-old hail-impacted roof remains functional, why tear it off if it looks ok from the ground and has 10 or 20 years of useful life remaining? Consider a practical hail damage scale when working with insurance companies:

Superficial damage – superficial flaws or defects (i.e., cosmetic) on a surface not visible from ground level and that do not impair material functionality or structural integrity.

Minor damage – minor and/or infrequent dents, scrapes, scratches, and burnish marks on a surface that minimally affect its appearance and that do not impair material functionality or structural integrity.

Severe damage – damage that significantly deforms or dislodges and impairs material functionality or structural integrity.

The Scam

Ambulance chasing (or barratry) is the unethical practice of lawyers pursuing clients recently involved in accidents or other traumatic events. Unfortunately, roofers have their version of this unscrupulous practice too. It works like this:

- A hailstorm is reported in the news.
- Soon after, a roofer performs a house-to-house canvass of a neighborhood, searching for the vulnerable.
- An unsuspecting senior tending the garden is the target.
- The roofer chalks dents in the downspouts and gutters for effect and tells a convincing tale; the senior takes the bait.
- In the garage, the roofer listens and takes notes as the senior provides personal identity information and files a claim with the insurance agent over the phone.
- A sign is posted in the yard that advertises the roofer.

Weather Checks

1. *Confirm Thunderstorm Activity*
   Check with the National Weather Service (NWS) or National Centers for Environmental Information (NCEI) storm event databases to determine if thunderstorms were in the area at the same time hail damage was reported. Detailed information about severe thunderstorms (e.g., location, time, wind speeds, hail occurrence and damage reports) that may be available in the storm data should only be used for anecdotal purposes.*
2. *Identify Thundercloud Type and Ceiling*
   Check with the Aviation Weather Center (aviationweather.gov) for METARs (aviation weather reports) that indicate the presence of cumulonimbus (or CB) clouds and ceilings above 13,000 ft (4 km), specifics necessary for hail.
3. *Collateral Damage*
   Check nearby vegetation, vehicles, buildings, and structures for visible signs of hail damage.

* NWS, NCEI or other storm database information should not be used in place of or to infer site-specific storm conditions or damage assessment.

Identify Roof Type and Assess Damage

1. *Asphalt Shingle* - Common mineral granule (sand and gravel) impregnated fiberglass mat shingle. Specify:
   - 3-tab shingle (basic)
   - Architectural (thicker)
   - Premium (custom, mimics slate, wood shakes, or other aesthetic look)
2. *Cedar Shake or Shingle* - Best used in drier climates. Shake appeal: rustic appearance, natural insulation properties, hand-split or taper-sawn aesthetic look. Shakes are thicker and more durable than cedar shingles, which are typically sawed smooth. Requires regular cleaning and treatment to prevent moss and algae growth. Can be fire-retardant treated.
   Specify:
   - Common grain (most affordable but shakes can curl)
   - Select (mix of straight and common grain shakes)
   - 100% straight grain (premium)
3. *Slate* - Naturally beautiful, durable, long-lasting metamorphic rock tile roof. A premium roofing option that is resistance to fire, water, and other weather elements.
4. *Clay Tile* - Shaped and fired clay or shale tiles. A premium long-lasting roofing option that is resistance to fire, water, and other weather elements. Common types: interlocking, non-interlocking, slate, scalloped, and barrel tiles.
   Specify:
   - Interlocking or non-interlocking
   - Slate
   - Clay tile type (flat, S-shaped, scalloped, barrel, etc.)
5. *Standing-seam Terne Metal* - Interlocking terne metal (steel sheet coated with lead-tin allow) panels with raised seams for weathertight and visually appealing roof. Known for corrosion resistance and ability to be soldered, making it suitable for complex roof designs and historic restorations.
6. *Built-up Roofs (BURs)* - Common commercial multi-layered, aggregate surfaced, flat or low-sloped roof system. Constructed with alternating layers of asphalt or coal tar (bitumen) and reinforcing fabrics like felt or fiberglass mats. Durable, long-lasting, weather, fire, and ponding water resistant.
6. *Rubber Flat Roof System (aka Single-ply Membrane)* - Popular commercial flat or low-sloped foam with membrane topcoat roofs. Flexible, durable, waterproof and cost-effectiveness if applied correctly. Common membrane types: EPDM (synthetic), TPO (polymer blend), PVC (plastic reinforced with polyester or fiberglass).
   Specify: EPDM, TOP, PVC or other type membrane.

Assess (Field Log)

A rite-in-rain all weather notebook is an essential tool for field work.

- Roof type
- Areas missing granules/aggregate
- Estimate % coverage missing granules/aggregate
- Material damage:
  Impact craters/indentions, burnish marks*, fissures (cracks or breaks) emanating from craters/indentions, punctures, and soft spots.
  Patterns (trajectories and spacing)
  Measurements (representative crater/indention depths and diameters)
  Slate, clay tile breaks (from larger hail)
  Photo-document
- Material weather and deformation (non-hail related):
  Shingle curl and cupping (caused by drying and UV exposure)
  Shingle buckle (caused by too tight spacing and thermal expansion)
  Shingle fungus and rot
  Gross loss of aggregate from shingles (caused by UV exposure and asphalt weathering)
  Dings/dents in light-gage metals (flashing, vent caps, gutters and downspouts, snow guards). Chalk and photograph.
  Rubber membrane weathering, discoloration, punctures, bubbles (caused by UV exposure, maintenance, and improper installation)
  Photo-document
- Consider Hail Impact Energy Table

*A burnish (or splatter) mark is a change in a surface's appearance caused by hail impact that results in a smoother, shinier, darker or otherwise discolored or deformed area.

Trajectory

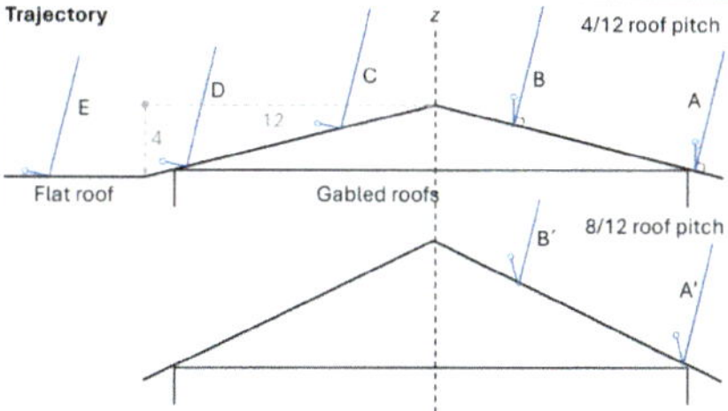

Severity of damage varies according to pitch and which side of the roof ridge (vertical dashed line z) the hail hits. Impact E on the flat surface has the greatest intensity, followed by impacts A and B, which are perpendicular to the vertical. Also, impacts A′ and B′ on the 8/12 pitched roof have less severity than strikes A and B on the 4/12 pitch. (After Noon, 2000.)

Hailstone Shape

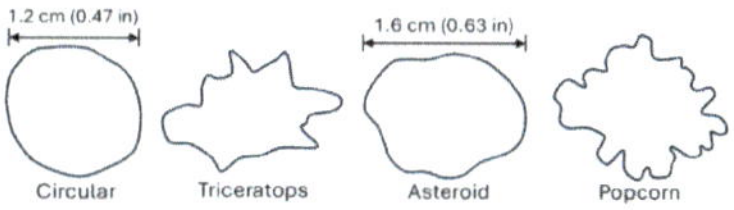

Note: Shape names are figurative.

Hail Impact Energy Table (Freefall and Crosswind comparison)

Size Relatable Object	Hail Size (cm)	Hail Size (in)	Terminal Velocity (v_T) (m/s)	Terminal Velocity (v_T) (mph)	Freefall Impact Energy/Area v_T (J/cm²)	Freefall Impact Energy/Area v_T (ft-lbf/in²)[4][5]	Crosswind 40 mph Impact Energy/Area v_T (ft-lbf/in²)[6]	Crosswind 60 mph	Crosswind 80 mph
US Quarter	2.54	1.00	22.4	49	0.38	1.80	3.00	4.41	6.44
US Half Dollar	3.18	1.25	25.0	55	0.59	2.82	4.30	6.09	8.63
Ping Pong Ball	3.81	1.50	27.3	61	0.64	4.01	5.80	7.96	11.00
Golf Ball	4.45	1.75	29.5	65	1.16	5.54	7.60	10.09	13.64
Golf Tee (length)	5.08	2.00	32.2	70	1.58	7.52	9.80	12.73	16.79
Tennis Ball	6.35	2.50	35.8	78	2.32	11.02	The following have similar strike intensity and chance of causing damage: • 1-inch hail with 60 mph wind • 1.25-inch hail with 40 mph wind • 1.5-inch hail in freefall w/o wind		
Baseball	6.99	2.75	38.0	82	2.80	13.33			
Std. Soup Can	7.62	3.00	39.3	85	3.30	15.70			

Notes: v_T is terminal velocity of free-falling hailstone without wind. Impact energy is computed based on spherical hailstone cross-sectional area, average ice density 90% that of water (or 0.9 g/cm³), the International Standard Atmosphere value for air density of 1.225 kg.m³ (0.00238 slugs/ft³), and terminal velocity. Cells highlighted yellow for freefall (left column) versus wind-enhanced (three crosswind columns) are for damage comparison. Impact Energy/ Area v_T < 2.5 ft-lbf/in² (0.53 J/cm²) is considered superficial and insufficient to cause material disfunction.

Estimating Hail Downward Impact Energy

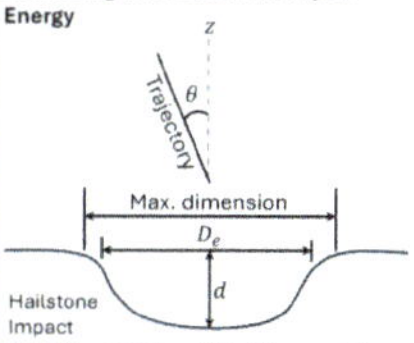

Hailstone Impact Energy: $KE/A = 4/3\,\rho D(v_T \cos\theta)^2$

KE is kinetic energy $= \frac{1}{2} m v_T^2$
v_T is terminal velocity
m is mass of ice $= V\rho$
V is ice volume
$\rho = 0.9\ g/cm^3$ is average density of ice
A is ice cross-sectional area $= 0.25\pi D^2$
D is diameter of spherical ice ball
θ is hailstone trajectory

FIGURE 9.11
Example quick reference guide for hail damage assessment.

Chalked downspouts on all sides of the house. Hail came from all directions? Upper winds in a thunderstorm can swirl with updrafts and hailstones may swirl inside a cloud as they form. However, once too heavy to stay aloft, the overall pattern of falling hail is more of a swath or trajectory, not a swirling pattern. The forensic engineer knows how to look for a directional pattern of dents in the roof system and distinguish it from older wear and tear. Further, roof surfaces have much larger areas than gutters and downspouts, and the trajectory favors the windward side of a pitched roof. Damage should be more prevalent and noticeable there.

A roofer with binoculars from the street (or an adjuster with a drone) can easily age a roof. And taking advantage of a hailstorm to replace an old roof or one with preexisting damage is one reason insurance premiums increase. The forensic engineer experienced in discerning cause and effect is best equipped to aid decision makers with replacements and establishing practical actuarial rates and premiums.

9.3 Building Condition Assessment

A Minnesota River flood caused damage to a rural church. The insurer hired a forensic engineer to perform a ***building condition assessment*** (BCA), a systematic evaluation of the building's physical state. The BCA involves inspecting building components to identify damage, deficiencies, latent conditions/issues, and necessary maintenance or repairs. Developing a punch list (or deficiency list) is ideal for this type of assessment. Table 9.1 is the engineer's punch list for this project. It includes recommendations, always appreciated by the insurance company.

9.4 Vibration Damage

Experience suggests two kinds of vibrational damage are most common: (1) that caused by operating machines and (2) that caused by construction activity. With the first, many types of buildings have mechanical and electrical rooms, and some, like high-rise hotels, condominiums, and casinos, have rooftop compounds that house a variety of equipment that operate at different frequencies, the combination of which can damage the enclosure. Vibration damage to existing infrastructure caused by heavy equipment operating at an adjacent construction site is an example of the latter. The following provides background in vibration mechanics for understanding how operating machinery can damage structures.

TABLE 9.1
Punch List

No.	Item	Location	Condition/ Comments	Recommendation
Building Exterior				
1.	Downspouts	All four corners of building.	Drain directly to foundation.	Divert runoff away from building with a. velocity breakers, or b. downspout extensions, or c. underground pipes to dry wells.
2.	Drainage grade	Perimeter of building all sides.	No drainage slope.	Regrade 5 ft (1.5 m) at 3%–5% all sides to slope runoff away from building.
3.	Roof gutter	Roof, north side, @ junction above building entrance.	Leaks water above outfall.	Check for clog/leak; repair as necessary.
4.	Deck support	Front right isolated pad foundation at south entrance to building.	Differential settlement.	Reestablish pad on NFS compacted soil as necessary to re-level deck corner.
5.	Basement wall	West side, stairwell down.	Wall heaved inward.	Trench excavation to foundation to expose wall. Rehabilitate block wall, Bituthene wrap wall, and backfill excavation with tamped NFS soil to foundation.
Building Interior – Upstairs				
6.	Fire extinguishers	All inside building.	Outdated certifications.	Recertify all extinguishers; replace inoperable as necessary.
7.	Door trim	Pastor's office.	Missing.	Install door trim to match.
8.	Ceiling leak	Upstairs coat closet.	Water stains from small roof leaks above ceiling.	Licensed and bonded roofer to check roof for leaks and repair as necessary.
Building Interior – Daylight Basement				
9.	Water/vapor barrier	Plastic sheeting outside of all exterior walls, below grade sections.	Deteriorated and/or failed.	Replace plastic with Bituthene wrap; backfill trenches with tamped NFS soil all walls to minimize potential for future heaves.
10.	Interior sheetrock	All interior basement walls east of and including kitchen and boiler room to east end of building.	Water damage and mold established in wall cavities and floorings in the rooms.	Remove bottom 4 ft (1.2 m) of sheetrock on both sides of stud walls and all floorings. Consult professional mold remediation specialist. Replace materials as necessary to reestablish functional thermal walls and floors.

(*Continued*)

TABLE 9.1 (*Continued*)

Punch List

No.	Item	Location	Condition/ Comments	Recommendation
Building Interiors – General				
11.	Lighting	NE storage room, hallway, and utility room.	Lights hanging with exposed wire or missing covers.	Repair/replace as necessary to code. (See item 16.)
12.	Baseboard heaters	All damaged/ corroded by water.	Check for rust and other damage.	Replace all damaged/corroded.
13.	Bathroom plumbing	Basement SE corner bathroom.	Septic backup; inoperable.	Licensed and bonded plumber to flush and replumb all lines/ fixtures as necessary to affect proper drainage/discharge.
14.	Boiler	Boiler room.	Pressure vessel leak in 18-year-old boiler.	Replace boiler with new.
15.	Window trim	Classroom 1 (adjacent boiler room).	Windows missing interior trim.	Install new trim to match.
16.	Electrical wiring	Basement rooms.	Rewire.	Licensed and bonded electrician to repair/rewire as necessary to code.

Notes: NFS is non-frost-susceptible; Bituthene (or comparable product) wrap is waterproofing.

Machine Vibration

Vibration, from Latin *vibrāre* "to shake," is the mechanical phenomenon of something that oscillates about an equilibrium point. Our vocal cords and eardrums, reed instruments, and mobile phones are examples of necessary or intentional vibrations. However, vibrational motions of motors and mechanical devices caused by imbalances in rotating parts, uneven friction, and deformations are undesirable and can be destructive.

Frequency, the number of times a body oscillates per unit time, is a fundamental parameter in machine design. Typically expressed in cycles per second (1/s), or Hertz (Hz), a machine has its characteristic ***natural frequency***[4] that presents as vibration with normal operation. Vibration forces are induced by rotating shafts, meshing gear teeth, roller bearing elements, vanes, blades, etc., and are characteristic of all operating machines regardless of the condition of their health.

Machine designers and engineers take extensive measures to prevent input frequencies from approaching natural frequencies to minimize the risk of damage. They recognize that destructive ***resonance*** can be induced over a range of frequencies, not just a single value. Conditions necessary to produce resonance are:

1. An external force of vibration must act on the body.
2. The body must have at least one natural frequency of vibration.
3. The external force of vibration must be similar to the object's natural frequency.

Resonance can produce vibrations strong enough to destroy the machine that produces them, or objects that experience them. For example, a large group of people walking across a pedestrian bridge can create extreme vibrations at the bridge's natural frequency and shake it. For this reason, soldiers marching across a bridge always break step when crossing. Buildings too have natural resonances. An earthquake sends out shock waves from its epicenter, and these waves vibrate the Earth as they pass. Ground vibrations that escalate to the natural frequency of a building will cause it to shake apart.

At machine scale, erratic control system operation and deformation of parts over time can cause destructive vibrations. A water heater connected to a power circuit that experiences voltage fluctuations or *electrical line noise*[5] can experience erratic operation. A rotary vane pump can experience shaft deformation as bearings wear (Fig. 9.12(a)). As shaft deformation progresses, vibration acceleration increases until it becomes destructive (Fig. 9.12(b)). With shaft deformation, vanes vibrate differently and complex finite element analysis is necessary to model the vane vibration mode shapes (Fig. 9.12(c)).

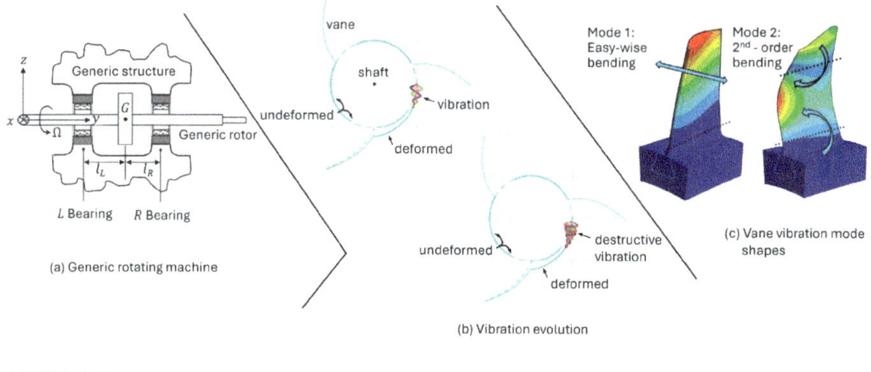

FIGURE 9.12
(a) Representation of a generic rotating machine operating on two bearings; (b) progressive shaft deformation leading to destructive vibration; and (c) finite element analysis of vane vibration modes[7].

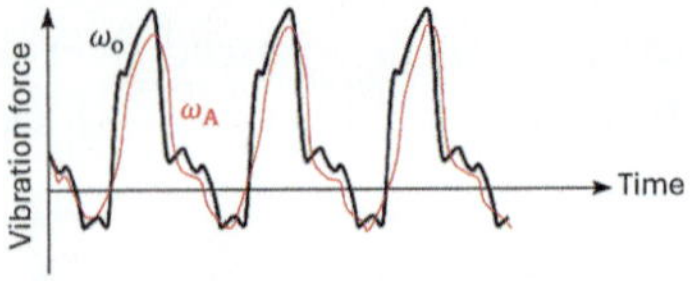

(a) Short term shock or vibration formation as input frequency approaches natural frequency of a periodic repeating wave form of a rotating machine component.

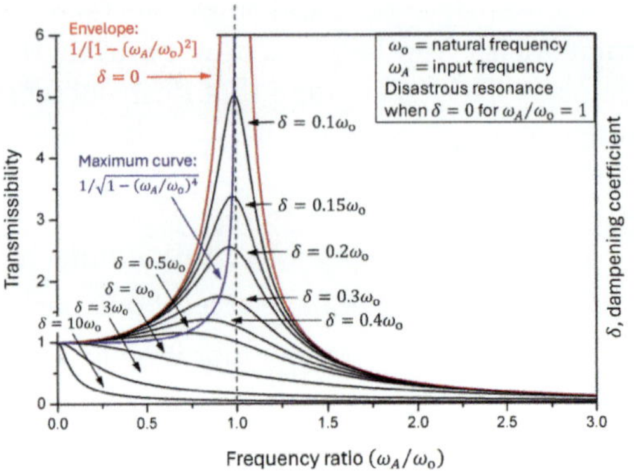

(b) Resonance transmissibility [8]

FIGURE 9.13
The importance of dampening when input frequency approaches that of natural frequency: (a) Short term shock or vibration formation as input frequency approaches natural frequency of a periodic repeating wave form of a rotating machine component; and (b) Resonance transmissibility[8].

Resonance transmissibility refers to the vibration force transmitted to that applied, or the ratio of output to input accelerations. Transmitted force is the vibration that is transmitted to the body or support structure of the machine, and applied force is the external one that generates the vibration force in the first place and causes it to be transmitted. In Figure 9.13(b), transmissibility illustrates how much greater the resonance acceleration is to that of the input. Transmissibility greater than 1 means amplification, and maximum amplification (i.e., disastrous resonance) occurs when the input frequency (ω_A) and natural frequency (ω_o) coincide, or $\omega_A / \omega_o = 1$. Also, as the frequency ratio approaches a value of 1 ***dampening***[6] decreases and vibration approaches destructive resonance.

Incident

Imagine a mechanical penthouse on the roof of a building with multiple rooms, each housing different types of machines and electrical equipment (Fig. 9.14). Elevator controllers, gas-fired water heaters, a circulation pump, an air conditioning unit, and a couple of large electrical panels, each operating at different frequencies and waveforms. Some of this equipment was mounted directly to bare concrete and others in metal cabinets bolted to the concrete floor. Further, some dampened, others not so well. You are standing on the bare concrete floor of the elevator control room with shoes and socks off. Wall cracks are visible. A passenger calls for the elevator, and the controller suddenly comes to life. Your feet feel the change in vibration, as

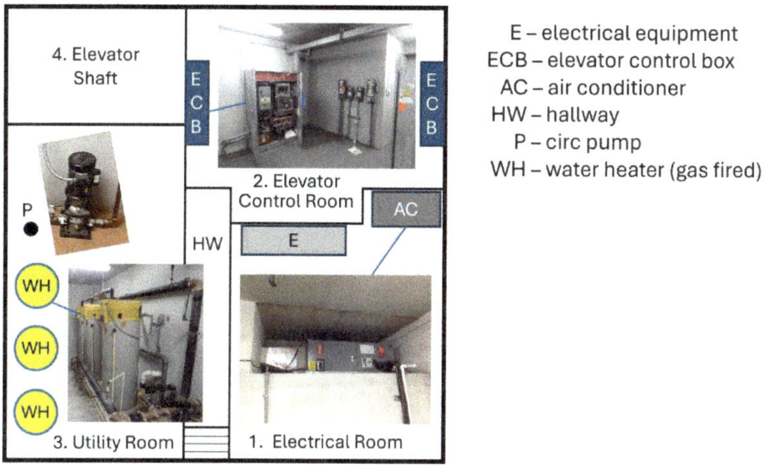

FIGURE 9.14
Layout of equipment in the four rooms of a building rooftop penthouse.

do your hands pressed on a wall. Then the circulation pump in an adjacent room kicks on, and again the vibrations change. The wall cracks begin to take on meaning, and you wonder whether the machines are causing the cracks? Turn to Case Study 9 for the rest of this story.

Notes

1. *Torch down* means a torch was used to melt and adhere modified bitumen sheets to the roof surface.
2. *Chinking* refers to using natural materials such as clay, mud, sand, and lime, moss, oakum, and natural fibers like straw or hemp to create a tight seal between logs, preventing air and moisture from entering the structure.
3. *Threshold hail* refers to the size of hail that can cause damage to a particular material. For example, the literature agrees that 1.25-inch (3.2 cm) hail is the recognized threshold for damage to asphalt shingles.
4. A machine with many components will generate a combination of frequencies, one from each moving part. With smooth operation, the combined overall frequency is the machine's *natural frequency*.
5. Two common types of *electrical line noise* are electromagnetic interference (EMI) and radio frequency interference (RFI).
6. Vibrations of an operating machine are *damped* by active insulation to prevent the vibrations from being transmitted to the foundation, walls, and adjoining rooms. For example, a transformer in an elevator control cabinet will be damped with rubber *shock absorbers* at its mounts to minimize transmission of vibrations to the floor beneath the cabinet.

References

1. Susan Buchanan (30 July 2010). *"Record Setting Hail Event in Vivian, South Dakota on July 23, 2010"*. Aberdeen, South Dakota: National Weather Service. Retrieved 26 June 2025.
2. Hail Basics (2006). National Severe Storms Laboratory, National Oceanic and Atmospheric Administration. Retrieved 26 June 2025.
3. Randall K. Noon (2000). *Forensic Engineering Investigation*. CRC Press, 375–377 pp. ISBN 978-0-429-24801-6.
4. Laura Shedd, Matthew R. Kumjian, Ian Giammanco, Tanya Brown-Giammanco and B. Ross Maiden (2021). Hailstone shapes. *J. Atmos. Sci.* 78:639–653. doi:10.1175/JAS-D-20-0250.1
5. Yimin Dai, Yao Deng, Hao Luo and Yanwen Long (2025). Experimental study on the effect of wind speed on hail impacts. *Appl. Sci.* 15(9):5120.
6. Rick Abbott. Wind Driven Hail Versus Hail Without Wind. Accessed https://abbottforensics.com 27 June 2025.
7. Meherwan P. Boyce (2012). Rotor Dynamics, In: *Gas Turbine Engineering Handbook*, 4th Ed. Butterworth-Heinemann, 215–250 pp.
8. File:Resonance.PNG. (10 May 2021). *Wikimedia Commons, the Free Media Repository*. Retrieved 18 September 2022 from https://commons.wikimedia.org/w/index.php?title=File:Resonance. PNG&oldid=559389523.

Case Study 9: Destructive Resonance

- Mechanical Equipment Penthouse
- Damage Assessment
- Vibration Analysis

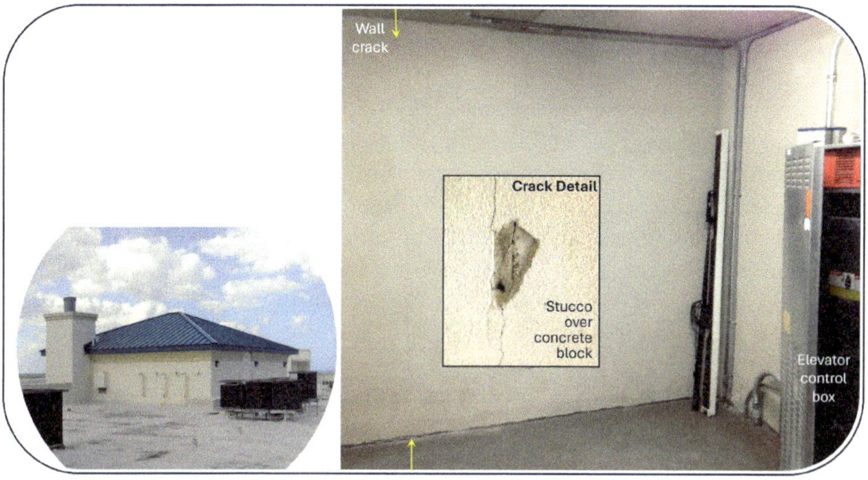

Message to students:

> Distress in a structure under light load is probably caused by something other than self-weight.

Penthouse Mystery

In the seventh year of a high-rise building, maintenance personnel reported abnormal cracks in walls of a rooftop mechanical (ME) penthouse. Coincident with the need for a building inspection, property management hired a forensic engineer to investigate the ME penthouse issue.

DOI: 10.1201/9781003623311-20

The ME penthouse was a single-story concrete block structure with stucco-finished walls, bare concrete floors, and a metal roof. Upon entry, four steps led down to a hallway with three doors: an electrical room, elevator equipment room, and utility room. The floor plan for the compound shows the layout of these rooms and the elevator shaft (Fig. CS9.1).

Various pieces of equipment were in the three rooms. Electrical panels inside cabinets, one on each side of a breaker box, were floor-mounted against the shared wall with the adjacent elevator control room. An air conditioner operated in an open cubbyhole above the cabinets. This unit conditioned air for the elevator control room. In the elevator control room were two floor-mounted control cabinets, one for each of two elevators that operated inside the adjacent elevator shaft. The utility room across the hall housed three gas-fired water heaters and a water circulation pump, all floor mounted. Photographs of this equipment are included in Figure CS9.1. Equipment operating parameters are summarized in Table CS9.1.

Generally, buildings experience three types of vibrations: periodic, random, and transient (Fig. CS9.1(a)). Periodic vibrations are repeating waveforms common to operating machines. Transient impulses that decay back to rest are associated with the elevator controllers and vehicles that come and go from the ground-level parking garage. Random waveforms such as earthquake vibrations never repeat. The operating equipment in the ME penthouse produced periodic and transient vibrations.

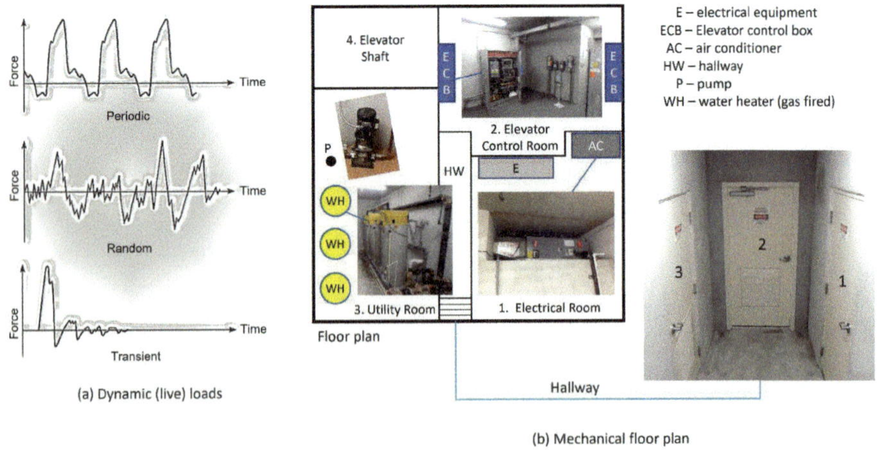

FIGURE CS9.1
Operating equipment dynamics and penthouse floor plan: (a) Dynamic (live) loads; and (b) Mechanical floor plan.

TABLE CS9.1

Equipment in ME Compound

Room	Equipment	Operating Parameters
1. Electrical Room	Bryant 1-ton, horizontal-mounted cabinet, air-cooling unit (1) Electrical panel (2) Breaker box (1)	208–230 V, 60 Hz operating frequency.
2. Elevator Equipment Room	Elevator control boxes (2) Main breakers (2 elevator, 2 elevator lights)	VFD (variable frequency drive) for improved load control. Positioning a load using a single-speed hoist may cause shock loading or load swings. A VFD allows for smooth operation by gradually slowing down a load. The elevator control autotransformers control the power supply to the electric circuits of elevators and lifts. 60 Hz.
3. Elevator Shaft	Otis elevator lifts (2)	60 Hz operating frequency.
4. Utility Room	AO Smith Cyclone XHE, gas-fired, 100-gal, 199,000 Btu water heater (3)	120 VAC, single phase, 60 Hz operating frequency.
	2.5 HP circulator pump (1)	60 Hz operating frequency.

Note: VAC is volts alternating current, HP is horsepower (1 HP = 746 watts), Btu is British thermal unit (1 kWh = 3412 Btu).

ME Penthouse Inspection

Numerous cracks were observed in the stucco walls of the three ME rooms. Although hairline cracks in corner joints are not a concern, other diagonal, stairstep, and vertical cracks in walls are. Stairstep cracks that show in stucco usually emanate from block mortar joints behind, and vertical cracks may translate from cracked blocks behind the stucco, as evident in the cover image crack detail. A vertical series of diagonal cracks in the mortar joint between two convergent walls in the utility room suggested differential wall movement. The number and variety of wall cracks in the ME penthouse indicated vibration damage.

A person standing in the penthouse rooms could feel floor and wall vibrations from the different equipment. Vibration testing indicated the elevator controllers vibrated at different frequencies, the gas-fired water heaters vibrated with consistency, and AC unit vibrations were highest. Elevator control room wall vibrations were greatest when the AC unit kicked on simultaneously with energizing of an elevator controller.

The transformers inside the elevator controller boxes (ECB) were dampened at their mounts in the cabinets, and the cabinets were bolted to the floor.

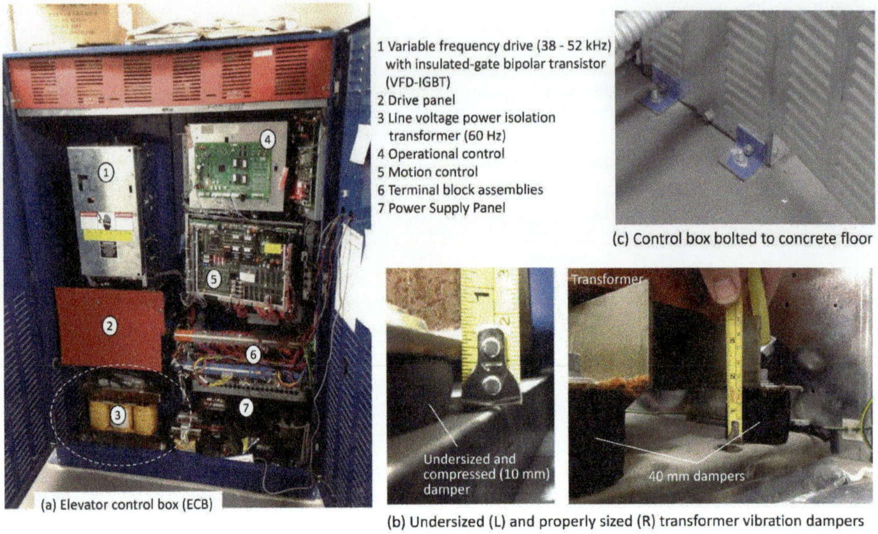

FIGURE CS9.2
(a) Elevator control panel components, (b) adequate and undersized transformer dampers, and (c) ECB fastened to floor (typical).

One could feel the vibration transfer from the cabinets to the concrete floor every time an ECB was energized when an elevator was called for. Similarly, vibrations were felt when the water heaters fired up and the recirculation pump energized. Wall cracks were measured, dated, and photographed as a baseline for crack monitoring over the next year.

Close inspection of the ECBs revealed some undersized rubber dampers (Fig. CS9.2(b)). Dampening of transformers is crucial to mitigating vibrations at the cabinet-floor interface. A damper modulates oscillations (bouncing), and its resistance to motion is proportional to how fast the motion occurs. Imagine that an energized transformer supported on four equal-height 40-mm relaxed rubber dampers might better reduce vibration to the floor than one that resides on compressed (stiff) 10-mm tall dampers. The AC unit in the electrical room was not dampened, and the water pump in the utility room was missing two of four fasteners in its plate mount to the concrete floor.

Vibration Analysis

Vibration analysis is a process that monitors vibration levels and signal patterns within a part, machinery, or structure to evaluate its overall condition and detect abnormal vibration events. Commonly performed on the time waveforms of vibration signals directly (*time domain frequency analysis*, or TDFA), or on a frequency spectrum (*frequency domain vibration analysis*, or FDVA) obtained by applying a *fast Fourier transform* technique on the time waveforms, analysis reveals the severity of abnormal vibrations with consideration for various parameters:

- root-mean-square (RMS)
- peak amplitude
- standard deviation
- kurtosis (a measure of *tailedness* of a distribution, or how often outliers occur)
- crest factor (a measure of how extreme the peaks are in a waveform)
- skewness

The RMS value is the most relevant measure of amplitude because it accounts for both the time history of the wave and energy content and is therefore most reflective of the destructive abilities of the vibration.

In Figure CS9.3, T is the period of the waveform, amplitude describes severity of vibrations, and peak amplitude is shock initiation of the elevator transformer. Peak-to-peak value indicates maximum wave excursion, useful for example as a measure of vibration displacement in evaluating a machine part's maximum stress. The peak value is useful as an indication of short-duration shocks. RMS superimposed on the elevator control signal identifies a double-amplitude vibration, suggesting need for tuning. In this case, replacing an undersized damper with a correct one alleviated the anomaly.

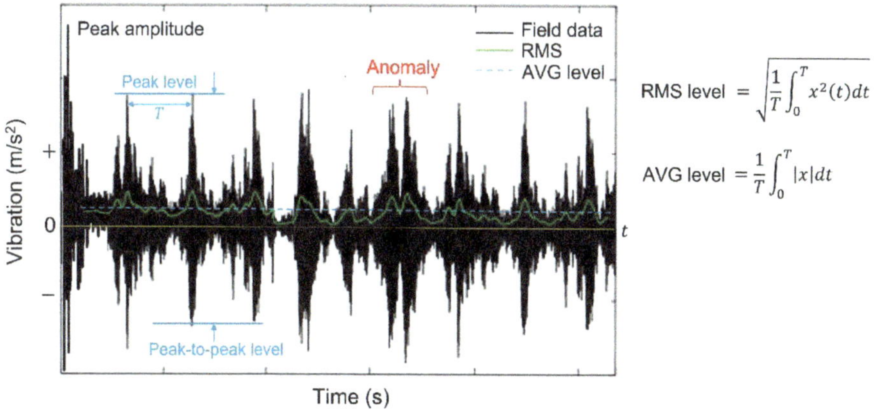

FIGURE CS9.3
Elevator control signal superimposed with RMS detection of anomalous double vibration. Peak amplitude is transformer actuation.

Machine condition vibration analysis can be measured with stud-mounted sensors (e.g., piezoelectric and micro-electromechanical) that quantify displacement, velocity, and acceleration, and *enveloped acceleration* (a measure of the high-frequency repetitive vibration signals). Instantaneous measurements of vibration acceleration, velocity, and enveloped acceleration can be accomplished with modern handheld multi-meter measuring tools capable of evaluating low-frequency (10 Hz to 1 kHz) and high-frequency (10 kHz to 30 kHz) machinery. A laser Doppler *vibrometer* can record vibration measurements of the surface of a rotating part. TDFA is typically used to monitor vibration levels; FDVA can detect cracks and other deformations that lead to diminished machine performance, damage, and/or destructive resonance. Machine frequency tuning of rotating machinery ensures that an adequate separation exists between natural frequencies and excitation sources in the prevention of damaging vibrations and disastrous resonance.

Equipment Remediation and Vibration Monitoring

Initial vibration analysis provided baseline values for operating equipment, detection of anomalies, and immediate remedies. An elevator control panel transformer's dampening was corrected, the AC unit was dampened, and the water pump mounting plate was fully fastened to the floor. Although these remedial actions reduced equipment-specific vibrations, vibration and wall crack monitoring over time was necessary to address impacts to the structure. The combined vibrations of multiple pieces of equipment had to be evaluated relative to structural impacts with quarterly crack monitoring as a phase I remediation plan. Monitoring results would then lead to phase II corrective action (e.g., equipment reconfiguration, additional dampening) and structural distress evaluation.

One might imagine the challenges of designing a mechanical room in which multiple different types of machines must operate. Design would be based on new equipment operating at optimal conditions with ideal dampening. But machines wear, parts sometimes deform ever so slowly, routine maintenance changes out parts, and large-item replacements can negatively impact maintenance schedules. Further, routine maintenance can sometimes lead to inconsistencies and incompatibilities when old parts are replaced by new *other* parts. (The inconsistent dampening of the elevator transformer in Figure CS9.2(b) is an example.)

10

Product Liability

CONTENTS

CONCEPTS	KEY TERMS	ADDITIONAL READING
Product Liability	Defective product	Exercise 19
Confined Space Work	Duty of care	Exercise 20
Regulatory Safety	Portable gas analyzer	
American Society for Testing and Materials	Elastomeric roof	Case Study 10
	Technical data sheet	
	Safety data sheet	
	Vertical platform lift	

Products are supposed to work for the benefit of users. When one does not, it has a bad design, a flaw, is cheaply made, inadequately safeguarded, or a combination of these. Consumers have recourse through the legal system when injury or damage occurs. This chapter is about ***product liability***, a realm of law designed to protect consumers. The forensic engineer is involved after a claim has been filed and the litigants take opposing positions regarding product design, fabrication, installation, or use. Through discussions about three different product liability situations, students will learn to distinguish a defective product, false advertisement, and improper installation. But first, we need to define product liability.

10.1 Product Liability

Product liability is a consumer protection concept. It refers to the legal responsibility that manufacturers, distributors, and installers must ensure products are safe for consumers. This concept holds businesses accountable

DOI: 10.1201/9781003623311-21

for injuries or damage caused by products that are defective or dangerous. Important aspects of product liability include:

- *Defective Products*: A product can be defective in design, manufacturing, marketing, or dysfunction (e.g., failure to warn, false advertisement).
- *Duty of Care*: Manufacturers and installers must exercise reasonable care in the design, manufacturing, distribution, and installation of products, and ensure they are safe for intended uses. Further, this liability can extend to wholesalers and retailers (i.e., sellers) depending on the situation.
- *Damages*: Manufacturers, installers, and sellers can be held liable for personal injury, property damage, and other losses associated with defective products.
- *Claims*: Claims can be based on negligence, breach of warranty, or strict liability* depending on the situation and jurisdiction.

When a product is found to be defective, a manufacturer is legally obligated to remedy the problem. They do so by exercising recalls, repairs, replacements, or refunds depending on the nature of the defect and the potential risk it poses to consumers.

10.2 Defective Generator and CO Poisoning

A licensed, bonded, and insured contractor was performing concrete placement and finish work at a remote Canadian construction site when a worker passed out and had to be medevacked to a hospital for hyperbaric oxygen treatment. Medical tests revealed the worker succumbed to carbon monoxide (CO) poisoning. The quick action by the work crew and emergency responders saved his life.

An insurance company hired a forensic engineer to investigate and determine how the worker was poisoned. The engineer traveled to the site, inspected construction conditions and machinery used to heat the tented workspace and power equipment (Fig. 10.1), and reviewed equipment maintenance records and detailed accounts provided by five co-workers.

Inspection of the heater and generator was performed by the engineer in the presence of an insurance representative, lawyers, and a military official. The 2-year-old 200,000 BTU propane portable torpedo heater and 3-year-old 5.5 kW 4.5 hp portable generator were found in good operating condition and without obvious malfunction. However, a ***portable gas analyzer*** (PGA)

* A manufacturer, installer, or seller can be held liable even if they are not negligent in causing the defect.

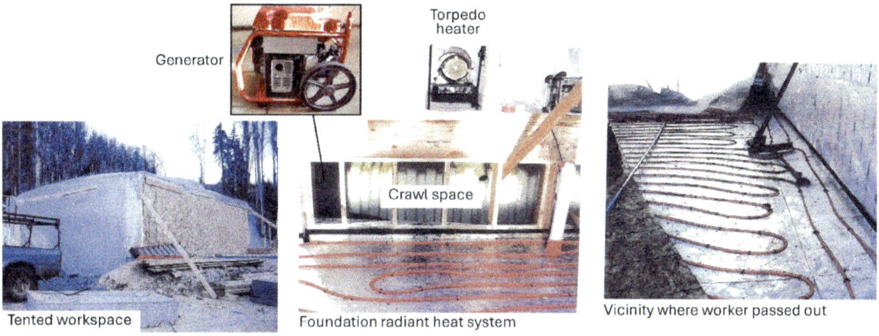

FIGURE 10.1
Tented enclosure and locations of the torpedo heater, portable generator, and worker at the time he passed out.

measured carbon monoxide (CO) emissions from the generator exhaust at 2820 g/hr, a high CO emission rate[1] and suggestive of an abnormal air-fuel mixture. Too much fuel or too little air can cause incomplete combustion and lead to higher CO emissions. This result prompted a review of the generator maintenance records.

Defective Carburetor

Records indicated a carburetor throttle valve problem in the 15th month of service and a rebuild kit repair. The generator carburetor was disconnected, bagged as evidence, and taken by the engineer to an independent small engine repair shop for diagnosis. Diagnosis revealed a damaged needle valve and a malfunctioning float assembly from an aftermarket kit. Defective carburetors in the subject generator were a known problem, as were issues with aftermarket repair kits. The repairman stated: *"They don't make them like they used to, and a worthless warranty. Use OEM always if available."* OEM is "Original Equipment Manufacturer," genuine parts made by the manufacturer.

Setting and Workspace Ventilation, Day of Incident

On the day of the incident, workers arrived at 7 AM, turned on the generator and heater, and began preparations for a winter concrete pour within a plastic-wrapped enclosure (Figs. 10.1 and 10.2(a)). Outside temperatures were between –32°C and –29°C (–26°F to –20°F).

The enclosure encompassed the foundation footprint for the pour and crawl space plenum on which the torpedo heater sat. Its power cord dropped down through a floor hatch and plugged into the generator located about 3 m (10 ft) away on the ground in the crawl space. With no forced ventilation of the enclosure and crawl space, the workspace constituted a confined space.

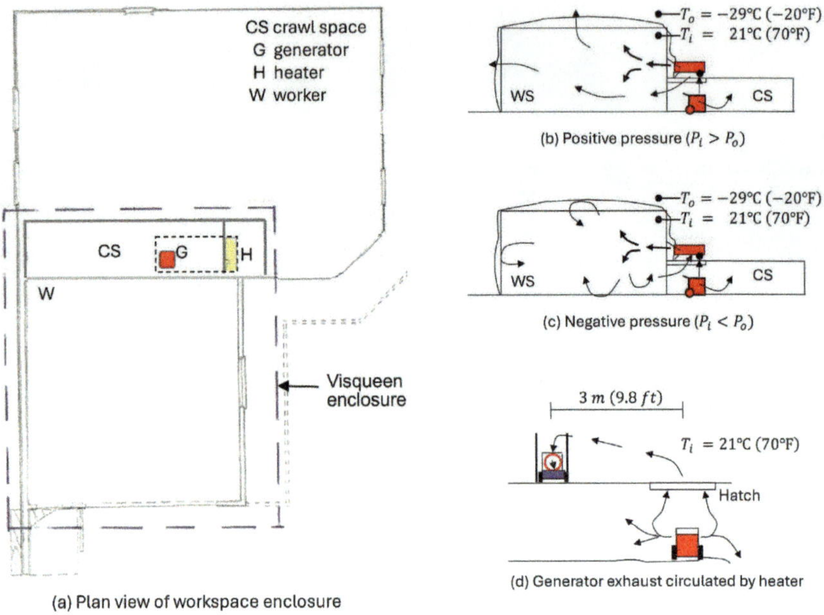

FIGURE 10.2
Proximity of generator and heater to CO-affected worker, and exhaust circulation scenarios.

A timeline of the days' concrete work (Table 10.1) indicated that Worker 1 (the CO-poisoned crew member) had worked inside the confined space for most of the 13-hour workday with only short breaks outside.

Generator exhaust contains a mix of gases: carbon monoxide, carbon dioxide, water vapor, and unburned hydrocarbons. Although some components

TABLE 10.1
General Timeline for Enclosure Work on the Day of the Incident

Date/Time	Description
7 AM	Crew arrived at site. Generator and heater turned on.
7–11 AM	Preparations for concrete pour.
11:30 AM	First concrete truck pour.
3:15 PM	Last concrete truck pour.
3:30 PM	Worker 1 working concrete alone in enclosure. Other crew members working at the site beyond the enclosure.
6 PM	Worker 1 felt nauseous and began taking fresh-air breaks outside enclosure.
7–8 PM	Worker 1 and two others completed work and tidied up for the night.
Approx. 8 PM	Workers noticed Worker 1 passed out in the corner of enclosure and rendered aid.
8:03 PM	Site foreman placed emergency call to project superintendent who then requested emergency medevac assistance.

like carbon dioxide and water vapor are denser than air, the overall exhaust mixture readily mixes with air through diffusion and convection in an open environment. However, in an enclosed or poorly ventilated area, heavier components like CO_2 and heavier-than-air hydrocarbons can accumulate near the ground, while lighter components like CO – its molecular weight being slightly lighter than that of air – will readily mix with air.

CO-Poisoning Hypothesis

In the workspace enclosure setting, generator exhaust initially rose being hotter than the surrounding air. As it cooled, it mixed with the surrounding air and dispersed. The torpedo heater, only 3 m (10 ft) away, drew exhaust up through the crawl space door (Fig. 10.2(d)) that was cracked open by the power cord, which enhanced dispersion within the enclosure. Crawl space air quickly became contaminated with exhaust, and replacement air drawn in through the seams of the plywood deck that comprised the sub-floor above the crawl space was not enough to sufficiently dilute the growing hazard. Further, the enclosure would have experienced a negative pressure environment[2] through the workday due to the much colder air temperatures outside. This afforded greater containment and circulation of inside air.

Portable Gas Analyzer for Confined Space Work

Having a portable gas analyzer available for confined space work is highly recommended and often required by safety regulations. The problem of poisoned air can be quantified with real-time monitoring or a simple mass balance analysis with measured gas concentrations in the field. A mass balance equation for computing CO concentration in an enclosed volume considers the volume of the enclosed space and the rate of change of CO based on sources and sinks:

$$\rlap{-}V\frac{dC}{dt} = E + QC_R - QC - R \qquad [10.1]$$

where

- $\rlap{-}V$ is the volume of the enclosed space $\left(m^3,\ ft^3\right)$
- dC/dt is the rate of change of CO concentration in the enclosed space $\left(ppm\,/\,hr\right)$
- E is the CO emission rate from indoor sources $\left(mg\,/\,hr,\ lb\,/\,hr\right)$
- Q is the replacement air flow rate (outdoor air or air from other sources) $\left(m^3\,/\,hr,\ ft^3\,/\,hr\right)$
- C_R is the CO concentration in the replacement air $\left(ppm\,/\,hr\right)$
- C is the CO concentration in the enclosed space $\left(ppm\,/\,hr\right)$
- R is the CO removal due to reactions or sinks $\left(mg\,/\,hr,\ lb\,/\,hr\right)$

Terms in Equation (10.1) are defined in Table 10.2.

TABLE 10.2

Description of Terms in Equation (10.1)

Term	Term Name	Description
$\forall \frac{dC}{dt}$	Accumulation	Rate at which CO accumulates or depletes within the enclosed space over time.
E	Source	Rate at which CO is generated within the enclosed space by generator, automotive exhaust, or fuel-burning appliances (e.g., furnaces, water heaters, stoves).
QC_R	Input from replacement air	Accounts for mass of CO entering enclosed space from replacement air (e.g., outdoor air, air from ventilation system).[a]
QC	Output from exhaust	Accounts for mass of CO leaving the enclosed space due to ventilation.
R	Sinks/reactions	Rate at which CO is removed from enclosed space through molecular reactions, filtration, and absorption by materials.[b]

Notes: [a]The concentration of CO in outdoor air can fluctuate due to traffic and other outdoor activity; CO in ventilation system air is likely to have a more uniform concentration. [b]This term is often negligible at new construction sites since CO in indoor air is stable, and without materials to absorb or infiltrate.

Using the subject enclosure for example, we can determine the CO concentration as a function of time, considering the initial concentration, ventilation rate (if any), source strength, and other parameters. The volume of the enclosed workspace was 204 m^3 (7200 cf), and the measured generator emission rate (E) was 2820 g/hr (or 2.82×10^6 mg/hr). A rule of thumb for approximating generator air flow rate (Q), the replacement air flow rate since no forced-air ventilation was provided, is $2.5 \times HP = cfm$. Therefore, the estimated replacement air flow rate for the 4.5 hp generator is 19.11 m^3/hr (675 ft^3/hr). The workspace had no means to filter, absorb, or otherwise diminish CO concentration, so we can assume no reactions and sinks ($R = 0$). With a PGA on site, one could have easily measured the initial CO concentration in the enclosure and replacement air and at any time thereafter. Further, for steady-state conditions (CO concentration no longer changing) and assuming fully mixed space, the accumulation term becomes zero, which simplifies the analysis.

Fortunately, regulations have since improved the situation and now require gas monitoring of confined spaces with government-funded projects.

Safety

There were several issues with the use of the torpedo heater and portable generator that related to a lack of workplace safety at the time of the CO poisoning incident. The following represents information and regulations in place in the United States at the time. Students need to know that confined space work is serious business.

The United States Consumer Product Safety Commission reviewed portable generator safety in 2006. A briefing of their findings produced some interesting statistics:[1]

- 5.0 kW to 5.9 kW generators were most involved with fatalities resulting from generator use.
- A 10 hp generator engine produced carbon monoxide at rates 450 to 1000 times that of modern idling cars.[2]
- The US averaged 51 generator carbon monoxide deaths per year between 2002 and 2005; annual generator deaths did not exceed 20 in previous years.
- Portable generators accounted for approximately 24% of carbon monoxide deaths in the United States annually.

The United States Department of Labor Occupational Safety and Health Administration (OSHA) defines proper installation and use of portable generators and heating devices at a temporary construction site in Title 29 of the Code of Federal Regulations (CFR) Part 1926. The standards discuss machine use in enclosures, with emphasis on ventilation and clearances (i.e., air quality and fire safety concerns). For example, in 29 CFR Part 1926.154 for temporary heating devices, the regulation stipulates that fresh air must be supplied to maintain health and safety of workers, and where natural means of fresh air supply is inadequate, mechanical ventilation must be supplied.[3]

The torpedo heater and generator had fixed safety labels on them. On the left side of the heater was a label that read:

> *WARNING Safety Requirements:* Improper use of this heater can cause serious injury or death from burns, explosion, electrical shock, and carbon monoxide poisoning. Heaters used in the vicinity of tarpaulins, canvas or similar enclosure materials shall be located a safe distance from such materials. The recommended minimum safe distance is 10 feet. It is further recommended that these enclosure materials be of a fire retardant nature. These enclosure materials shall be securely fastened to prevent them from igniting or from upsetting heater due to wind action.

Excerpts from warning labels on the portable generator stated:

> Failure to read and follow these instructions can result in death, bodily injury and/or property damage.
>
> Read and follow operating instructions before running generator.
>
> Engines emit carbon monoxide. DO NOT run in enclosed area.
>
> Keep exhaust gas at least 5 feet from any combustibles or structure.

These safety rules and warnings were also stated in the owner's manuals for the machines.

OSHA regulations also mandate clearance requirements for small engines near combustible materials. For example, 29 CFR 1926.154 requires the sides of temporary heating devices to be at least 12 inches (30.5 cm) from combustibles, and 10 feet (3 m) from enclosure coverings.[4] Furthermore, said coverings must be securely fastened to prevent contact with the heater and ignition due to wind action. The torpedo heater was positioned on the crawl space plenum in immediate proximity to draped visqueen and woodwork.

Carbon Monoxide Levels

Potential effects of CO on humans are well documented by health organizations and government agencies concerned with occupational health and safety. A threshold for CO exposure is the ***permissible exposure limit*** (PEL), defined by OSHA as 50 ppm as an 8-hr time-weighted average value. A CO levels chart (Table 10.3) gives perspective to the PEL within a range of exposure levels and associated physical symptoms. The chart is a compilation of CO threshold values as set by the various US and international organizations.

The emergency room blood draw from Worker 1 measured his blood CO concentration at 156 mg/L (equivalent to 156 ppm). This was after pure oxygen treatment with a ventilator during medevac.

TABLE 10.3

Carbon Monoxide Levels, Regulatory Thresholds, and Symptoms

Carbon Monoxide Level (ppm)	Threshold	Symptoms
0	Safe level	
6	WHO 24-hr TWA	
9	ASHRAE, EPA, NAAQS, WHO 8-hr TWA	
25	ACGIH 8-hr TWA	
35	WHO 1-hr TWA NIOSH 10-hr TWA NAAQS 1-hr TWA	Physical symptoms[a] after 6–8 hours
50	OSHA 8-hr TWA (PEL)	
70–149	UL 1–4 hr alarm[b]	
150	UL 10–50 min alarm	Physical symptoms in 1–2 hours. Life-threatening after 3 hours.
200	NIOSH 15-min STEL OSHA 15-min TWA	Physical symptoms after 2-3 hours
400+	UL 4-min alarm	Physical symptoms after 45 minutes. Unconscious within 2 hours. Fatal in 2-3 hours.

(Continued)

TABLE 10.3 (***Continued***)

Carbon Monoxide Levels, Regulatory Thresholds, and Symptoms

Carbon Monoxide Level (ppm)	Threshold	Symptoms
800		Physical symptoms within 20 minutes. Fatal within 1 hour.
1600		Physical symptoms within 5-10 minutes. Fatal within 25-30 minutes.
12,800		Fatal within 1-3 minutes

Notes: [a]Physical symptoms may include headache, fatigue, dizziness, drowsiness, and nausea.
[b]Alarm means a detector's audible sound. Workers in confined spaces typically wear CO detectors with audible alarms set to go off if a preset threshold value is exceeded. WHO is the World Health Organization; TWA is time-weighted average; ASHRAE is the American Society of Heating, Refrigerating and Air Conditioning Engineers; EPA is the US Environmental Protection Agency; NAAQS is the US National Ambient Air Quality Standards; ACGIH is the American Conference of Governmental Industrial Hygienists; NIOSH is the US National Institute for Occupational Safety and Health; OSHA is the US Occupational Safety and Health Administration; PEL is permissible exposure limit; UL is Underwriters Laboratories (a global safety certification organization); and STEL is short-term exposure limit.

10.3 Failed Elastomeric Roof System (False Claim)

Rubber Roofing Industry and ASTM Standards

Rubberized roofing emerged in the early 1960s as a synthetic alternative to traditional roofing materials like tar and gravel built-up roof (BUR) and asphalt for low-slope and flat roofs. Developed by DuPont in the late 1950s, ethylene propylene diene monomer (EPDM) gained popularity in the 1970s due to an oil embargo[3] that significantly increased the cost and reduced the quality of asphalt-based roofing materials. EPDM became the leading material in the single-ply roofing market during the 1980s. Continued evolution of the rubber membrane industry led to polyvinyl chloride (PVC), thermoplastic polyolefin (TPO), and other acrylic- and silicone-based coatings since the 1990s.

Today, rubberized roof coating is a $2B industry globally. Multiple manufacturers produce a variety of product types for a broad selection of rubber roof coatings in the US market. Consumers have many choices and much information to challenge decision-making. Generally, water-based products are cheaper than solvent-based, and 2025 unit costs range from $0.45 to $3.50 per square foot. Prized for their waterproofing, flexibility, and resistance to UV radiation, weathering, and punctures, these coatings are a popular choice for flat foamed roofs, and manufacturers tout 20- to 50-year roofs for all-weather applications.

The ***American Society for Testing and Materials*** (ASTM) is a globally recognized organization that develops and publishes standards for a wide range of building materials and products. Table 10.4 summarizes ASTM standards for common rubberized roof coatings. The tests set minimum performance standards for key properties such as adhesion, elongation, permeance, tear and weathering resistance, low-temperature flexibility, and tensile strength. Low-temperature flexibility testing is especially important for the application of water-based coatings in cold climates. Water-based elastomeric membrane products are a more budget-friendly option commonly used on well-drained flat roofs that do not experience ponding water. Solvent-based coatings typically do not include water as an ingredient, provide greater resistance to ponding water, and are better suited for cold climates.

Incident

A two-story condominium complex in a northern state was built with a tar and fiberglass flattop roof in the mid-1980s. The roof was resurfaced with an elastomeric membrane in August 2011. An evaluation of commercially available rubberized roof products was conducted, elastomeric products were selected, and the manufacturer specified quantities of primer and topcoat for the new roof cover. The elastomeric system was purchased and applied per manufacturer instructions. Eight months later, after the first winter, a roof inspection revealed membrane deterioration. Detachment, lesions, and black splotches on the surface were reported and photographed. Membrane deterioration continued through the next year. The manufacturer was contacted and sent documentation of the distress conditions. Initially, the manufacturer suggested algae was the cause of membrane degradation and then implied improper installation.

TABLE 10.4

ASTM Standards for Rubberized Coatings Used in Roofing

Standard	Type Coating
ASTM D6083	Liquid-applied acrylic
ASTM D6694	Liquid-applied silicone
ASTM D6947	Liquid-applied elastomeric urethane polymer
ASTM D4637	Liquid-applied EPDM
ASTM D6878	Liquid-applied TPO polymer
ASTM D4434/D4434M	Liquid-applied PVC
Other Relevant Standards	**Test**
ASTM D4798	Accelerated weathering test for various roof coatings
ICC Testing Application Standard 143-95	Low-temperature flexibility test for white elastomeric roof coatings (crucial for roof coatings in cold climates)

Note: ICC is the International Code Council.

A forensics expert was subsequently hired by the condominium association to investigate membrane performance and failure. The manufacturer subsequently hired their own experts to evaluate and cold test the membrane products. Although some opinions differed, the experts reached the same conclusion that the products failed cold testing and did not perform well. Further, a salesperson for the manufacturer stated in an email to the customer that their products were tested down to -60°F (-51°C) and passed, and a couple of reviews told of sub-freezing failures of the same products.

When asked to produce cold testing data to support the -60°F (-51°C) claim, the manufacturer did not reply to the plaintiff's engineer. The manufacturer was then subpoenaed for this information, and through their attorney, formally replied that they did not have the cold testing data. The case quickly settled with the manufacturer found responsible for selling a product under false pretenses and liable for a new roof.

Guidance

The investigating engineer may find the following guidance useful:

1. PVC, EPDM, and TPO are the three most common roof coating types for low-slope and flat roofs.
2. Silicone, acrylic, and polyurethane coatings bond well to spray polyurethane foam (SPF) and gravel-impregnated bitumen underlayment, and some BUR. Synthetic roof coatings do not bond well with tacky asphalt substrate.
3. Read reviews and ratings, an anecdotal yet sometimes trendy basis for product acceptance and approval.
4. Obtain a ***Technical Data Sheet*** (TDS) and ***Safety Data Sheet*** (SDS)[4] for each product of consideration.
5. Verify that ASTM performance standards are met for adhesion, elongation, permeance, tear resistance, weathering resistance, low-temperature flexibility, and tensile strength. A good TDS will cite the ASTM standard next to the property value.
6. Check the SDS for percent water per unit volume. This differentiates water-based from solvent based, and higher water content leads to greater thinning as a coating cures.
7. Compare material characteristics, properties, and claims. (This is where a spreadsheet is useful.)
   - **Characteristics**: Adherence to specific substrates and recommended membrane mil thickness are examples of two important *characteristics* to consider.
   - **Properties**: Service temperature rating and accelerated low-temperature flexibility are examples of important *properties* to consider for cold climate applications.

- **Claims**: A claim that a water-based product works well in cold climates should be considered suspect until proven otherwise. (The higher the water ingredient percentage, the greater the potential for membrane thinning during curing and in-service brittle fracture.) (One can always ask a manufacturer for their cold test results to support such a claim.) Another type of suspicious claim relates to longevity. Check the warranty against a claim of ≥ 25 years of useful life for a rubber roof coating. (Experience suggests that most rubber roof coatings significantly weather and deteriorate before 20 years of service. Unfortunately, *performance evaluation* is not a criterion in an SDS.)

8. Membrane thickness scale:
   - 45 mil (0.045 in or 1.14 mm): Thin for roof applications.
   - 60 mil (0.060 in or 1.52 mm): Minimum thickness for roof applications. Balances cost and performance.
   - 75 mil (0.075 in or 1.90 mm): Thicker option for reinforced membranes. Polyester scrim or fabric embedded between layers (or passes) is a common reinforcement of EPDM products.
   - 90 mil (0.090 in or 2.29 mm): Thickest standard option for maximum durability and longevity.

10.4 Vertical Platform Lift Accident (Improper installation and design flaw)

Church service just ended, and a disabled parishioner was headed down in the lift for Sunday brunch with family and friends. Then came the fall.

A CHURCH hired an ELEVATOR COMPANY to install a ***vertical platform lift*** (lift)[5] to accommodate parishioners confined to wheelchairs. The lift was operated for three years without incident, before one fateful day when a man in a wheelchair rolled backward out of the lift during descent (Fig. 10.3) and was seriously injured. A personal injury lawsuit ensued, and a forensic ENGINEER was hired to investigate the lift incident.

The ENGINEER reviewed case file information, lift plans and specifications, and relevant design codes, conducted interviews, and visited the incident scene twice to document lift static and operational conditions. Lawyers and the CHURCH'S maintenance man attended the first visit. The lawyers, a representative from the ELEVATOR COMPANY, the state's ELEVATOR INSPECTOR, and an EXPERT ELECTRICIAN were present during the second operational demonstration. During the initial visit and after introductions, the ENGINEER stated: "Please do not communicate with me during

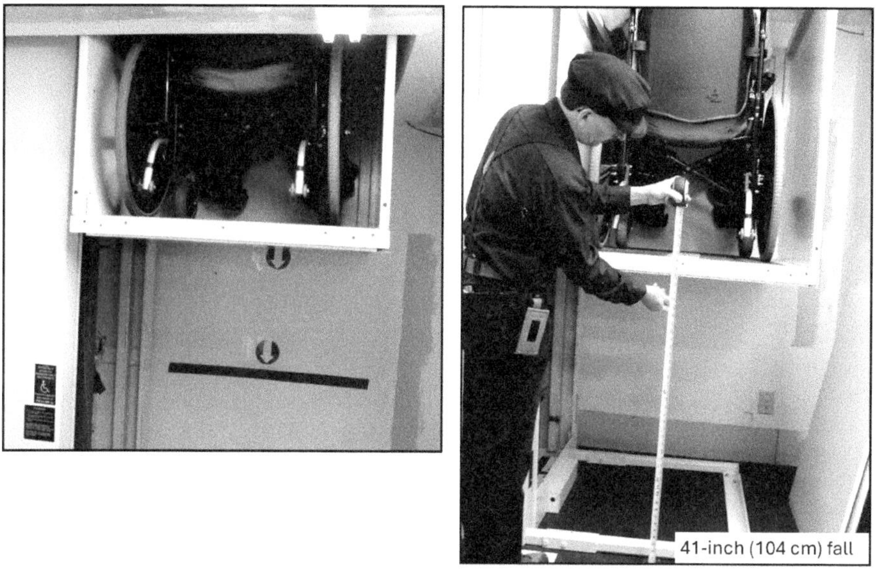

FIGURE 10.3
Vertical platform lift.

my investigation and allow me to ask the questions should I have any. I will not be answering any questions from any of you. Thank you for understanding this request." Lift conditions were documented and photographed during both visits.

Investigation Findings

The following summarizes important aspects of the ENGINEER'S investigation:

1. From the Pastor's deposition:
   - The CHURCH received a grant and held fundraisers for the Lift Accessibility Project. Approximately $38,000 was raised for the project, and approximately $5000 in residual funds remained after lift installation. Further, the cost to install a hoistway by an independent contractor was $4400. This price included materials, labor for a two-person crew, and shipping of a fire-rated door.
   - A member of the CHURCH council and building committee was charged with purchasing the lift and overseeing its installation. This person coordinated the lift purchase, oversaw its installation, and told the Pastor that lift permitting was not required.

2. From ELEVATOR COMPANY disclosures:
    - CHURCH decisions about lift and installation scope were based on limited funding.
    - At least two products were suggested: (1) a flex-lift limited use/limited application elevator and (2) the standard porch lift. Hoistway, pit, and flip-down ramp components were excluded.
    - State inspection was discussed.
    - Two employees of the ELEVATOR COMPANY installed the lift with temporary power for functional checks. A *"lift in wait of its hoistway,"* no one could recall if temporary power was disconnected or later reestablished. At the time of installation, the lift was subject to ASME A18.1-2003 and 2002 NEC (or NFPA 70-2002) codes. (ASME is American Society of Mechanical Engineers, NEC is National Electrical Code, and NFPA is National Fire Protection Association.)
    - The Installation and Service Manual for the porch lift specified: (1) installer responsibility for code compliance (section 286); (2) proper [electrical] wiring according to enclosed schematic and NEC compliance (section 289); (3) platform gate installation (section 297); (4) proper wiring of call/send controls and interlocks (sections 298-299); (5) proper base frame fastening (section 300); (6) installer final checklist (section 302); and (7) lift maintenance requirements (section 304).
    - The CHURCH was provided the Installation and Service Manual (ISM) with the stipulation that the hoistway must be completed before the installer can complete several operational checks and assist with state certification.
3. Plans and specifications provided by a contractor to the manufacturer were for a standard porch lift. Drawings showed a 34-inch (86 cm) inside clearance for the platform lift (the minimum allowed width for a wheelchair), a 42-inch (107 cm) platform lift height, and the lift centered with the upper landing gate in accordance with code requirements. However, notes on the drawings stipulated: "….hoistway enclosure with doors and disembark gate configuration by others."
4. The hole cut in the upper-level floor of the CHURCH for the hoistway severed joists, which compromised floor support in the area around the opening. (None of the interviewees could recall who cut the hole.)
5. Lift installation was not completed. No hoistway, lift door, bottom lift gate, or lockout mechanism was installed, nor was the lift properly wired, placarded, and permitted for operation.

6. First-floor interior work created a *lift room* (Fig. 10.4(A)). However, the doorway to this room was not handicapped-accessible. A 1-inch-tall threshold impeded wheelchair access.

FIGURE 10.4
Lift with installation problems highlighted.

7. Lift deficiencies included:
    - A misaligned gate at the top landing.
    - Operational space more than code allowances (Figs. 10.4 and 10.5).
    - Incomplete anchoring and fastening of the lift frame (Fig. 10.6).
    - Nonfunctional gate interlock, call switches, slack cable switch, and audio alarm.

    The controller, field wiring, and power supply were indicative of a temporary installation. The ELEVATOR INSPECTOR's report cited 14 structural and electrical gross safety violations.
8. Other issues identified:
    - Pinch points where the platform lift traveled through the upper floor (Fig. 10.4(B)).
    - The upper landing gate (Fig. 10.4(C)) was not the type of gate specified in the porch lift ISM provided by the ELEVATOR COMPANY. Also, the gate end posts were larger and of a different configuration than that shown in the manual.
    - The gate strike plate was significantly worn, suggesting a used gate had been installed.

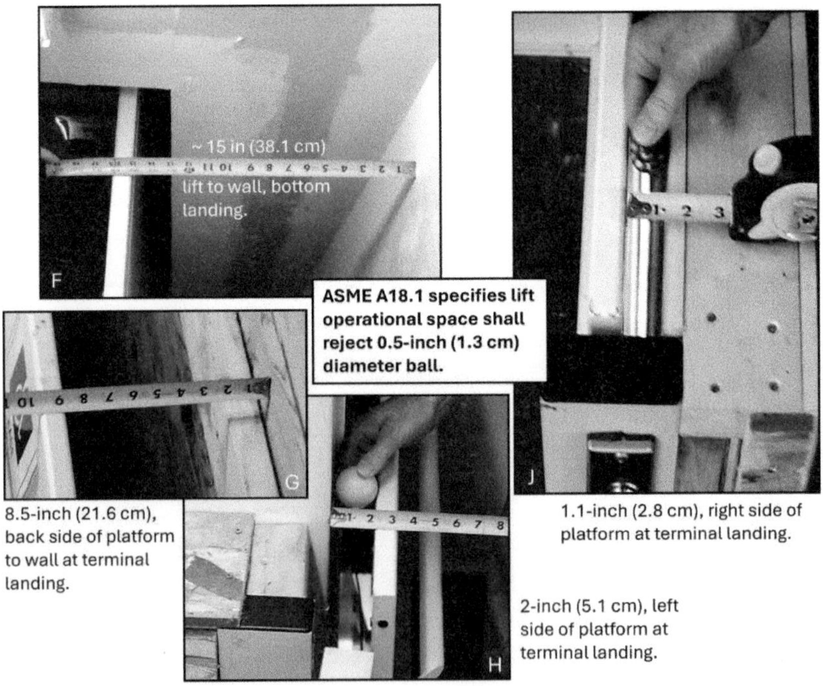

FIGURE 10.5
Lift operational space code violations.

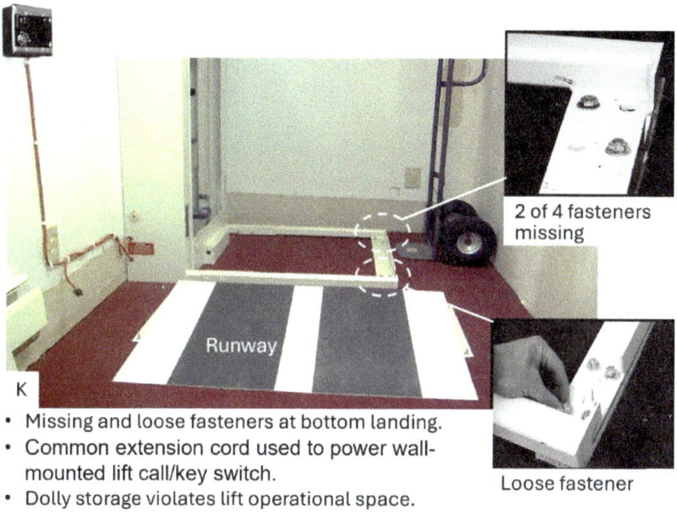

FIGURE 10.6
Problems with base frame installation, wiring, and storage at the bottom landing.

- The right-side gate post housed the electro-mechanical interlock and call switch that were not connected for service.
- The manual gate lock hole had been drill-enlarged for easier access with an Allen wrench.

9. From EXPERT ELECTRICIAN deposition:
   The lift tower was opened for inspection. The controller was not set for proper lift operation as factory jumpers were still in place. An exposed wire protruded from a 4-square box located beneath the controller. Sheathing of an electrical cord passing through the tower frame had been cut by the sharp metal edges of a tight hole that lacked a grommet sleeve. Main power to the lift and lower landing call/key switch was provided by a common extension cord (Fig. 10.6(K)). The slack cable switch was improperly wired, rendering it nonfunctional. This switch is an important safety feature; it sets the brakes should the lift's suspension cable break. The suspension cable tag was not filled out as an inspection requirement.
10. Review of the Department of Labor and Workforce Development records indicated that the CHURCH lift had not been inspected, registered, and certified for operation. The state ELEVATOR INSPECTOR confirmed this violation of state regulation.
11. From CHURCH building committee depositions:
    - The CHURCH was to complete the installation of the hoistway and fire-rated doors before the lift would be permanently wired, and the lift was only temporarily wired to accommodate hoistway construction.

- A CHURCH member performed lift electrical work as a cost-saving measure.
- The CHURCH operated the lift by unconventional means for three years before the accident. An Allen wrench kept at the upper landing was used to enter/exit the platform lift because the upper landing gate was not properly wired for operation. Although the key was secured in the sacristy to ensure operation with "appropriate instruction," the CHURCH allowed the lift to be operated by various people that assisted wheelchair occupants. Further, the injured party was not formally trained in lift operation.
- The lift could stop intermittently and without alarm when the pressure operational switch was released.

An example of a state-approved vertical platform lift is shown in Figure 10.7.

FIGURE 10.7
Example state-approved vertical platform lift. The hoistway is the outer enclosure.

Outcomes

A mediated settlement was agreed to by the parties. Significant mediation outcomes were:

- The CHURCH was responsible for ensuring completion and certification of the lift installation.
- The CHURCH had few answers to important questions about who made what decisions regarding lift installation and wiring.
- The ELEVATOR COMPANY was irresponsible in designing and installing a *customized partial lift* and should have depowered the lift after their initial lift checks had been completed. It provided an ISM to the CHURCH that illustrated a complete installation, and should have followed up with platform lift completion, inspection, and certification assist.
- The lift had a design flaw in the pressure operational switch. Intermittent stops without an alarm or lift door are an accident waiting to happen, and it did with the injured party.
- The lift must either be completed by professional contractors and inspected and certified by the state's ELEVATOR INSPECTOR at CHURCH expense or be permanently decommissioned. The lift was not to be used until one of these options was met.

Cooper: TARS, what's your honesty parameter?

TARS: 90 percent.

Cooper: 90 percent?

TARS: Absolute honesty isn't always the most diplomatic, nor the safest form of communication with emotional beings.

Excerpt from Interstellar
2014 Paramount Pictures movie

Notes

1. Modern generators with closed-loop electronic fuel injection and integrated catalyst can achieve CO emissions well below 500 g/hr. In 2016, the US Consumer Product Safety Commission (CPSC) proposed a rule to limit CO emissions from portable generators based on engine type: One-cylinder, Class II SI (spark ignition) engines should not exceed 150 g/hr. and two-cylinder, Class II SI engines should not exceed 300 g/hr. By 2028, California has a zero-emission standards plan for newly manufactured portable generators.
2. Cold and warm air have different air pressures due to their differing densities. Cold air is denser and exerts higher pressure than less dense warm air. This is

because cooler air molecules pack closer together and move more slowly which increases density and pressure, while warmer air molecules move faster and spread out, which decreases density and pressure.

3. The 1970s OPEC (Organization of the Petroleum Exporting Countries) oil embargo on nations that supported Israel during the Yom Kippur War of 1973 led to a global energy crisis. This embargo, along with production cuts and a sharp increase in oil prices, caused severe economic disruptions in the United States and other affected countries.

4. SDS (Safety Data Sheet) replaced the term MSDS (Material Safety Data Sheet) with full implementation in 2015. The change was an upgrade to the Globally Harmonized System (GHS) for chemical classification and labeling. The SDS format is more standardized and comprehensive than the previous MSDS, with 16 sections instead of the 9 with MSDS.

5. A *vertical platform lift* is a type of elevator designed for a single wheelchair. Comprising a lift inside an enclosed *hoistway*, it operates much like a common elevator.

References

1. Consumer Product Safety Commission (2006). Staff review of portable generator safety, 26 Oct 2006.

2. H. Frey, et al. (2003). On-road measurement of vehicle tailpipe emissions using a portable instrument, *J. Air Waste Mgmt. Assoc.*, V. 53.

3. G.S. Earnest, et al. (1997). Carbon monoxide poisonings from small, gasoline-powered, internal combustion engines: Just what is a "well-ventilated area"? *Am. Indust. Hyg. Assoc. J.*, 58, 787–791.

4. CCH (2003). *OSHA Standards for the Construction Industry*, 1st ed., CCH Inc., Chicago, 597 p.

Case Study 10: Side Dump Rupture

Expert Team:

- Forensic Engineer
- Structural Engineer
- Heavy Machinery Mechanical Engineer

Incident

A heavy equipment service company regularly operated side dump trucks for road construction and mining projects. One side dump failed in its third year of service during routine hauling and off-loading. The chassis ruptured during a dump when the operator heard a loud pop. The truck was immediately taken out of service, and the truck manufacturer was contacted. An insurance claim was filed, and an investigation was launched to determine the cause of the failure. Failure analysis involved a forensic engineer, a structural engineer experienced in materials testing and stress analysis, and a mechanical engineer familiar with heavy equipment design and operation.

DOI: 10.1201/9781003623311-22

Investigation

The investigation comprised the following tasks:

- Establish the side dump trailer history.
- Review manufacturer design specifications, recalls, technical bulletins, and warranty notice history.
- Inspect various similar side dumps in the fleet.
- Evaluate the side dump for fatigue and deformation.
- Static and dynamic stress testing.

History, Design, and Operation

Light-duty side dump trailers of this type have a useful life of 7 to 10 years, depending on maintenance and upkeep. Fifty 2004 side dumps were sold by the manufacturer. The original owner of the subject trailer was a heavy equipment leasing company that sold the dump truck after one year of service. Unbeknownst to the buyer, the manufacturer had issued a warranty notice to owners for repair of a defect in that model year trailer. A request was sent to the manufacturer for information about the warranty notice without response. No recalls were issued for the 2004 model year trailer.

The investigation team inspected other similar side dumps within the company fleet, and those of another friendly company. Similar side dumps of model years 2003, 2005, and 2006 were inspected. No signs of chassis fatigue were observed. No other 2004 model year side dump was available for comparison.

The steel trailer was fabricated with two rail beams joined with perpendicular cross beams and structural tubing ends to form the mainframe, which was augmented with brace plates and diagonal cross-bracing (Fig. CS10.1). This weldment was supported on triple axles with dual tires on each side, springs, and an industrial suspension system. The steel tub was mounted

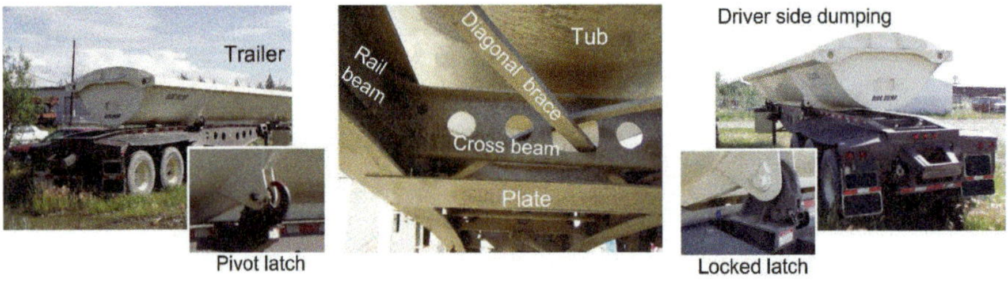

FIGURE CS10.1
Steel frame side dump trailer: rail beams braced with cross beams, brace plates, and diagonals.

to the mid-span of the frame with two locked latches on the passenger side, and a pair of pivot latches on the opposite side for driver-side off-loading. A pair of opposing hydraulic lifts mounted under the center of the chassis were used to maneuver the tub during off-loading.

Side dump operation for gravel hauls was routine but occasionally problematic with full or *sticky loads* (cohesive and frozen soil can stick to the tub) when the trailer would tip and then rebound once the load dumped. Sticky loads sometimes caused the trailer to tip and lift the tires off the ground on the opposite side of the off-load before rebounding. A loader was sometimes used to push on the center upper rim of the tub to assist with difficult dumps.

Mine hauls were more abusive with off-road transportation of rock and ore at open-pit mines. Bumpy roads with frequent potholes, slopes, and curves, and frequent excessive loads challenged trailer welds and mounts. Dump trucks were serviced every six months or as needed.

Evaluation of Fatigue and Deformation

Both forensic and structural engineers evaluated the trailer for signs of fatigue and deformation. The tongue (front) section of the chassis had been racked and twisted clockwise such that the rail beams were permanently bent to the right (Fig. CS10.2(a), (b), and (c)). The left rail beam exhibited a

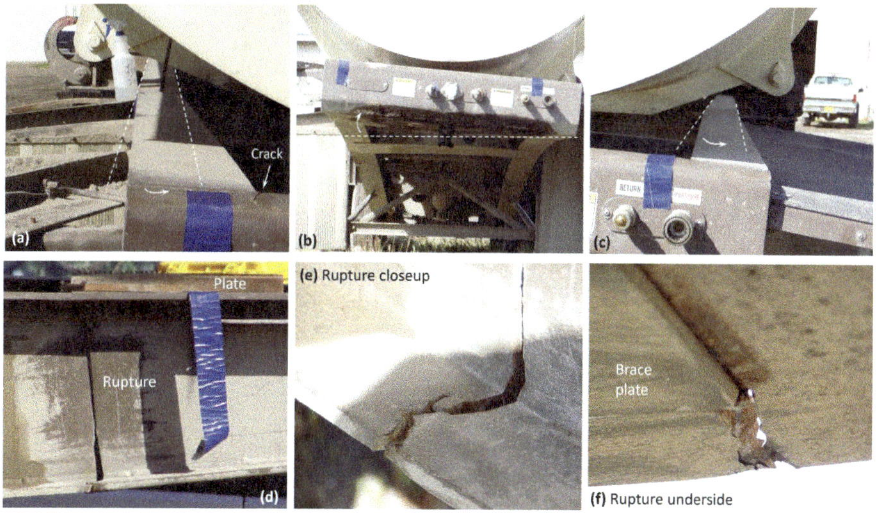

FIGURE CS10.2
Chassis torque and rupture: (a) bent left rail beam, (b) tongue torsional deformation, (c) bent right rail beam, (d) left rail beam rupture behind reinforcement plate, (e) side view of rupture, and (f) underside view of rupture at reinforcement plate.

tear emanating from the bottom flange upward through the web section (Fig. CS10.2(d) and (e)). This rupture was located at the trailing edge of top and bottom brace plates (Fig. CS10.2(d) and (f)). A 2.5 cm (1 in) long fatigue crack across the bottom flange of the right rail beam was also coincident with the trailing edges of the cross bracing. The proximity of the rail beam rupture immediately aft of the pair of reinforcing plates suggested a potential design flaw in the positioning of the reinforcement plates.

Another major problem with the trailer was poor quality welds. Numerous instances of weld fatigue were observed in the frame, some at critical junctures and mounts. For example, incomplete fusion was observed at the tub mounts of the two hydraulic lifts, mounts for the diagonal cross brace members that were pulling out, and along the weld that joins the right rail beam with the tongue hydraulic control tube (Fig. CS10.3). One cross beam to rail beam connection exhibited a fast low-voltage weld line with splatter and a missing bead at the bottom (Fig. CS10.3(c)); an adjacent diagonal bead across the rail flange should have filled the gap. In other instances, poor welds exhibited porosity and blowholes (poor gas shielding during welding), cracks and splits, and warpage (irregular heating during welding). Many of the frame cracks and welds exhibited rust and soil particulates inside the cracks, indicating they developed before the frame ruptured.

Two other observations related to a history of tub-to-frame contact (Fig. CS10.4(a)) and a deteriorating trailer hitch-pin assembly (Fig. CS10.4(b)). The tub had been contacting the trailing edge of the tongue beam over time.

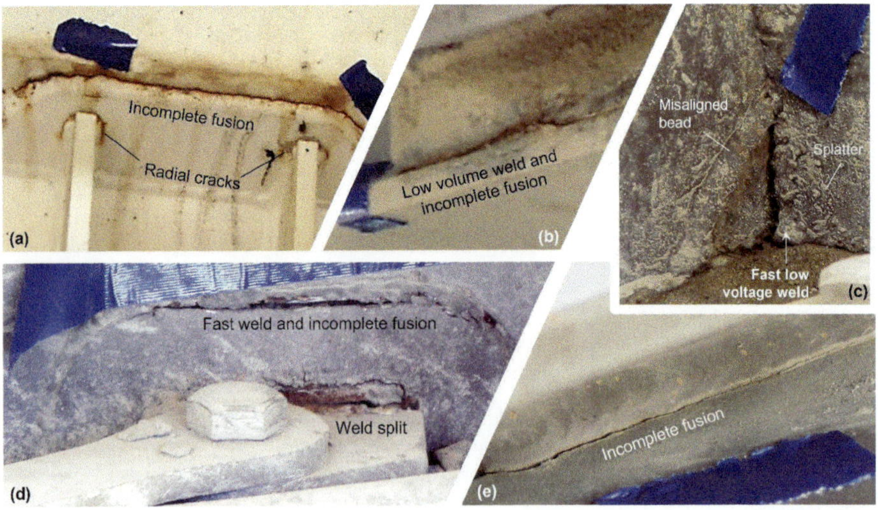

FIGURE CS10.3
Examples of poor welds at: (a) a tub hydraulic lift mount; (b) flange support tab; (c) rail beam–crossbeam juncture; (d) diagonal cross bracing mount; and (e) rail beam-tongue beam butt weld.

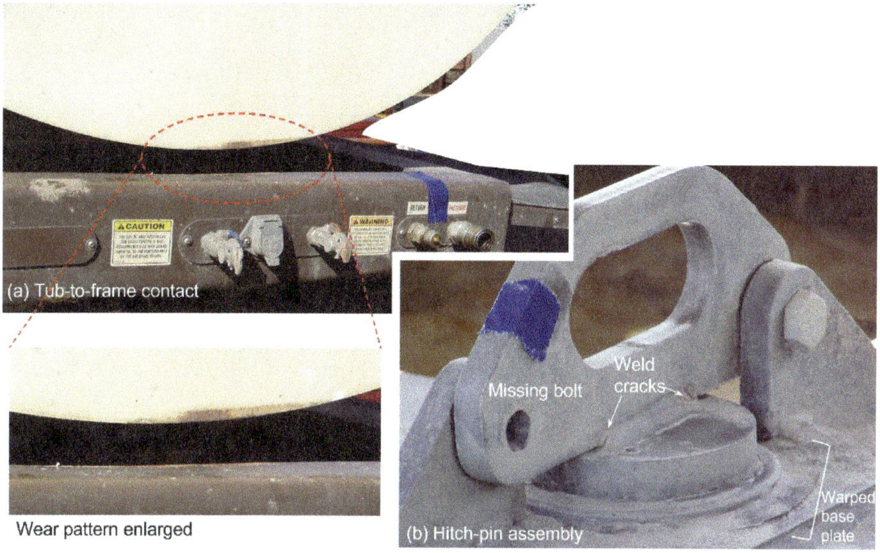

FIGURE CS10.4
(a) Trailer tub-to-frame wear pattern and (b) distressed hitch-pin assembly.

Rust patterns that developed after paint wore away attested to light repetitive impacts during off-loading. A deformed hitch pin assembly exhibited a radial weld crack around a portion of the base plate that was warped. Light rust on cracks in the butt weld that joined the hitch plate and pin suggested these cracks had developed over time. The hitch pin was also missing one of its two bolt fasteners. Since the assembly was galvanized and the radial crack did not exhibit rust, the origin of the base plate warp was uncertain.

Stress Analysis

Metallurgical analysis was deemed unnecessary as the documented distresses were attributed to design and fabrication issues. The trailer chassis was the focus of stress analysis, particularly the proximity of reinforcing plates to the rail rupture.

The structural engineer combined finite element analysis (FEA) with static and dynamic testing using Abaqus FEA software to simulate cyclic loading of the trailer frame. Important input parameters included frame member and reinforcement plate configurations, dimensions, and moments of inertia, steel properties (carbon content and yield stress), relative beam and reinforcing plate positions, tub operational loads (20 to 30 tons), and weldment reactions. The trailer was designed for 25-ton loads; 30 tons represents an extra-heavy load of large rock ore.

Von Mises stress (or equivalent stress) is a value used in engineering to predict the failure of ductile metals under complex loading conditions. The concept is based on the *maximum distortion energy theory*, which states that yielding (or failure) occurs when the distortion energy per unit volume of material reaches a critical value. When a chassis component is subjected to various loads (tension, compression, shear, bending, and torsion) during loading, hauling, and off-loading, it experiences complex stress states that are difficult to evaluate with a single value. Von Mises stress simplifies a complex stress state into a single scalar value, which represents the overall stress state at a particular point in the chassis. For example, Figure CS10.5 shows von Mises stress results in the vicinity of the rupture for a 20-ton load. Note that the maximum stress value of 66.4 MPa (9.63 ksi) occurs in the rail beam beneath the edge of the cross brace. Modeling during design would locate the cross brace a little further back over the beam to strengthen this area of the beam.

Comparing this von Mises stress value to the material's yield strength (the stress level at which the metal begins to permanently deform) allows the engineer to evaluate yielding for various load scenarios. If the von Mises stress exceeds the yield strength, the material is considered to have deformed. Also, repetitive stressing of insufficiently supported locations will fatigue and eventually lead to chassis rupture.

Static stress testing of the frame for driver-side dumping – the nominal operating mode of the trailer – confirmed the left rail beam would experience the greatest tensile and torsional stresses in the rupture area. Further, the right rail beam would experience compression and torsion stresses in the same general area on the opposite side, but to a lesser degree.

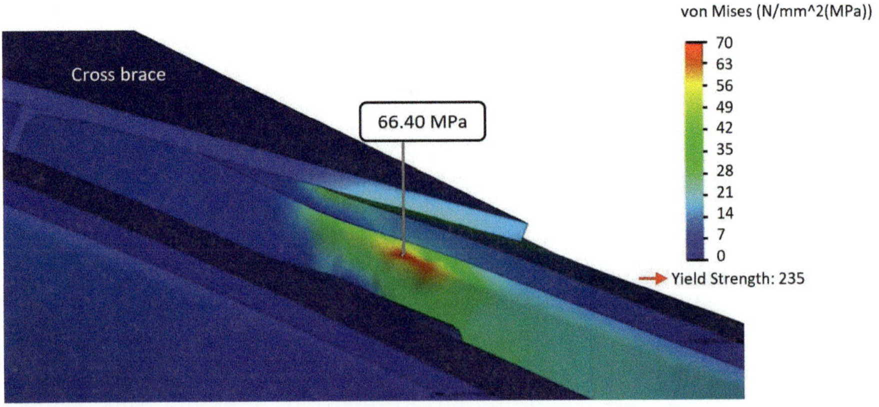

FIGURE CS10.5
Example von Mises stress analysis of trailer chassis at cross brace.

This explained the proximal top and bottom pair of mid-span reinforcement plates. However, dynamic simulations suggested the plates would ideally be positioned approximately 6.5 cm (2.56 in) back from their welded position. The 12.5 mm (~ 0.5 in) plate thickness was sufficient for reinforcement.

Fabrication Quality

A heavy machinery expert was consulted to review and opine about the trailer design and distress. The opinion of this expert was that

- Cross-brace placement should be in accordance with design plans and specifications, and weld-fabrication in accordance with American Welding Society (AWS) D1.1/D1.1M standards for fabrication of structural steel.
- The frame design was good, including cross brace thickness and configurations. Assuming design plans correctly specified locations of mid-span brace plates, their positions slightly offset or elsewhere located would be a fabrication error. Misread or misinterpreted plans, inattention to detail, and/or complacency are among the common causes of fabrication errors.
- The overall weld quality of the trailer frame was poor. Weld quality is a function of materials compatibility, temperature, and weld process. Signs of hydrogen embrittlement (filler metal with too much hydrogen), fast and low-temperature welds, and careless welding were evident. Figure CS10.3(c) demonstrates all three problems at a critical juncture of two structural frame beams, a bottom corner where access may have been difficult. Hydrogen embrittlement is often the cause of cracked or split welds. Incomplete fusion (i.e., a discontinuity between weld and base metals that is open to the surface) of hydraulic lift and diagonal cross brace mounts attest to poor welds at critical supports.
- Improper weld heat treatment was observed. A combination of single-pass and multi-pass butt welds was observed. Generally, multipass welds are necessary to properly join larger, more substantial members of the steel frame. Single-pass welds were used to fix light-load members and appurtenances to the main frame. Although heat treatment is not a concern with single-pass welding, it is crucial with thicker multi-pass welds. Proper preheating and sequencing of weld passes helps ensure that a consistent *interpass temperature* (i.e., the temperature at which successive weld runs are deposited) is maintained for a strong weld. Improper control of weld temperature can

lead to inconsistent grain structure in the metals and rapid cooling, resulting in a weak bond. The mount pullouts and some of the weld cracks and splits observed were indicative of poor heat control and weak welds.

- The warped base plate of the hitch-pin assembly was a fabrication defect. Weld material hid the defect until repetitive trailer racking exposed it. The dislodged bolt was not missing long as the hitch had not yet begun to deform the singular remaining fastener.

Summary

The side dump left rail beam rupture was caused by improper placement of a pair of reinforcement plates just ahead of a mid-span critical stress zone in the frame. A cyclical load simulation of the trailer frame using FEA coupled with dynamic stress analysis indicated the reinforcement plates should have been welded approximately 6.5 cm (2.56 in) further back from their fabricated position. The reinforcement plates were of sufficient thickness and size to strengthen the mainframe.

Routine driver-side dumping imparted nominal clockwise torque on the trailer frame. Occasionally, heavy or sticky loads would torque the frame enough to lift the left side wheels off the ground. Properly reinforced trailers are designed to elastically rebound without frame deformation under these operating conditions. Operating conditions of the trailer were considered normal for gravel hauls, but sometimes excessive with rock ore hauls. Repetitive excessive loads were a contributing factor to the rupture.

Numerous poor-quality welds were incidental to frame rupture. Although they would eventually contribute to a reduced useful life of the trailer, the permanent deformation of the frame and reinforcement diagonals was a consequence of the instantaneous rupture of the frame.

An agreement was worked out between the trailer owner and manufacturer for a replacement trailer, reputations were preserved, and litigation was avoided.

11

Water Treatment Systems

CONTENTS

CONCEPTS	KEY TERMS	ADDITIONAL READING
Water Quality Management	Pathogens	Exercise 21
Water Treatment Technology	Waterborne disease	Exercise 22
Industrial Process Water	Vulnerability assessment	
Forensic Chemistry	Syndromic surveillance	Case Study 11

> When the well is dry, we learn the value of water.
>
> **Paraphrased, Benjamin Franklin**

Water, essential for all bodily functions, enables digestion, nutrient transport, joint lubrication, waste flushing, and temperature regulation. It also supports community health, farming and industry, and sanitation. Safe water prevents illnesses and ensures basic survival. Lack of it means contaminated water, which can lead to deadly waterborne diseases, infections, and death. Further, clean water is essential for industrial cooling, processing, and cleaning. Thus, protection of water systems is a high infrastructure priority.

11.1 A Hundred Years of Water Quality Management

Drinking Water and Waterborne Diseases

The 1912 *Public Health Service Act* transformed a service for sick mariners into a comprehensive federal agency dedicated to protecting and improving the health of the entire population of the United States. This landmark

DOI: 10.1201/9781003623311-23

law established non-enforceable guidelines for ***waterborne disease*** (pathogens[1]) in interstate carriers. Formal monitoring of waterborne pathogens in the U.S. water supply system began in 1971, when the CDC (Centers for Disease Control), EPA (Environmental Protection Agency), and CSTE (Council of State and Territorial Epidemiologists) launched the Waterborne Disease and Outbreak Surveillance System (WBDOSS) for national tracking. Continuously refined by new science and laws since:

- 1974 Safe Drinking Water Act
- Quantitative Polymerase Chain Reaction (qPCR) pathogen-detecting technology in the 1990s
- EPA rules IESWTR and LT1ESWTR[2] in 1998 and 2002

Community water systems served 95% of the U.S. population in 2021[1]. A hundred years of basic monitoring of water contamination and systematic, national-level surveillance with WBDOSS has tracked waterborne disease outbreaks (Table 11.1). An interesting observation about the data is the increase in outbreaks during the first quarter of this century, largely attributed to aging pipes and distribution systems. Corrosion can leach lead and copper into water, while cracks or breaks in pipes can allow outside contaminants to intrude into the clean water supply. Small water systems, which serve most U.S. communities, often lack the financial and technical resources to maintain infrastructure and consistently meet all health standards.[2]

TABLE 11.1

Drinking Water Outbreaks, Illnesses and Deaths in the U.S. 1920–2021[3],[4],[1]

Community and Non-Community Systems				Community Water Systems	
Time Period	No. Outbreaks[1]	Cases of Illness	Deaths	Outbreaks per Year	Illnesses per Outbreak
1920–1930	255	102,024	669	17.2	513
1931–1940	275	93,306	320	12.8	748
1941–1950	313	53,935	61	9.6	467
1951–1960	111	12,491	10	4.1	247
1961–1970	131	46,399	20	3.9	1023
1971–1980	324	78,155	6	12.3	483
1981–1990	254	63,549	6	12.4	289
1991–2000	173	432,733[2]	60[2]	6.1	6929[2]
2001–2010	167	15,673	--	13.4	781
2011–2021	340	37,733	122	17.2	1210

[1] Excludes recreational water and private systems. Non-community water systems serve primarily transient populations (i.e., institutions, industries, camps, parks, hotels, and service stations). Community water systems serve most of the U.S. population.

[2] 403,000 cases, 50 deaths of immunocompromised people, and approximately 90% of illnesses per outbreak are attributed to the 1993 Milwaukee, Wisconsin *cryptosporidium parvum* outbreak.

Community Water Treatment Technologies

Community water treatment is a multi-stage process comprising removal of physical contaminants, followed by disinfection, and advanced treatment for dealing with specific chemical issues (Fig. 11.1). Physical contaminants are

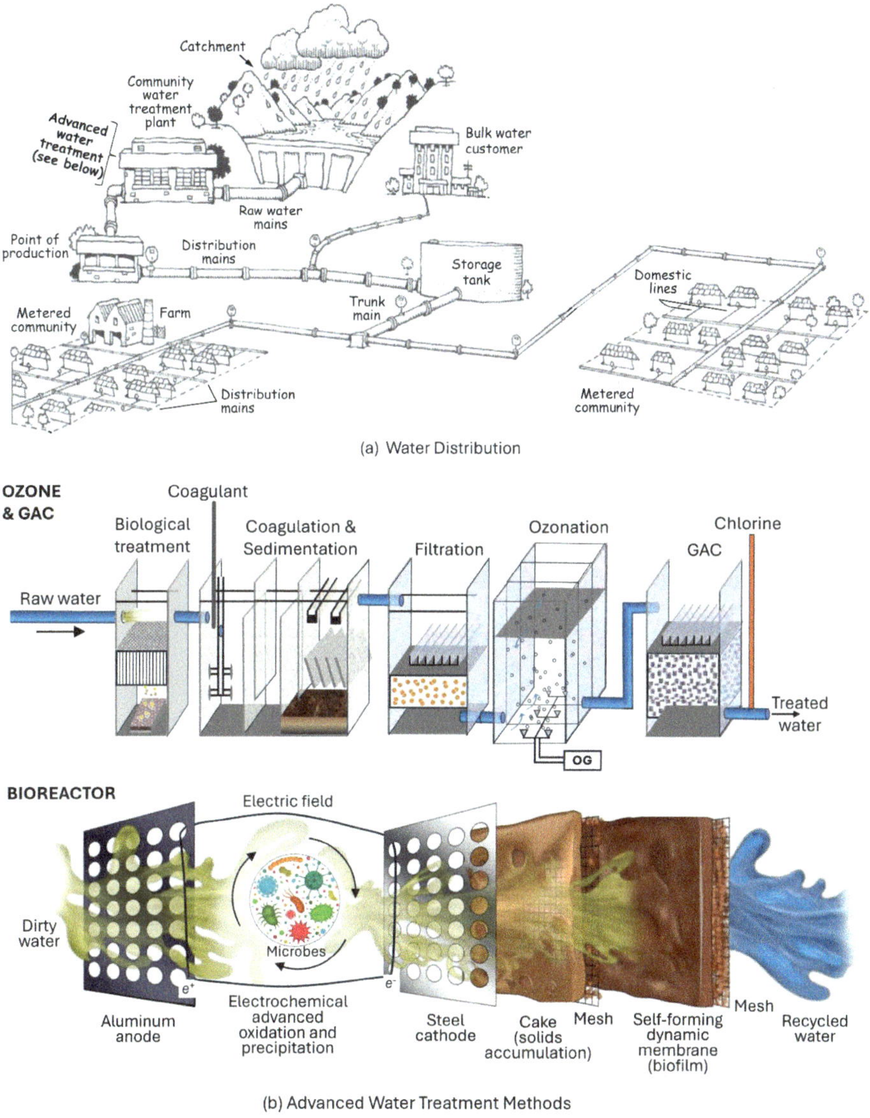

FIGURE 11.1
(a) Example water distribution from source to users and (b) illustrations of a couple of advanced water treatment methods. OG is an ozone generator, and the *bioreactor* is adapted from its original source[5] for industrial process water application here.

removed by coagulation/flocculation, sedimentation, and filtration. Filtration can be basic with sand and gravel media, or with membranes (MF, UF, NF, or RO)[3]. Disinfection with chlorine, ultraviolet light, or ozone kills pathogens (bacteria, viruses, and parasites). Emerging technology like advanced *smart membranes* (AFMs, MBRs, and catalytic)[4], ion exchange (IX), and granular activated carbon (GAC) are used to combat specific chemical issues like PFAS and pharmaceuticals and improve water reuse for complex industrial and municipal needs. Water treatment requires lots of energy, chemicals, monitoring and maintenance, and the more intricate is a treatment method, the greater the need to manage dosing and byproducts, and prevent biofouling, corrosion, and scaling.

Industrial Process Water

Industrial process water is used in manufacturing for cooling, heating, rinsing, dilution, and as an ingredient. Its treatment reduces scaling and corrosion, extends the useful life of equipment, and enables recycling of water within a facility, thereby balancing operational needs with regulatory compliance.

Process water typically requires specific quality levels that can only be achieved by treatment. Pretreatment, filtration, and chemical addition (e.g., coagulants, flocculants, softeners, scale inhibitors) are treatment processes that remove contaminants and impurities and control minerals. Deionization, reverse osmosis, or distillation are advanced treatments that remove impurities to achieve ultra-pure water quality.

11.2 Water Treatment Issues

The American Society of Civil Engineers (ASCE) grades U.S. infrastructure quality every few years. According to the ASCE 2025 Infrastructure Report Card, grades for drinking water and wastewater sectors are C– and D+, respectively. U.S. drinking water supply, estimated at 39B gallons/day, is withdrawn from surface waters (61%) and groundwater (39%) and distributed through approximately 2.2M miles of pipe. Approximately 18.2B gallons/day is withdrawn from surface (74%) and ground sources for industrial use. Utilities replace between 1% and 5% of their pipelines annually[6].

From source to facility, water endures a gauntlet of challenges before use. Host to microbes and parasites, tainted by wastes, pesticides, and fertilizers from agriculture, and industrial chemicals, water must be decontaminated, disinfected, and neutralized before use. Also, aged and failing infrastructure can contaminate drinking and processing water through pipe corrosion, breaches allowing pathogen/chemical intrusion, biofilm buildup, and inadequate treatment (Table 11.2).

TABLE 11.2

Common Problems with Drinking Water Distribution Systems

Aging Pipes and Corrosion

- *Lead/Copper*: Old lead and copper pipes and fixtures corrode, leaching Pb and Cu into water, especially if water is acidic.
- *Biofilm*: Bacteria colonize older rusty iron/steel pipes, creating biofilms that harbor microbes and cause leaks, leading to contamination.

Breaches and Leaks

Deteriorated pipes can crack or break, allowing contaminated groundwater, soil, and pathogens to seep in. Materials that degrade and contribute to contamination in older water systems include lead pipes and solder, zinc-coated galvanized steel, cast iron, and copper pipes.

Source Water Issues

Infrastructure designed to treat a particular source water may not protect against other contaminants from agriculture (pesticides, fertilizers) or industrial waste (heavy metals, solvents) that seep/migrate into the source.

Inadequate Treatment and Distribution

- *Disinfection Byproducts (DBPs)*: Chlorine used to kill germs can react with organic matter, forming harmful byproducts like trihalomethanes and haloacetic acids.
- *Incomplete Treatment*: Systems might not remove emerging contaminants like PFAS (forever chemicals) or nitrates from agricultural runoff. System upgrades with more sophisticated techniques (RO, UV, Ozone, GAC) often come after serious outbreak events.
- *Human Error*: Proper dosing, monitoring, and routine testing (turbidity, pH, temperature, conductivity, organics) are crucial for optimizing water treatment. Modern monitoring with real-time sensors and SCADA (Supervisory Control and Data Acquisition) for automated control reduces the potential for operator error and equipment malfunction.

Notes: RO is reverse osmosis; UV is ultraviolet light; Ozone is ozone gas (O_3); GAC is granular activated carbon filters.

Treatment failures can lead to disease outbreaks (Milwaukee, WI, 1993), public health crises (Flint, MI, 2014 and Newark, NJ, 2016), boil advisories, and supply interruptions. As the U.S. continues to grapple with funding infrastructure rehabilitation, news will continue to highlight the failures. Nevertheless, water infrastructure resilience, and water conservation and recycling are high-priority considerations moving forward.

11.3 Forensic Chemistry in Water System Investigations

Chemistry in water system investigations identifies contamination sources, age, and pathways using advanced analytics like *chromatography* and *spectroscopy* to detect, quantify, and trace pollutants. Forensic chromatography uses separation techniques like Gas Chromatography (GC) and High-Performance

TABLE 11.3

Important Roles of Forensic Chemistry in Water Treatment Investigations

Source Identification	
Pinpointing the origin of contaminants in water uses a multi-pronged approach that combines lab analysis (like GC-MS/LC-MS for chemicals, qPCR for microbes) with fingerprinting techniques (isotopes, genetic markers, chemical ratios), and field assessments to trace sources from industrial, agricultural, or sewage origins.	
Age and Impacts	
Water transport models (like MIKE, AQUATOX, SWAT) are mathematical tools used to simulate how dissolved contaminants move, disperse, and change in water. They use physics (flow, advection, dispersion) and chemistry (degradation, sorption) to predict contaminant spread for management, water quality impacts, and remediation[7].	
Contaminant Characterization	
Advanced lab techniques (like GC-MS, ICP-MS) are used for identifying specific harmful substances (metals, organics, TPH, VOC[5]) and their concentrations.	
Compliance and Fraud Detection	
Testing water against permit limits (like National Pollution Discharge Elimination System, a federal regulation for the discharge of pollutants from point sources (like pipes) into U.S. waters) and investigating improper discharges and infrastructure sabotage (deliberate leaks).	
Environmental Forensics	
Chemistry using GC-MS and isotope analysis is used to build a scientific narrative (composition, source identification and differentiation (*chemometrics*), age, fate, and transport) to help prove cause and effect in environmental damage and support legal cases[8].	

Liquid Chromatography (HPLC), often coupled with Mass Spectrometry (MS) as GC-MS/LC-MS, and Ion Chromatography (IC) to differentiate and identify contaminants. Forensic spectroscopy uses detection and identification techniques like Fourier-Transform Infrared (FTIR), Raman, Ultraviolet Visible Light (UV-Vis), Inductively Coupled Plasma (ICP), and Fluorescence. These nondestructive techniques are used to help differentiate between accidental spills, faulty infrastructure, or deliberate contamination. They transform raw chemical data into chemical fingerprints and actionable intelligence. Key roles of forensic chemistry in water investigations are summarized in Table 11.3.

11.4 Case Briefs

Three water crisis events are next summarized. The role of forensic chemistry in cause and origin determination with the Newark case brief is of particular interest. A short list of industrial mishaps (HDPE pipes, equipment failure, and sabotage) is then presented to further student interest in the subject. Also, Case Study 11 *Water Mistreatment* completes the chapter with an in-depth study about industrial chiller corrosion and breach of contract.

Milwaukee Cryptosporidium Outbreak, April 1993

Waste-tainted Lake Michigan water supply and inadequate filtration over a 16-day period at one of two treatment plants led to the largest recorded waterborne disease outbreak in the U.S. The most likely source of contamination was a sewage treatment plant outlet two miles upstream in Lake Michigan. Coincident with the contaminated inflow to the south Howard Avenue Water Treatment Plant (HAWTP) was filtration with inoperable streaming-current and turbidity monitors. Nearly half of the people that received drinking water from this plant became ill, resulting in approximately 403,000 confirmed illnesses and 69 deaths (mostly immunocompromised individuals).

Cause: Source water contamination and treatment mismanagement.

Investigation. A comprehensive investigation comprising the scope below provided multiple lines of independent evidence for the *cryptosporidium* outbreak determination:

1. Review and comparison of north and south (HAWTP) water plants drinking water quality data for January to April 1993.
2. Review of HAWTP policies, procedures, and functionality for the same period. (HAWTP was closed on April 8, three days after the beginning of the outbreak.)
3. Surveillance of testing labs and healthcare facilities/providers throughout the greater Milwaukee area for indications/diagnoses of waterborne disease.
4. Laboratory and clinical confirmations of *cryptosporidiosis* infection.
5. Survey of households in the greater Milwaukee area.
6. Data compilations and statistical analyses.

Outcomes. Approximately US$508M was spent on repairs, upgrades, and water quality monitoring improvements to the plants. Water treatment incorporated activated charcoal filtration, ozone disinfection, and SCADA real-time monitoring. Also, the HAWTP intake pipe was moved further out into Lake Michigan for higher-quality supply water.

Flint, Michigan Water Crisis, 2014–19

In spring 2014, the Flint Water Treatment Plant (Flint WTP) switched their drinking water supply from Detroit's system to the Flint River as a cost-saving measure. With inadequate corrosion control, lead leached from pipes into drinking water and caused widespread poisoning, especially in children. Residents complained about discoloration, foul-smelling, off-tasting water, and health issues like skin rashes, itchy skin, and hair loss. Lead levels in residential water exceeded the federal level of 15 parts per billion (ppb) in 17% of homes tested in a 2015 Virginia Tech study. Also, a 2015 independent pediatric study reported that "… nearly 9,000 children were supplied lead-contaminated water for 18 months."[9]

State and citizen action groups formed, including the NRDC (Natural Resources Defense Council), a major U.S.-based international nonprofit environmental advocacy group. The coalition sued the city of Flint and Wisconsin with a Safe Drinking Water Act lawsuit in 2016 after city, state, and federal agencies failed to act on the crisis.

Cause: Source change mismanagement. In 2013, the state of Wisconsin took control of indebted Flint, and the governor appointed an emergency manager to oversee and cut costs. The switch to a cheaper water supply (the Flint River) was an interim cost-saving measure until a new water pipeline from Lake Huron could be built. The Flint WTP was ill-equipped to adequately treat the Flint River water.[10]

Flint River History. The Flint River's history in Flint, Michigan is a century-long story of industry, pollution, and crisis. A water source for Native Americans, logging, and auto manufacturing, the river became an industrial sewer leading to the infamous 2014 Flint water crisis. Efforts are ongoing to restore the river's health and the riverfront for community use.

Outcomes. Flint citizens and action groups were successful in some civil suits, but the state Attorney General's office intervened and terminated all criminal cases. One civil case settled for US $626M, but no empowered officials were indicted. In 2024, the EPA issued a new rule that reduced the federal action level for lead to 10 ppb. (The EPA acknowledges that "there is not safe level of exposure to lead.") The rule also mandated the replacement of all lead service lines in the U.S. within ten years. Workers completed replacement of 11,000 lead pipes in Flint in July 2025. [9]

Newark Safe Drinking Water Case, 2016–24

The Newark Watershed Conservation and Development Corporation (NWCDC), a nonprofit organization, was established in 1973 by a Newark Municipal Council ordinance. NWCDC's creation was meant for water stewardship, with responsibility for managing Newark's freshwater assets. In 2010, corruption was exposed by the Newark Water Group – a citizen action group – in its efforts to prevent Newark from privatizing its water system. NWCDC became embroiled in governance failures and a massive fraud and kickback scandal. The New Jersey Office of the State Comptroller investigated and found that from 2008 through 2011, the NWCDC recklessly and improperly spent millions of dollars of public funds with little to no oversight by either its board of trustees or the City of Newark. An independent audit discovered that over a million dollars of taxpayer money went missing[11]. NWCDC was dissolved, and several people were convicted and sentenced to lengthy federal prison terms.

In March 2016, environmental officials announced that drinking water in multiple Newark Public Schools contained elevated lead levels above the 15 ppb federal standard. After repeated unsuccessful attempts to resolve the issue without litigation, the Newark Education Workers Caucus and the NRDC filed a citizen suit against the City of Newark and New Jersey

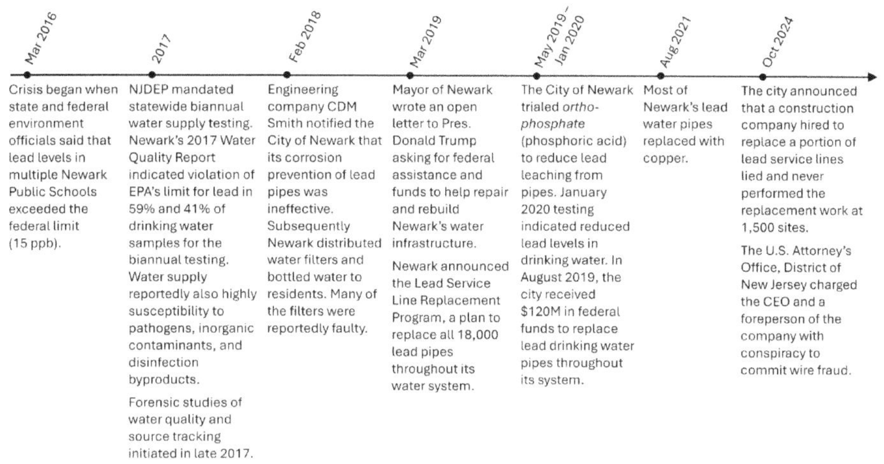

FIGURE 11.2
Newark water crisis timeline. (NJDEP is New Jersey Department of Environmental Protection.)

state officials in June 2018. Their intent was to secure safe drinking water for Newark residents; no money damages were sought[12]. The saga played out over eight years (Fig. 11.2). With federal assistance, Newark's lead pipes were replaced with copper, but not before another fraud and corruption scandal was uncovered in 2024. Legal actions continue as of this writing.

Forensic chemistry studies revealed treatment failures and the challenges of managing two water sources (Wanaque & Pequannock watersheds) with legacy lead infrastructure. Analysis revealed that the city's water lacked sufficient corrosion control that allowed lead to leach. The studies focused on the complex interplay of water source, treatment, and old lead service lines, and played a role in uncovering the fraud in pipe replacement efforts.

Causes: Records show that the mismanagement and corruption at the NWCDC (Newark Watershed Conservation and Development Corporation, managers of Newark's Water Safety Plan) occurred around the same time that the water's acidity levels started increasing, which accelerated the corrosion of pipes that caused lead to leach into the water. Additionally, city officials later admitted that some records of water tests were "lost" during the period of corruption at the NWCDC.

Forensic Study. Newark's drinking water distribution system used two geographically distinct sources (Fig. 11.3):

- Pequannock Water Treatment Plant (WTP), located ~ 25 miles NW of Newark, served customers in the West Ward and portions of the North, Central, and South Wards; and
- Wanaque WTP, located ~25 miles NNW of Newark, served the East Ward and the remaining portions of the North, Central, and South Wards.

FIGURE 11.3
Newark water system boundary, the Wanaque gradient in blue, and the Pequannock gradient is the unshaded section (obtained from City of Newark).

These two plants had different source waters, corrosion control strategies, and water parameters (pH, flow rate, etc.)[13].

In 2016, Newark's water and sewer system included more than 1000 miles of pipes more than 100 years old. Approximately 23,800 lead service lines (LSLs) – about 60% of the city's total service network – were identified within the system[11].

In 1992, under the EPA's new LCR, the action levels for lead and copper were 15 ppb and 1.3 ppm (or 1300 ppb), respectively, and biennial sampling was the norm. In accordance with federal regulation 40 CFR 141.86,

- Community Water Systems (CWS) that were above the action level for lead or copper were required to monitor semiannually.
- Systems that exceed 10 ppb for lead but were below 15 ppb were required to monitor for lead annually and copper triennially.

- Systems below 10 ppb for lead and 1.3 ppm for copper in consecutive monitoring periods were eligible for reduced monitoring.
- Any water system that utilized corrosion control treatment (CCT) was also required to conduct a CCT study if trigger levels (15 ppb for lead and 1300 ppb for copper) were exceeded.

Beginning on January 1, 2017, the NJDEP moved all large CWSs to standard monitoring (100 samples every 6 months). Sampling of the Newark water system is summarized in Tables 11.4 and 11.5. Calculated results from existing data for each independent source are shown in Figure 11.4.

TABLE 11.4

Newark Water System LCR Compliance Data, 1992–2021. (Annotated from original source [16].)

Sample Year	90th Percentile for Lead (ppb)	No. Samples	
2021 (1st)	6.6	116	
2020 (2nd)	13 (Elevated)	136	
2020 (1st)	13.9 (Elevated)	140	
2019 (2nd)	**31**	495	
2019 (1st)	**57**	348	Biannual sampling: required 100 samples every 6 months (2017–2021)
2018 (2nd)	**47.9** — Exceedances (Newark water crisis period) (2017–2019)	246	
2018 (1st)	**17.8**	126	
2017 (2nd)	**26.7**	153	
2017 (1st)	**27**	129	
2015	10 (Elevated)	52	
2012	9	53	
2009	11	52	Reduced triennial sampling (2003 -16): required 50 samples once every 3 years (2003–2015)
2006	7.9	52	
2003	8.3	57	
1998	12.4 (Elevated)	102	Biannual sampling: required 100 samples (1992–1998)
1992	**26.4** (Exceedance)	229	
1992	EPA LCR established 15 ppb Pb and 1300 ppb Cu ALs)		

TABLE 11.5

Newark Water System LCR Compliance Data, 2005–15. (Distribution across wards and water sources.)[16]

Sampling Period	LCR 90th Percentile for Lead (ppb)	Total Samples (*N*)	Non-LSL Samples (Cu with Pb solder) (*N* [%])	North Ward (*N*)	Central Ward (*N*)	South Ward (*N*)	West Ward (*N*)	Pequan-nock (*N*)	Wa-naque (*N*)
2013–15	10	52	16 (30.8%)	13	1	10	2	25	27
2010–12	9	53	14 (26.4%)	13	1	9	2	24	27
2007–09	11	52	13 (25%)	13	2	9	2	25	27
2005–07	7.9	52	13 (25%)	15	1	8	2	25	27

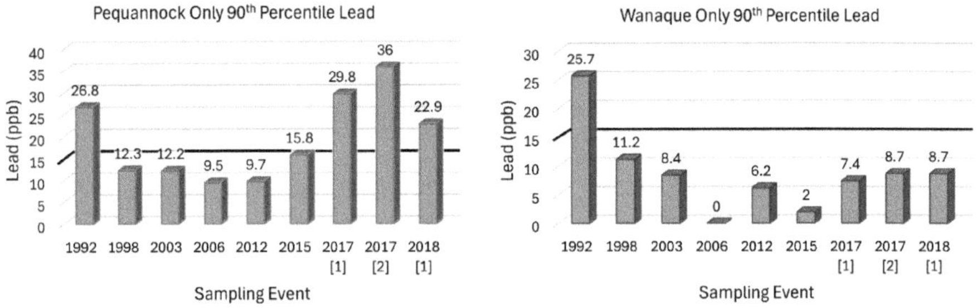

FIGURE 11.4
Newark water system LCR compliance sampling data, 1992–2018, for Pequannock (left) and Wanaque (right) sectors. (Amended from original source [16].)

Some observations about this information follow.
Table 11.4:

- Newark first exceeded the action level for lead in 1992 (90th percentile = 26.4 ppb) shortly after the LCR was implemented in 1991.
- Limited data were available for six years thereafter.
- Newark tested below the action level (90th percentile = 12.4 ppb) during the 1998 monitoring period, and was subsequently placed on reduced triennial sampling, requiring only 50 samples for testing once every 3 years.
- Newark remained on reduced triennial sampling from 2003 to 2016.

Table 11.5 shows the distribution of Newark's LCR compliance data sampling across wards and water sources across four triennial monitoring periods between 2005 and 2015.

- Approximately half of the samples were taken from each of the two water systems.

- Sampling distributions heavily favored the North Ward (Pequannock system) and South Ward (Pequannock and Wanaque systems). It is unclear why few samples were collected and tested for the East, Central, and West wards.
- Also, 25% or more of the sample locations were not from a lead service line (LSL).
- Each of the four triennial monitoring periods was below the 15-ppb action level.

Figure 11.4 shows the independent 90th percentile for lead results for the Pequannock and Wanaque systems for 1992 to 2018.

- Both systems would have exceeded the action level in 1992.
- The Pequannock system generally had higher lead levels thereafter and would have exceeded the action level from 2015 to 2018.
- The Wanaque system would not have exceeded the action level after 1992.

Corrosion Control Treatment

Both Newark WTPs implemented corrosion control using different treatments in the mid to late 1990s[13]. Based upon a 1990s city Corrosion Control Study, sodium silicate was chosen as the corrosion inhibitor for the Pequannock WTP. The study determined at the time that sodium silicate treatment in lead pipes fared better than orthophosphate or blended phosphate. Zinc orthophosphate was used as the corrosion inhibitor for the Wanaque WTP during the same time. Zinc orthophosphate became the new corrosion inhibitor for the Pequannock WTP in May 2019[14],[15].

Corrosion Control Treatment (CCT) can reduce but not necessarily eliminate lead leachate into drinking water. More successful at reducing soluble lead than particulate lead, CCT also does not control other conditions unrelated to corrosivity (e.g., plumbing age and disturbances, prolonged stagnation, low water use, and high water flow). LCR revisions focus heavily on CCT, sometimes allowing small CWSs to substitute CCT for LSL replacement when the action level is exceeded[16].

Ineffective CCT as a Potential Cause of Increased Lead in Drinking Water

In 2018, EPA and consultant CDM Smith characterized Newark pipe scales using x-ray diffraction analysis, scanning electron microscopy (SEM), and energy dispersive spectroscopy (EDS). Their findings indicated that silicate complexation with lead was not occurring, possibly resulting in higher soluble lead levels. A silica crust (SiO2) found on all pipes likely did not act as a barrier against lead release from the pipe, as it was porous[16].

Tetravalent lead (Pb(IV)) compounds can form scales on pipes in highly oxidizing waters with elevated oxidation reduction potential (ORP[6])[14].

Pb(IV) scales are effective at reducing lead corrosion when appropriate water chemistry (high pH and high ORP) is maintained. However, Pb(IV) scales will be disrupted, resulting in release of particulate lead, if water chemistry changes (e.g., pH is lowered or ORP is lowered due to change in disinfection)[16]. Also, the newly exposed surface is vulnerable to lead leaching from water with low pH.

Should redox conditions change, highly insoluble PbO_2 (plattnerite) reduces back to a more soluble Pb(II). Alternatively, if ORP is maintained but pH decreases, then previously stable PbO_2 can convert to more soluble Pb(II) carbonate compounds and soluble Pb, which increases both particulate and soluble water lead levels[13].

The EPA-CDM Smith study determined that LSL scales primarily consisted of hydrocerussite, cerussite, and plattnerite. Plattnerite is a Pb(IV) compound while hydrocerussite and cerussite are divalent Pb(II) carbonate-based scales. Hydrocerussite is stable at pH $\geq$ 9; cerussite is stable at pH 8 to 8.5. Since both Pb(II) and Pb(IV) scales were found on Newark's lead pipes, reduction of pH within Newark's distribution system may have been the primary cause for elevated lead levels by increasing the solubility of the scales[13].[7]

Sodium silicate is listed as a corrosion inhibitor in the EPA LCR. However, many studies have shown that silicate CCT is ineffective for controlling lead release in water systems. When sodium silicates dissolve in water, pH increases. Alternatively, orthophosphate controls lead solubility through the formation of hydroxyl-pyromorphite or other poorly soluble lead phosphate minerals. Generally, a pH increase will decrease lead solubility, but the specific role of silicate alone to increase pH is unknown[15].

It is possible that when Newark began using sodium silicate in the 1990s for the Pequannock system, the pH adjustment was enough to form the hydrocerussite, cerussite, and plattnerite scales, preventing large soluble lead release. However, when pH decreased, the scales could not be maintained, which resulted in soluble and particulate lead release[16].

A short list of additional water system failures is presented below. Research these case investigations for some interesting findings and outcomes.

Richmond, VA Water – Wastewater Calamity (2024–25)

A January 2024 storm overwhelmed the Hopewell Regional Wastewater Treatment Facility near Richmond, Virginia, with multiple untreated sewage overflows caused by faulty wiring, broken equipment, and lack of maintenance. Local waterways were contaminated with millions of gallons of raw sewage. Then in early 2025, a power failure at Richmond's WTP caused by faulty wiring and IT system failures resulted in loss of water service and a boil-water advisory for the metro area. Regulatory agencies issued violations citing management errors and lack of preventative maintenance and preparedness.

Oxidative Degradation of HDPE Pipes from Exposure to Drinking Water Disinfectants (2009 and beyond)

Significant studies showed that drinking water disinfectants like chlorine and chlorine dioxide cause oxidative degradation (OxDeg) in HDPE pipes, resulting in embrittlement, premature cracking, and reduced useful life. OxDeg consumes pipe antioxidants (additives in plastic pipes that prevent degradation from heat, oxygen, and chemicals) and forms reactive carbonyl groups on pipe surfaces that make them brittle and susceptible to failure. Investigators used FTIR to detect carbonyl formation and OIT (Oxidation Induction Time) tests to measure antioxidant depletion. Research findings about HDPE susceptibility to chemical migration and degradation products into water.

Corrosion Failure of Tube Heat Exchanger (2015)

An industrial shell and tube heat exchanger at a desalinization plant failed due to tube corrosion. The exchanger was used for heating desalinized basin water during winter. Desalinated water treated with RO technology contained up to 800 ppm chloride, and pH was adjusted using hydrochloric acid (HCl). Exchanger tubes were made of 316 stainless steel (316 SS), and tube temperatures reached as high as 80°C during heating[17].

The corroded tubes were subjected to chemical analysis using optical emission spectrometry to verify 316 SS composition. Electrochemical corrosion testing (cyclic anodic polarization with a computerized potentiostat in a solution containing 1000 ppm chloride) was then performed on the tubes to assess their susceptibility to pitting. Microscopic evaluation defined the corrosion mechanism for different chloride levels between 800 and 1000 ppm (operating conditions) [17].

Microscopic examination revealed tube pitting corrosion was greatest with water at higher chloride levels and temperature. Test results led to recommendations to control chloride at less than 300 ppm in basin water by optimizing the RO process and controlling pH using sulfuric acid instead of HCl (which increases chloride content in the water).

Sylvan Township, Michigan Security Breach and Vulnerability Assessments (2024)

On July 8, 2024, Sylvan Township, Michigan, experienced a security breach at its WTP, leading to vandalism, control systems tampering, and a precautionary "Do Not Drink" advisory for residents. The vandal damaged a chemical scale used for measuring water additives, poured cement down drains, and tampered with control systems, electrical connections, and infrastructure components. Research post-incident investigations of physical and cyber vulnerabilities and formal vulnerability assessments for critical infrastructure.

11.5 Vulnerability Assessment, Syndromic Surveillance, and AI for Protection of Community Water Systems

Future protection of our vital water systems includes vulnerability assessments, syndromic surveillance, and artificial intelligence (AI). A ***vulnerability assessment*** (VA) of a CWS identifies threats (natural, malevolent, cyber), weaknesses in infrastructure (pipes, treatment, storage), and mitigation strategies to ensure water quality and reliability. Further, *risk and resilience assessment* and *indicator-based metrics* (land use, infrastructure, service gaps) with consideration for climate change are tools used to evaluate the impact of potential failures on water reliability, quality, and public health. Required by law (Bioterrorism Act (2002) and America's Water Infrastructure Act (2018) since 9/11, VAs are integral to the protection of critical infrastructure.

Syndromic surveillance in water systems uses real-time health data (like ER visits for diarrhea, gastrointestinal issues, vomiting) to spot potential waterborne outbreaks early. Acting as an early warning system before lab confirmations and incorporating it with environmental data (like water quality sensing information) has potential for early detection of contamination and improved prevention (Fig. 11.5).

AI is an upcoming technology in water treatment systems that uses machine learning for real-time monitoring, predictive maintenance, and process optimization, potentially leading to better water quality, increased efficiency, and cost savings. Further, AI can improve system sustainability by analyzing sensor data to adjust chemical dosing, forecast demand, detect leaks, and create virtual *digital twins*[8] for simulating complex scenarios. This technology has the potential to transform complex water operations and help utilities manage resources more intelligently.

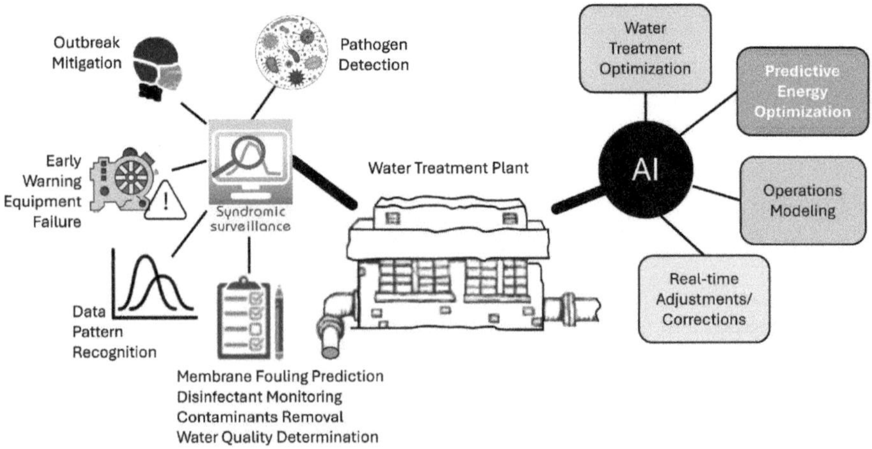

FIGURE 11.5
AI and syndromic surveillance in water treatment monitoring.

Notes

1. A pathogen is a bacterium, virus, or other microorganism that can cause disease. Some familiar pathogens are *E. coli, Salmonella, Legionella, Giardia,* and *Cryptosporidium.*
2. The EPA's Interim Enhanced Surface Water Treatment Rule (IESWTR) and Long Term 1 Enhanced Surface Water Treatment Rule (LT1ESWTR) regulate microbial pathogens in drinking water. IESWTR requirements pertain to large systems (10k+ people) and those of LT1ESWTR to smaller systems (<10k people). The focus of the regulations is strict filtration and turbidity monitoring for removal of pathogens while balancing disinfection byproduct risks.
3. MF is microfiltration, UF is ultrafiltration, NF is nanofiltration, and RO is reverse osmosis.
4. Smart membranes integrate nanomaterials, catalysts, and electrical charges to remove tough contaminants (viruses, pharmaceuticals, and microplastics). AFM is automatic microfiber membrane, MBR is membrane bioreactor, and GAC is gram-activated carbon.
5. TPH is total petroleum hydrocarbons; VOC is volatile organic compounds.
6. ORP (Oxidation-Reduction Potential) in water measures its ability to cleanse itself or break down contaminants, indicating water quality and sanitizer effectiveness (e.g., chlorine above 650 mV typically means well-sanitized water). A positive ORP (e.g., +200 to +700 mV) means the water is oxidizing and has strong sanitizing power (good for killing germs) but also promotes scaling. A negative ORP (e.g.,–50 mV to –800 mV) indicates it's more reducing, meaning the ORP is busy breaking down waste or contaminants, making sanitizers less effective.
7. A statistically significant change in pH (±0.01) can alter lead pipe scaling. According to a CDM Smith report (ref. 13), "… the average pH ranged from 7.6 to 8.0 from 2005 to 2012 with a few sustained periods above pH 8.0. After 2012, pH dropped to an average range of 7.5–7.7." Also, pH decreased from 8.08 to 7.33 from 2015 to 2016. Variability/trends in water acidity could not be inferred for years prior to 2005 as the pH data was scant.
8. A digital twin is a dynamic, virtual replica of a physical process or system. Synchronized with real-time data from sensors to mirror its real-world counterpart's performance, conditions, and behavior throughout its lifecycle, the replica enables analysis, simulation, and optimization for improved decision-making and predictive maintenance.

References

1. WBDOSS Annual Surveillance Reports, http://www.cdc.gov/healthywater/surveillance/ index.html. Retrieved 14 Dec 2025.
2. USEPA (2023). Report on the Environment Trend Summary Protocol, EPA/600/R-23/321, www.epa.gov/roe.

3. American Water Works Association (2006). Manual of Water Supply Practices-M48, Waterborne Pathogens, 2nd Ed.
4. Jasen M. Kuhn, et al. *Surveillance of Waterborne Disease Outbreaks Associated with Drinking Water – U.S., 2015–2020.* Surveillance Summaries, 14 Mar 2024, 73(1);1–23.
5. J.M.J. Millanar-Marfa et al. (2022). Advanced wastewater treatment and membrane fouling control by electro-encapsulated self-forming dynamic membrane bioreactor. *npj Clean Water* 5:38.
6. D. Olson et al. (2025). *2025 Report Card for America's Infrastructure.* Reston, VA: American Society of Civil Engineers, 2025. Accessed: 21 Dec 2025. [Online]. Available: https://infrastructurereportcard.org/
7. P. Talukdar, B. Kumar, and V.V. Kulkarni (2023). A review of water quality models and monitoring methods for capabilities of pollutant source identification, classification, and transport simulation. *Rev Environ Sci Biotechnol.* 22:653–677.
8. D.L. Saber, D. Mauro, and T. Sirivedhin (2005). Applications of forensic chemistry to environmental work. *J Ind Microbiol Biotechnol.* 32(11-12):665–668. doi: 10.1007/s10295-005-0013-z.
9. M. Denchak, *Flint Water Crisis: Everything You Need to Know,* July 1, 2025. https://www.nrdc.org/stories/flint-water-crisis-everything-you-need-to-know. Accessed: 23 Dec 2025.
10. Flint Water Advisory Task Force. Flint Water Advisory Task Force Final Report, March 2016. https://flintwaterstudy.org/wp-content/uploads/2016/03/Flint-task-force-report_2438442_ver1.0.pdf, retrieved 27 Dec 2025.
11. State of New Jersey, Office of State Comptroller. *Investigation Report: Newark Watershed Conservation and Development Corporation*, February 19, 2014. https://www.nj.gov/comptroller/news/docs/newark_watershed.pdf. Retrieved 22 Dec 2025.
12. NRDC, *Newark Safe Drinking Water Case Documents,* February 12, 2021. https://www.nrdc.org/resources/newark-drinking-water-case-documents. Accessed: 23 Dec 2025.
13. CDM Smith. (2019). Pequannock WTP corrosion control review and recommendations. Final City of Newark, Lead and Copper Rule Compliance Study.
14. CDM Smith. (2019). Wanaque gradient corrosion control review. City of Newark Lead and Copper Rule Compliance Study Wanaque Gradient.
15. CDM Smith. (2021). Pequannock gradient corrosion control treatment effectiveness evaluation and pipe loop study results. Final Report, City of Newark.
16. S. A. Stratton et al. (2023). The lead and copper rule: Limitations and lessons learned from Newark, New Jersey. *WIREs Water.* 10:e1620. DOI: 10.1002/wat2.1620
17. I.M. Ghayad, Z.A. Hamid, and N. Gomaa (2015). A case study: Corrosion failure of tube heat exchanger. *J. Metallurgical Eng* (ME). 4:57–61. doi:10.14355/me.2015.04.007

Case Study 11: Water Mistreatment

- Water Quality Management
- *Chiller Corrosion*
- *Beach of Contract*

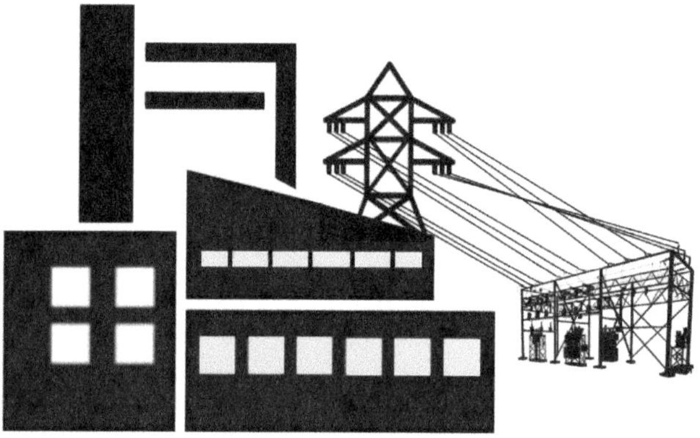

Chillers provide process cooling of critical power plant equipment such as turbines, generators, electrical rooms, and other machinery that generate substantial heat during operation. By circulating a chilled fluid (i.e., closed-cycle recirculating water in this case) through heat exchangers, chillers absorb and remove excess heat, preventing overheating and optimizing mechanical performance. However, customizing process water chemistry for specific use is crucial to prevent fouling, scaling, and corrosion of chiller equipment.

Recommended process water for power plants is typically low in minerals and suspended solids and treated to control both corrosion and biological growth. Closed-cycle recirculating systems using treated groundwater, municipal, or advanced-treated recycled water are preferred for their efficiency and reduced environmental impact. Regular monitoring and testing, and compliance with environmental regulations are essential for optimal equipment operation, sustainability, and residuals discharge.

DOI: 10.1201/9781003623311-24

Background

On March 28, 2016, a state university (University) issued an *Invitation for Bids* (IFB) with plans and specifications detailing a scope of work for repair or replacement of aged mechanical units at the University's arena energy plant (the Project). The scope of work was to provide chillers, cooling towers, pumps, air handling units, fan coil units, fans, and other items required for the Project.

Chillers and the Contracted Parties

On October 6, 2016, the University's Procurement Director awarded the $3.34M contract to a domestic commercial plumbing company (the Contractor) with experience in the type of work identified in the IFB. Shortly thereafter, the Contractor purchased three chillers, each composed of a helical rotary compressor, evaporator, and condenser, from a building technology and energy solutions company (the Manufacturer) to initiate the Work. Figure CS11.1 portrays an example of a water-cooled chiller system at an industrial plant and chiller function. Of note are the copper barrel tube sheets in the evaporator and condenser, the focus of scaling and corrosion for this case study.

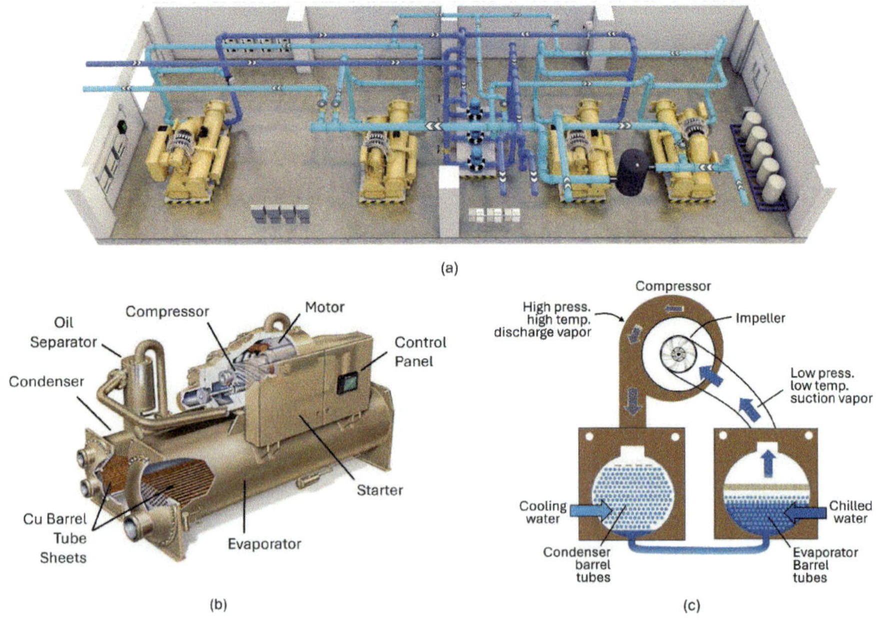

FIGURE CS11.1
(a) Example industrial layout of a water-cooled chiller system and (b) chiller components[1], and (c) illustration of water circulation through copper barrel tube sheets.

The Manufacturer (and chiller operating manual) recommended the services of a qualified water treatment specialist to determine process water treatment requirements for the Project. This was because chillers require chemical water treatment for proper operation, and untreated or improperly treated water can cause scaling, erosion, corrosion, and algae or slime buildup within the equipment.[2] Consequently, on June 8, 2017, the Contractor entered into a purchase order agreement (the Agreement) with a specialty chemical services company (the Specialist) to provide the required chemical treatment for the chillers. The Agreement specified:

1. Water treatment equipment, one year of chemical supplies, and monthly analytical service for the University equipment in accordance with the IFB specifications.
2. Expressed warranties to the Contractor.
3. The Specialist must obtain and maintain comprehensive Commercial General Liability insurance (CGL insurance), including property damage and contractual liability coverage, and name the Contractor as an additional insured (the Policy).

The Specialist identified three specific chemical products intended for use in the water treatment of the chiller systems as part of the Agreement. However, the dosage of each chemical product and instructions for their integration into the process water were not specified in the Agreement.

⚠ **CAUTION**

Process Water Treatment!

The use of untreated or improperly treated water in a chiller may result in scaling, erosion, corrosion, algae or slime. It is recommended that the services of a qualified water treatment specialist be engaged to determine what water treatment, if any, is required. Chiller manufacturer assumes no responsibility for equipment failures which result from untreated or improperly treated water, or saline or brackish water.

The chiller systems went online in July of 2017. On August 8, 2017, the Specialist began adding a single chemical corrosion inhibitor to the condenser, but not to the chilled water loop portion of the system. Notably, the Specialist did not integrate any other chemical product at that time, excluding a key biocide required per IFB specifications. While acknowledging that silica is a key water quality parameter that can be involved in corrosion processes when present in the service water, the Specialist chose not to test for it in the initial start-up of the system. Contractual relationships and important contract documents are summarized in Table CS11.1.

TABLE CS11.1

Contracted Parties, Contracts, and Subcontractors

Party	Role/Contracts	Description
University	Project owner and IFB	Chiller system bid solicitor
Trane	Chiller manufacturer	Equipment provider
Plumbing Company	Contractor	Equipment and chiller systems installer
Chemical Service Company	Specialist	Chiller process water treatment subcontractor to Plumbing Company
Agreement	Contract between Contractor and Specialist	Contractual obligations
CGL Insurance	Specialist insurance policy with Contractor as added insured	Contractor hold-harmless for water impacts on chiller system
Corrosion Investigation		
Corrosion Consultant	Subcontractor to Contractor	Chiller corrosion investigation
First Insurance Company	Specialist's insurer	Required for contracting
New Replacement Equipment		
Chem, LLC	Replacement Chemical Specialist	Water treatment subcontractor to Plumbing Company for new replaced chillers

Corrosion Problem Detection

On October 15, 2017, the Specialist serviced the chiller system and noted that a corrosion inhibitor was not added to the chilled water loop portion, commenting that *"current conditions may lead to corrosion."* The Specialist subsequently added a corrosion inhibitor to the chilled water portion of the system on October 30, 2017, nearly three months after starting work on the Project.

Shortly thereafter, on November 1, 2017, the University performed eddy current testing (ECT) on the equipment installed for the Project. ECT is an electromagnetic testing method used in nondestructive testing to detect and characterize surface and subsurface flaws in conductive materials. Performed in accordance with American Society for Testing and Materials (ASTM) Standard E2884-22, *Standard Guide for Eddy Current Testing of Electrically Conducting Materials Using Conformable Sensor Arrays*, the test revealed that corrosion may be occurring on the chiller's internal condenser barrel tube sheets. The tube sheets are covered with a steel cover (or *head*), and within each cylinder (evaporator and condenser) are copper pipes configured in layers (or *sheets*). Process water is circulated through the copper pipes and interacts with the steel heads of each cylinder. The ECT revealed excessive corrosion in the chillers, particularly on condenser tube sheets and the heads.

Aware of the substandard water treatment needed to protect the chillers, and after notification to the University of the ECT findings, the Contractor retained a private corrosion consultancy (the Consultant) to investigate the conditions of the newly installed mechanical equipment.

On January 29 and 30, 2018, chillers 1 and 2 were taken off-line and the condensers inspected. The Consultant determined that the barrel tube sheets were showing signs of *tuberculation*[1] caused by premature corrosion, and that the steel heads were also corroding (Figure CS11.2).

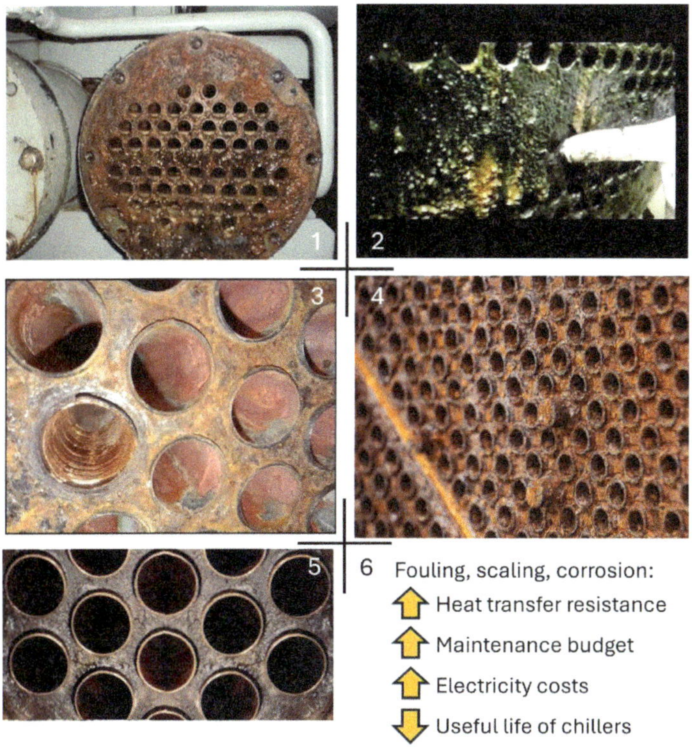

FIGURE CS11.2

Example chiller unit corrosion. Image 1: Front end of condenser with head removed[4]. Image 2: Barrel tube sheet microbial corrosion[5]. Image 3: Barrel tube tuberculation[6]. Image 4: Tube galvanic corrosion[7]. Image 5: Undamaged copper barrel tubes (contrast with Image 3). Image 6: Economic impacts of fouled/corroded chillers.

Claim

The University issued a formal written claim to the Contractor about the damaged chillers on July 23, 2018. This served as *"official notice that the chiller condenser barrel tube sheets are rejected and either need to be repaired or replaced to the satisfaction of the chiller manufacturer such that all warranties related to the chillers shall be provided in accordance with the contract documents and also meet the University's expectations of the chillers' operation and performance life expectancy"* (the Claim).

On July 31, 2018, the Contractor, through its legal counsel, provided a *Notice of Tender of Claim*[2] (the Tender) to both Specialist and its insurer, First Insurance Company (FIC), to cover the costs of replacement. The Tender highlighted the Specialist's obligation to defend, indemnify, and hold harmless the Contractor from loss and liability of any nature or kind arising out of or existing because of the Specialist's work under the Agreement. The Agreement provided that the Specialist *"...shall be responsible for and pay all damages sustained by the Contractor, including but not limited to all consequential damages and liquidated damages."* Furthermore, the Agreement required the Specialist to abide by the following:

> [Specialist] shall defend, indemnify and hold Plumbing Company, its affiliated companies, its officers, directors, employees, agents and customers harmless, with counsel selected by Plumbing Company, from loss and liability of any nature or kind arising out of, or existing because of the infringement or alleged infringement of any patent for or on account of the manufacture, sale of use of any goods furnished hereunder.

On August 2, 2018, the Specialist, through its counsel, denied the Tender and claimed it satisfied the Agreement and IFB specifications by providing *"the specified chemical treatment equipment"* and *"have fulfilled the requirement for (1) one-year supply of chemicals and analytical services."* Additionally, the Specialist affirmed its denial of the Tender because the University's Claim rejected the chillers themselves, which the Specialist did not purchase or install. Further, the Specialist claimed that its Corrosion Consultant determined that the water treatment program and alleged monitoring of the chillers did not cause corrosion. On August 9, 2018, FIC sent a letter to the Contractor's counsel denying the Tender because it believed that the Specialist's ongoing operations did not cause or contribute to the alleged property damage.

In February 2019, the University renewed its demand to restore or replace the chillers in accordance with the Manufacturer's recommendations, including, among other things, cleaning the entire Project, implementing an acceptable treatment plan, and testing the water treatment system to ensure proper operation. By March 15, 2019, the Contractor confirmed it would

- Fully replace the existing chillers (replacement was cheaper than repairs).

- Factory test the existing mechanical system in accordance with the Contract.
- Provide all test results to the University before installation of the replacement equipment.

On April 2, 2019, the University approved the replacement of the chillers after extensive investigation and cost analysis performed by the Contractor.

Lawsuit

A breach of contract lawsuit was filed by the Contractor against the subcontracted Specialist in April 2019. What follows describes the next three years of pre-trial disclosures, chiller system testing, COVID-19 pandemic delays, legal motions and technical arguments made, and the judge's final decision. Testimony by the Plaintiff's expert water quality engineer was crucial to the outcome of this case.

Breach of Contract

On April 5, 2019, the Contractor's (the Plaintiff) attorney filed a legal complaint (the Complaint) with the state's circuit court against the Specialist and related entities (the Defendant). The Plaintiff alleged (1) breach of contract, (2) breach of express warranty, and (3) declaratory relief for defense and indemnification. Specifically,

1. *Breach of Contract*
   a. The Agreement is a valid and enforceable contract between the Plaintiff and the Defendant.
   b. The Defendant failed to perform under the Agreement, including but not limited to the duty to defend and indemnify the Plaintiff against the Claim.
   c. As a result of the Defendant's breach, the Plaintiff has been damaged in an amount to be proven at trial.
2. *Breach of Express Warranty*
   a. The Defendant expressly warranted that the goods, articles, and work covered by the Agreement would comply with the requirements of the IFB specifications, drawings, special conditions, samples or other description provided by the Plaintiff.
   b. The Defendant expressly warranted that the goods furnished under the Agreement would be of good material and workmanship.

c. The Defendant expressly warranted that all goods, articles, and work under the Agreement would be merchantable and fit and sufficient for the purpose intended.
d. The Defendant's failure to provide goods, articles, and work in accordance with the requirements of the specifications, drawings, special conditions, samples or other description provided by the Plaintiff constitutes a breach of its express warranty.
e. As a direct and proximate cause of the Defendant's breach, the Plaintiff has been damaged in an amount to be proven at trial.

3. *Declaratory Relief for Defense and Indemnification*
   a. The goods furnished by the Plaintiff pursuant to the Agreement were not fit for the particular purpose for which the goods were required.
   b. The Plaintiff is entitled to reimbursement for attorneys' fees already paid in defending against and including payment for the future defense of the Claim the Contractor placed on the Specialist's insurers, including but not limited to defense against any litigation brought against it relating in any way to corrosion of the chillers due to the chemical treatment provided by the Defendant and any future appeals.
   c. The Plaintiff is entitled to indemnification per the Claim against the Specialist's insurers for the allegations against the Contractor made by the University.

Pre-trial Disclosures

On May 17, 2019, the Plaintiff and Defendant issued *Requests for Production of documents.*[3] Document gathering and disclosures continued in April and July 2019, while the Plaintiff processed the chiller replacement orders, equipment coordination, and shop drawing review procedures. Also, a water quality chemist and engineer (the WQ Expert) was retained to determine why the water treatment led to precipitation and corrosion.

Then in late July through early August 2019, the damaged chillers were removed and replaced with equipment of the same make and manufacturer. The damaged equipment was transported to a secure storage location. Meanwhile, the Contractor retained a new water treatment chemical services provider, CHEM, LLC (new Replacement Specialist) for the new systems. The Replacement Specialist began its water treatment program of the new units on August 9, 2019, which included

- A stronger scale and corrosion inhibitor than was used in the original chillers.
- Use of bromine, a non-oxidizing biocide to mitigate biofouling.

The University then performed ECT on the replacement chillers during October 15–17, 2019, and periodically thereafter through 2021; no abnormal conditions were reported.

Corrosion Testing

On December 17, 2019, the Replacement Specialist obtained additional corrosion samples from the failed chiller units at the University's storage location. Corrosion was evident on evaporator and condenser tube sheets of each of the damaged chillers.

Corrosion samples were analyzed using scanning electron microscopy with energy-dispersive X-ray spectroscopy (SEM/EDS). The SEM/EDS analyses detected iron (Fe), oxygen (O), carbon (C), silicon (Si), and varying amounts of other elements. Of particular concern was silicon, a well-known precipitant that causes scaling. The quantity of silicon detected in the samples was abnormally high, as shown in Figure CS11.3. Silica, a compound of silicon and oxygen, precipitates out of the water and deposits on surfaces, forming a *scale* or coating. Scale buildup on barrel tube surfaces can significantly reduce chiller efficiency and lifespan.

COVID-19 Pandemic Delay

On March 5, 2020, the state Governor declared a state of emergency in response to the emerging public health threat posed at that time by COVID-19. The Order mandated, except for emergency and time-sensitive matters, that all in-person appearances for civil, family, and, to the extent possible, criminal dockets were limited starting March 17, 2020, until or unless otherwise ordered. The Presiding Judge of each circuit court developed a specific plan

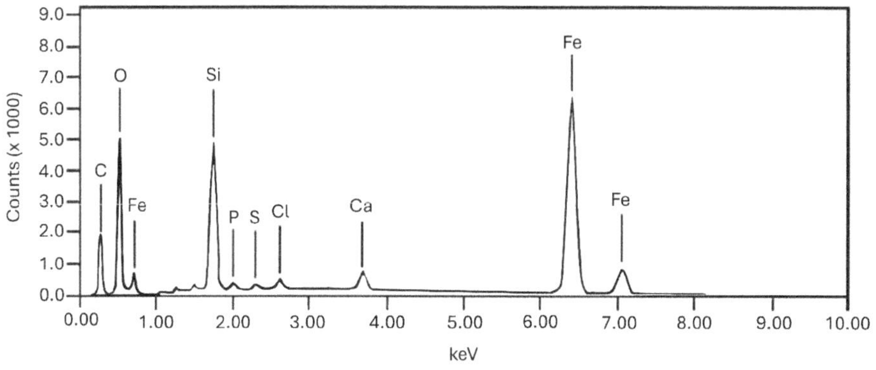

FIGURE CS11.3
Example of a scanning electron microscopy with energy dispersion test results from a representative chiller corrosion sample.

to significantly reduce the need for in-court appearances while the Order was in effect and postpone court proceedings as appropriate. All ongoing trials were postponed to a date after April 30, 2020. The state initially had planned to have trials resume on May 1, 2021; video depositions were allowed during that time. The trial for this case was rescheduled for November 1, 2021.

Trial

Depositions

Plaintiff and Defendant depositions were taken prior to trial. Depositions are formal interviews under oath that allow the opposing attorneys to gather information and determine what was known and who knows what about the case. The lawyers deposed employees who worked at the power plant, managers, and principals of the parties involved:

- General Manager (Chemical Services Company), on February 22, 2021.
- Project Manager (Chemical Services Company), on February 24, 2021.
- Technical Service Representative (Chemical Services Company), on February 26, 2021.
- Associate Principal (Plumbing Company), on July 28, 2021.
- Executive Vice-President (Plumbing Company), on August 19, 2021.

Remarkably, the Contractor's WQ Expert was not subpoenaed for deposition. The Defendant chose to move to trial without the WQ Expert's testimony at great risk.

On September 3, 2021, the Defendant filed a motion for *summary judgment* (a court order that decides a case without a full trial) prior to trial commencement, with a formal hearing scheduled for October 6, 2021. The Plaintiff (or Contractor) submitted documentation to rebut the Defendant's (or Specialist's) motion and urged the Court to consider:

> ...contractual [plain] language interpretation of the Specialist's duty to defend and indemnify the Contractor for its resulting losses, including but not limited to the costs that the Contractor incurred to repair the chillers, as well as attorney fees and costs that the Contractor incurred defending against the University's claim relating to the corrosion, and the Specialist's obligation to name the Contractor as an additional named insured under its CGL policy.

At the October 6 hearing, based on the Court's review of documents filed by the Plaintiff and Defendant, the pre-trial motion for summary judgment was denied. The trial remained scheduled to begin November 1, 2021.

Prevailing Arguments

The trial commenced with direct examination and cross-examination of deposed witnesses. Information gleaned from witness testimonies and pre-trial depositions was used during the trial to question the litigants, and in several instances the Defendant was inconsistent with answers provided during the trial. The most significant testimony was provided by the deposed Plaintiff's employees and WQ Expert. The following narrative highlights key communications (notes, emails, and orally in deposition) by the Defendant and technical arguments presented by the WQ Expert, whom the defense chose not to depose. This content was important to judicial understanding and the outcome of this case.

1. Defendant communications and interpretations:
   - August 3, 2017, Defendant: *"Chemical treatment program initiated for the condenser system. Our [product] was put online. This was the chemical identified to be used in the IFB and was in the approved submittal. We've used this product throughout the university for a number of years."*
   - October 31, 2017. Chillers opened for initial tube brushing and eddy current inspection. University personnel expressed concern that tube sheets looked worse than they should have after only three months of operation. Specialist was asked how it would address this issue. Specialist replied: *"We decided to be proactive and add other products to our chemical program."*
   - November 3, 2017 email from the chiller manufacturer to Plaintiff's management executives that documented their water treatment recommendations to the Specialist. Excerpts: *"Using untreated or improperly treated water in these units may result in inefficient operation and possible tube damage"*; *"The use of untreated or improperly treated water in a chiller could result in scaling, erosion, corrosion, algae or slime"*; and *"The services of a qualified water treatment specialist should be engaged to determine what water treatment, if any, is required."* These reiterated statements highlight issues not sufficiently addressed by the Defendant.
   - Email from the chemical provider to Defendant's staff dated December 7, 2017 pertaining to chiller #3, stating the material build up on the unit was primarily organic matter. Specifically: *"the iron and organic material are indicative of iron-reducing bacteria presence and their byproduct."*
   - December 2017, Defendant: *"Removed [chemical product 1] and put online [chemical product 2] another more concentrated organo-phosphonate-based scale and corrosion inhibitor and biocide. Increased dosage to continue to passivate condenser system."*

- February 2018, Defendant: "*Removed [chemical product 2] and put online [chemical product 4] a concentrated version of original chemical 1, continuing to feed biocide as well. We're currently on this chemical feed program.*"
- Defendant email to Plaintiff on June 7, 2018 acknowledging that biocide was not included in the water treatment regimen until December 2017, and that the regimen dosages were frequently modified.
- June 18, 2020 interrogatory asking Defendant to describe in detail and with specificity dosing and dosage rates used for each chemical used and how the dosage rate was determined. Defendant responded: "*...dosage for the [inhibitor] was 240 ppm.*" Hence, the Specialist acknowledged overdosing the CS-4618 chemical at a level significantly higher than the industry standard. The Defendant further stated: "*...no changes were made to the chemical treatment program until after the corrosion issue was brought up by the university.*" This implied that chiller corrosion existed before changes to the water treatment regimen, and indicated equipment fouling was due to improper chemical treatment.

2. The Plaintiff's WQ Expert effectively defended the technical evidence obtained during the disclosure process and at trial:
   - **The Specialist failed to provide complete and proper water treatment for the mechanical chiller units installed at the University power plant in July/August 2017.** Chillers are complex machines that require chemical water treatment and maintenance to operate properly. With improper treatment and maintenance, equipment will foul, corrode, lose efficiency, and may experience other problems. The Specialist provided substandard water treatment to the original chillers, which resulted in equipment damage (i.e., corrosion, scale, and biofilm).
   - **The Specialist failed to exercise a standard of care in selection and dosing of chemicals in treating process water.** Experienced water treatment contractors typically implement a combination of corrosion inhibitors and biocides for chiller equipment. In this case, the Specialist overdosed the process water with an incompatible inhibitor. The industry standard target dosage for a compatible corrosion inhibitor is on the order of 2.0 parts per million (ppm). Recalling the Defendant's June 18, 2020 interrogatory, the Specialist overdosed the process water with an incompatible corrosion inhibitor at 240 ppm, well above the industry standard. Inappropriate chemicals and/or incorrect dosing of circulating water will corrode or otherwise damage equipment.
   - **The Specialist did not demonstrate an understanding of the source water chemistry, in this case, the Public Water Purveyor's[4] Supply.**

The Specialist used a phosphonate-based inhibitor one month after the chiller systems went online. It is widely known within the industry that phosphonate-based inhibitors are ineffective at preventing corrosion in water with high concentrations of silica. The Specialist knew (or should have known) that the municipal supply water contained appreciable silica and should have selected an appropriate corrosion inhibitor for silicate water such as sodium metasilicate, sodium silicate, and other silicate formulations.

Water Quality Expert Elaboration

The aqueous chemistry of silica (SiO_2) is regulated by several coupled processes such as dissolution/precipitation, complexation to cations and anions in the aqueous phase, and cation-anion complexation at the particle-water interface. "*Properly treated water would prevent silica compounds from precipitating out of solution and depositing on the surface in amounts required for under-deposit corrosion*[5]*, and the corrosion patterns observed on the tube sheets are consistent with under-deposit corrosion that may manifest itself as galvanic corrosion. Under-deposit corrosion is caused when a precipitated material (even a non-corrosive material) deposits on a surface which leads to the formation of an anode-cathode pair known as a differential aeration cell*[6]*. For this to occur in the case of the condenser being studied, a compound had to precipitate out of the treated water solution and deposit on the surface of the tube sheet and the condition had to be conducive to corrosion. The Defendant, the original water treatment contractor, monitored the water treatment starting at approximately the same time the units were brought online. However, the parameters they monitored were lacking. For example, no direct measurement of silicate or silica was provided even though the Defendant knew or should have known of the high silica levels in the municipal water supply used as cooling water. Therefore, the Defendant's selection of treatment chemicals was flawed from the beginning of the service Agreement. The layered structure of the corrosion products indicates that corrosion occurred over time and that the precise corrosion reactions changed due to changes in water chemistry or localized changes [no chemical feed versus excessive chemical feed] in the conditions under the deposit. The corrosion rate of the chiller steel is sensitive to the pH of the water in the system and any process excursions resulting in temporary acidic conditions would have caused accelerated corrosion during that time period where improper treatment was practiced.*"

- **Defendant failed to implement microbial protection of the equipment.** Industry practice includes biocide compatibility in the prevention of chiller biofouling. The Specialist did not add biocide until after three months of operation. Bacterial growth leads to microbiologically induced corrosion (MIC) of equipment. MIC can result in metal deterioration due to the metabolic activity of various microorganisms. Several bacterial types can participate in MIC, such as sulfate-reducing bacteria (SRB), acid-producing bacteria (APB), and iron-reducing bacteria (IRB). These microorganisms and their metabolic products can form a biofilm layer on metallic components, and metabolic activity can facilitate accelerated metal corrosion. Further, the Plaintiff's WQ Expert refuted defense expert testimony that MIC impacts were the result of improperly installed equipment that caused the observed pitting in the steel chillers.

Water Quality Expert Anecdote

Original chiller #3 did not experience excessive corrosion because it operated only half of the operating time as the other two units. This is important because the buildup could not be explained solely due to galvanic corrosion, as purported by the Defendant's experts. The foulants were predominantly composed of organic matter with some iron and silica present in far less amounts. (Recall the December 7, 2017 email from the chemical provider to the Defendant.) These facts, first deemed inconsequential by the Defendant's experts, later contradicted their causal opinion during trial.

- ***Corrosion investigation and comparative water treatments reveal flawed initial chemical regimens that resulted in equipment damage.*** After the new chillers were brought into service, a replacement water treatment specialist contractor (CHEM, LLC) was retained to manage water quality. CHEM, LLC's treatment of the same water in the new chillers did not result in corrosion and fouling. This demonstrated that proper water treatment works. Also, the Plaintiff's Corrosion Consultant's investigation report found that after "*...the newly added [non-oxidizing] biocide was added along with bromine tablets and a scale and corrosion inhibitor designed for higher silica levels, the microbial count trended with improved control.*" This attested to the biofouling problem in the original chillers when the Defendant managed water treatment.
- The WQ Expert concluded at trial: "*It is my opinion that the Defendant was negligent as a water treatment specialist and caused fouling in the chillers that resulted in irreversible damage. The chemicals used by the Defendant as initially applied caused fouling issues because the treatment methods used did not adequately address the chemistry incompatibility of the municipal water supplied to the chillers. The Defendant failed to understand that the municipal water supply would lose its disinfectant residual in the chillers and result in the immediate growth of microbiological organisms within the equipment. Additionally, the fact that Defendant did not apply a compatible corrosion inhibitor specifically to address the elevated silica levels in the municipal water supply further aggravated the situation with under-deposit corrosion that developed over time.*"

In Summary, the Plaintiff's employees provided evidence of water mismanagement and their WQ Expert demonstrated how the Defendant willingly and knowingly breached contractual terms as outlined in the Complaint: (1) Breach of Contract, (2) Breach of Express Warranty, and (3) Declaratory Relief for Defense and Indemnification.

Judgment Rendered

The Judge agreed that the evidence presented supported the Plaintiff's Complaint in that the Defendant (1) breached their Contract with the Plaintiff, and (2) failed to provide complete and proper treatment for the mechanical chillers (condenser and evaporator components). Furthermore:

- The Court confirmed that water treatment for chillers is a complicated science, with chemicals interacting with one another, and that complex chemical interactions can have an effect on the ability of the treated water to inhibit or cause corrosion.
- The Plaintiff's testifying parties consistently agreed that the many investigations demonstrated that the corrosion was caused by inadequate chemical treatment provided by Defendant, and not by other causal factors related to the equipment or its installation.
- The Defendant clearly did not understand the significance of the source water chemistry to realize their treatment was insufficient and detrimental to the mechanical equipment.

Based on the proceedings of the Trial, the Judge ruled in favor of the Plaintiff and found the Defendant liable for Breach of Contract on all counts as presented in the Plaintiff's Complaint.

Epilogue

Cooling power plant machinery is a systems engineering endeavor that requires complex water quality management. Machinery is metal and water corrodes. An experienced environmental engineer certified as a water quality specialist is prized.

> It is always cheaper to do the job right the first time.
>
> **Philip Bayard Crosby**
> *Businessman,*
> *Quality Management Philosopher*

Notes

1. *Tuberculation* is the formation of small, bumpy mounds of corrosion products inside and on the ends of the barrel tubes. This buildup increases tube/pipe roughness and hinders water flow and can potentially cause backpressure in the system. Although copper is generally resistant to corrosion, water chemistry, bacteria, and electrical currents can contribute to the formation of tubercles.[3]
2. A *Notice of Tender of Claim* is a formal notification to an insurance company of a claim and request that the insurer take over the defense and/or indemnify the insured. Serving as notification about a potential lawsuit or legal claim, it provides the insurer an opportunity to either accept the defense and/or indemnify the policyholder or deny coverage.
3. A *Request for Production* is a legal tool to obtain documents and other tangible items from the opposing party during the discovery phase in a lawsuit. In the essence of fairness, it is used to obtain relevant information, authenticate records, and prevent surprise evidence at trial.
4. A *public water purveyor* is a public or private entity that supplies potable water to the public through a public water system. These systems are typically regulated by state health departments or the U.S. Environmental Protection Agency.
5. *Under-deposit corrosion* is a form of localized corrosion that occurs beneath or around deposits on a metal surface.
6. A *differential aeration cell* is a type of corrosion cell that forms when there's a difference in oxygen concentration at different points on a metal surface in contact with an electrolyte. This oxygen variance leads to a potential difference and a flow of current, resulting in corrosion at the area with lower oxygen availability.

References

1. *Trane, Series R Helical Rotary Liquid Chillers, Manual* RLC-PRC020-EN, June 2006.
2. *Consensus on Operating Practices for the Control of Feedwater and Boiler Water Chemistry in Modern Industrial Boilers*, The American Society of Mechanical Engineers, New York, NY, 1994.
3. *Guidelines for Control of Flow-Accelerated Corrosion in Fossil and Combined Cycle Plants*, EPRI Document 3002011569, the Electric Power Research Institute, Palo Alto, CA, 2017.
4. Fernandez, R. (2024). Heat exchanger condenser. https://khia.belzona.com/images/khia_ images/rfernandes/Main/rfernandes_3124_1.JPG.
5. Frazer, G. (2010). Microbiological corrosion of chiller tube sheet. https://i.ytimg.com/vi/4O5_lY_ 4Yo0/hqdefault.jpg.
6. Corroserve (2022). Galvanic corrosion. https://www.corroserve.com/wp-content/uploads/ 2022/07/Galvanic-corrosion-scaled.jpg.
7. (2023). Chiller tube replacement. https://5.imimg.com/data5/SELLER/Default/2023/11/358087223RJ/VQ/CS/34100340/chiller-tube-replacement-500x500.jpeg.

Index

Page numbers followed by f, t, and n indicate figures, tables, and notes, respectively.